Environmental Science and Engineering

Environmental Science

Series editors

Rod Allan, Burlington, Canada
Ulrich Förstner, Hamburg, Germany
Wim Salomons, Haren, The Netherlands

More information about this series at http://www.springer.com/series/3234

Songlin Feng · Weiguang Huang
Jun Wang · Mingquan Wang
Jun Zha
Editors

Low-carbon City and New-type Urbanization

Proceedings of Chinese Low-carbon City Development International Conference

Springer

Editors
Songlin Feng
Shanghai Advanced Research Institute
Chinese Academy of Sciences
Shanghai
People's Republic of China

Weiguang Huang
Shanghai Advanced Research Institute
Chinese Academy of Sciences
Shanghai
People's Republic of China

Jun Wang
Shanghai Advanced Research Institute
Chinese Academy of Sciences
Shanghai
People's Republic of China

Mingquan Wang
Shanghai Advanced Research Institute
Chinese Academy of Sciences
Shanghai
People's Republic of China

Jun Zha
East China Architecture Design
and Research Institute
Shanghai
People's Republic of China

ISSN 1863-5520 ISSN 1863-5539 (electronic)
Environmental Science and Engineering
ISSN 1431-6250
Environmental Science
ISBN 978-3-662-45968-3 ISBN 978-3-662-45969-0 (eBook)
DOI 10.1007/978-3-662-45969-0

Library of Congress Control Number: 2014959133

Springer Heidelberg New York Dordrecht London

Printed on acid-free paper

Springer-Verlag GmbH Berlin Heidelberg is part of Springer Science+Business Media
(www.springer.com)

Contents

Research on Urban Climate Problem-Solving-Oriented Planning Approaches in Xiamen

Bo Yu, Dai-feng Song, Shu-yu Liu and Shao-sen Wang

Abstract Urban development is a key factor of urban climate issues. And the quality of urban development is largely determined by urban planning approaches. Based on modern quality management theory, by means of expert interview and questionnaire survey, basic information and key issues of urban planning approaches in Xiamen, which are followed by optimization proposal, are collected and analyzed from aspects of process, procedure, and organization structure, from the perspective of integrating urban climate problem-solving and urban planning.

Keywords Urban climate · Planning process · Organization structure · Xiamen

1 Foreword

At present, China's cities are preparing to enter a period of low energy consumption, low emissions, high efficiency, and harmony with nature. The city of Xiamen has initiated the program 'Beautiful Xiamen,' which is aiming at 'resource conservation' and 'eco-friendly.' However, the constantly increasing number of extreme temperature and smog days indicate that urban climate problems in Xiamen are exacerbating. Studies based on urban climatology have demonstrated that urban heat island effect, air pollution, urban ventilation, and other urban climate issues directly result from unhealthy urban development. To a large extent, the quality of urban development is affected by urban planning scheme. William Edward Deming proved that the quality of product exactly depends on working approaches by red bead experiment (Edwards Deming 1994). At present, 'producing a plan' remains one of the most important tasks of planners. Logically, the relationship between planning approaches and the quality of 'plan' cannot against the above-mentioned

B. Yu · D. Song · S. Liu (✉) · S. Wang
School of Architecture and Civil Engineering, Xiamen University, Xiamen 361005, People's Republic of China
e-mail: liu_shuyu@msn.com

S. Feng et al. (eds.), *Low-carbon City and New-type Urbanization*,
Environmental Science and Engineering, DOI 10.1007/978-3-662-45969-0_1

principle. In this sense, the analysis and optimization of urban planning approaches in Xiamen should be recognized as an important starting point for its urban climate problem-solving.

2 Situation Analysis

2.1 Planning Process

Based on related laws and regulations, process of compiling the master plan of Xiamen is summarized into that shown in Fig. 1. Some steps such as preliminary study, compilation of master plan framework, compiling of urban master plan, environmental impact assessment (EIA), supervision, and inspection might integrate studies on urban climate issues. But there is no subject research about urban climate involved in the present planning system.

The process of compiling the detailed plan of Xiamen is different from that of other cities. According to Sect. 16 of Chap. II of 'Technical Regulation for Urban Planning Management in Xiamen (2010),' 'The compiling of regulatory plan of Xiamen consists of two phases, the outline and statutory map. The outline should include layout of public facilities and infrastructure, landuse-capacity and other contents. The output have to get publicity in accordance with statutory procedures,

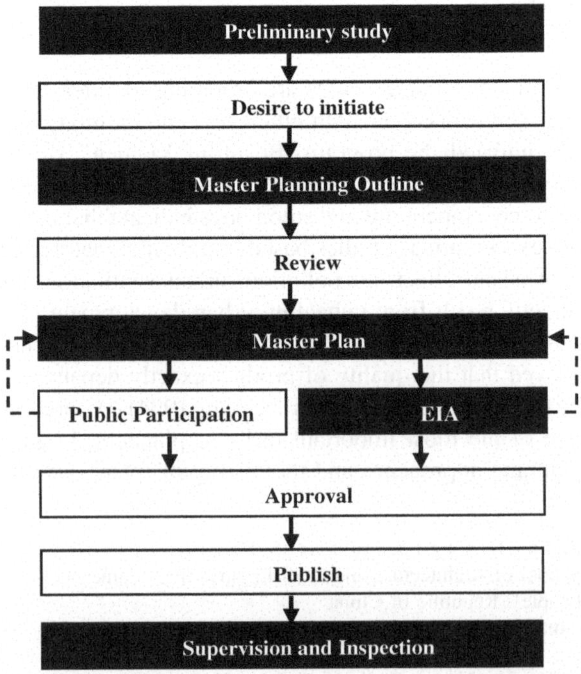

Fig. 1 Schematic of process of compiling the master plan of Xiamen. (*Source* Made by authors. According to the Articles 7 and 8 of Standing Committee of the National People's Congress (2002), Articles 13 and 16 of Ministry of Housing and Urban-Rural Development (2005), and Article 1.4 of State Environmental Protection Administration. *Note Solid arrows*—inevitable links, *dotted arrows*—possible links; *textbox with black background*—steps involving urban climate)

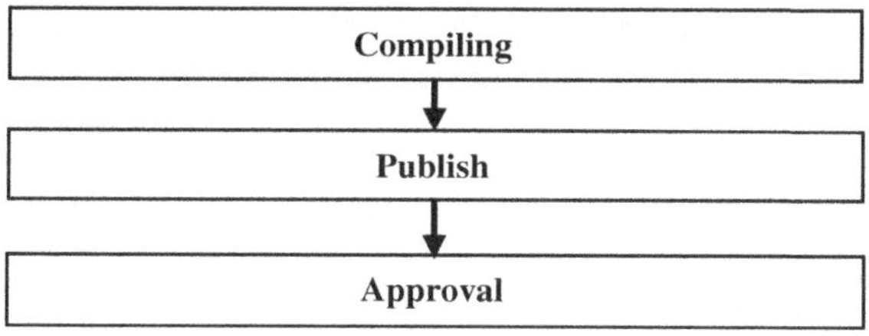

Fig. 2 Schematic of process of compiling regulatory plan outline of Xiamen. [*Source* Made by authors. According to the Articles 15 and 16 of Xiamen Planning Bureau (2010)]

it can be executed only after being approved (Fig. 2). Statutory maps of certain area operational projects should be based on the output of outline, it should be triaged by the city Planning Bureau then reported to city government and be reviewed in accordance with legal procedure. It will be published and initiate after being approved (Fig. 3). After the lot is transferred, the statutory maps should be recorded in the Urban Planning Bureau Information System and used as the basis for examining subsequent development.'

'When the lot is transferred, the statutory maps should be regarded as boundary condition for subsequent design. This condition must be strictly enforced and shall not be adjusted in land development' (He 2009). That is so-called hard bundling type transfer. In practice, as a condition of land transfer, statutory map is the result of compromise between planning department and the developers. If land transfer does not go well, the map will be modified for several times in a long period. Although its flexibility will be improved, its binding force will be reduced in turn.

However, the guidance effect of planning environmental impact assessment (PEIA) in urban planning is not sufficient. As an important part of the urban planning, 'China's EIA of urban and rural planning is still in its infancy (Dai et al. 2010).' Firstly, PEIA was executed late. The 'Environmental Impact Assessment Law of the PRC' went into effect on September 1, 2003, and the 'Planning Environmental Impact Assessment Ordinance' went into effect on October 1, 2009.

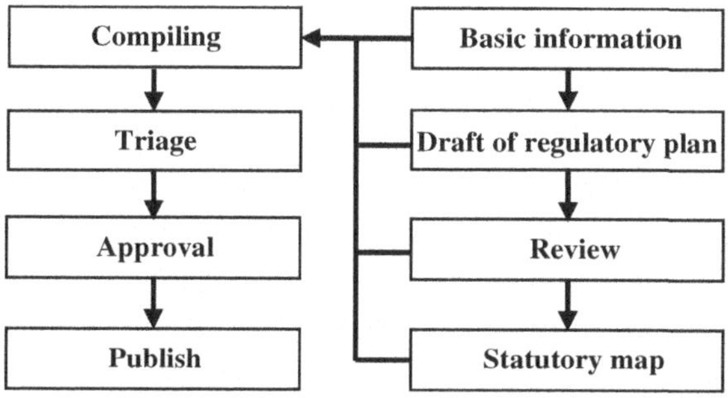

Fig. 3 Schematic of process of compiling statutory map of Xiamen. [*Source* Made by authors. According to the Articles 15 and 16 of Xiamen Planning Bureau (2010) and He (2009)]

Secondly, PEIA intervened too late in the planning process. Typically, it takes a long time to finish PEIA. Plans probably have been approved before the PEIA can make optimization. It can only make a partial adjustment instead of overall optimizing, even sometimes have no role in optimizing. Finally, the expression of output of PEIA is not intuitive enough to be understood by planners. It was unable to guide the spatial layout, urban morphology, or other aspects.

2.2 Procedures

In this regard, there are still many problems in Xiamen planning practice; for example, subject researches are not comprehensive and output of which is not included in statutory maps, climatic factors are not included in urban land-use suitability assessment and ecological function regionalization, and urban climate factors are rarely considered in PEIA. Next let us take a look at each of these issues in more detail.

Subject researches on heat island, wind environment, air pollution, and other urban climate issues still have to be increased. In the level of master plan, results of existing researches on heat island are hardly applied because they lack actual reference. In the level of detailed plan, due to the lack of technical team and limited funds and time, there is no condition to carry out studies on heat pollution of specific projects. As public do not have a strong awareness, studies on wind environment are rare, and the output that can directly guide the urban planning and design is lacking. The guidance of atmospheric pollution studies in PEIA on urban spatial layout is limited.

Results of studies on urban climate are not included in statutory maps. The impact of constructive activities on urban climate is considered to only a slight degree in the compiling of statutory maps. For example, there are many requirements about project location, land-use function, site area, floor area, building density, green rate, building height, and other aspects in the two statutory maps that 'Kaiyuan Industrial District B-14 Lot Statutory Map (2014a)' and 'Weibo Paper Lot Statutory Map (2014b)' announced by Planning Bureau in August 2014. But these requirements are not based on climate issues.

In terms of master planning, urban climate issues did not exert impact on the compiling of urban land-use suitability assessment, ecological function regionalization, and other integrating plans. The compiling of urban land-use suitability assessment and ecological function regionalization is carried out in the phase of EIA of master plan. According to 'Environmental Impact Assessment of Urban Master Plan in Xiamen (2010–2020),' based on slope, geological conditions, topography, land-use status, and a lot of other factors, urban land use is divided into three categories that 'suitable build land use, buildable land use and unsuitable land use' (Fig. 4). Ecological function regionalization is mainly based on geological conditions, ecological environment (water, vegetation, etc.), economic patterns, and other factors. The area including land and waters is divided into dozens of

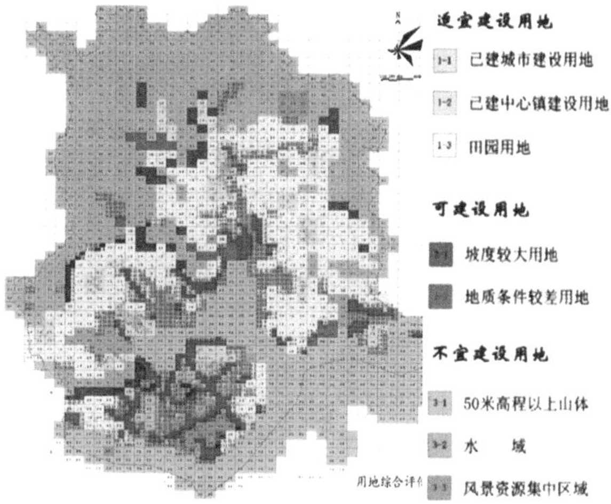

Fig. 4 Urban land-use suitability assessment map of Xiamen. [*Source* Shi (2010)]

ecological function areas such as Gulangyu and Mount Wanshi scenic ecological function area and Mount Caijianwei forestal ecological function area. Each ecological functional area will play a part in conserving water, absorbing pollutants, controlling flood, or other functions. It can be seen that the current compiling of urban land-use suitability assessment and ecological function regionalization is mainly based on the consideration of economic benefits and certain environmental principles. Urban climate factors are not involved in it.

The evaluating indicators of EIA about the impact exerted by urban spatial morphology on urban climate are inadequate. According to the 'Planning Environmental Impact Assessment Guidelines (Trial),' which is published in 2003, environmental evaluation indicators of climate issues mainly include energy consumption and flood control, the starting point of which lies in the reduction of greenhouse gas emissions and reduction of climate change disasters (Table 1). Urban ventilation and reduction of heat island are not involved.

2.3 Organization Structure

Experience of developed countries shows that establishment and improvement of interagency cooperation is the key to deal with all kinds of ecological problems. However, the cooperation among most of planning departments and other sectors in China is still in the exploratory stage, and there are no clear provisions about it in the planning system. In Xiamen, public sectors that are participating in the compiling of regulatory plan include the Education Bureau, Sub-district Administrative Office, and Community. Meteorological Bureau and Environmental Protection Bureau are not included. And in other sectors, there are no specific professional organizations or persons who are responsible for contacting with Planning Bureau.

Table 1 The expression model of environmental goals and evaluating indicators of regional planning

Environmental issues	Goals	Evaluating indicators
Biodiversity science	Protect and extend biological diversity	Come up to international/national protection standards
	Protect and expand the special habitats and populations	
Water	Prevent water pollution destroying natural ecological system	Reaching standard of water quality of rivers, lakes, and coastal waters
	Reduce the emission of water pollutants, water environmental function areas up to standard	Eutrophication level of lakes
		Quality and quantity of drinking water sources
	Keep balance between consumption and supply of groundwater	Water supply source guaranteed efficiency
		The scale and efficiency of centralized sewage treatment
		Control of emission of industrial water pollutants
Solid waste and soil	Reduce pollution and ensure the quality and quantity of soil	Cultivated area
		Green area
	Waste minimization (reuse, compost, energy utilization)	Control area and quantity of water and soil erosion
		Usage and management of chemical fertilizer and pesticide
		Innocuous disposal of domestic garbage
		Hazardous waste disposal (hazardous wastes and general industrial solid wastes)
Air	Reduce the emission of air pollutants, air environmental function areas come up to standard	Number of days the air quality meets safety standards
		Control emissions of air pollutants
		Reduce proportion of emissions of air pollutants
		Information of whether motor vehicle exhaust emission reaches standards
Acoustic environment	Reduce noise and vibration	Reaching standard rate of traffic noise
		Proportion of categories I and II noise functional area (quality of regional noise)

(continued)

Table 1 (continued)

Environmental issues	Goals	Evaluating indicators
Energy and mineral resources	Use energy efficiently	Proportion of centralized heat supply
	Increase the proportion of clean energy	Electricity supply
	Reduce the consumption of mineral resources	The use of gas
	Increase the reuse of materials	Fire coal
Climate	Reduce greenhouse gas emissions Reducing the climate change disaster	Energy consumption Flood prevention
Cultural heritage and landscape	Protect historical buildings, monuments, and other important cultural properties	Proportion of buildings and monuments on the endangered list and historical significance, cultural connotation, and amusement value (delights, popularity, etc.)
		Aesthetic value (landscape aesthetic feeling degree, strangeness, integrity, etc.)
	Value and protect geographic and topographic landscape (such as mountain landscape, canyon landscape, seacoast landscape, karst landform, wind erosion landform)	Scientific value

Source HJ/T (2003). *Note* Goals and evaluation indicators should be determined based on type, level and development status and environmental status of certain area

As an important tool to protect the urban ecological environment and promote sustainable urban development, 'Environmental Protection Master Plan of Beautiful Xiamen' cannot coordinate well with the master plan.

On the one hand, there is no effective communication mechanism between the environmental protection master plan and master plan. Environmental protection master plan is difficult to exert impact on the master plan, which is largely due to the compiling of environmental protection master plan which is behind the master plan. So it will be difficult to embody the significance of environmental protection master plan. In addition, planning regulations implemented by Environmental Protection Bureau are different from those implemented by Ministry of Housing and Urban-Rural Development (MOHURD), and two types of laws are different and even contradictory in terms of standards, goals, and other aspects. Therefore, it is difficult to integrate environmental protection master plan and master plan.

On the other hand, the environmental protection master plan is mainly compiled by engineers and its output is lack of thoughts about urban planning, so it is not likely to be applied to urban planning. In the process of compiling the environmental protection master plan, Planning Bureau is responsible for industries and facilities and Urban Development and Environment Institute is responsible for the researches of environmental issues. Engineers of Urban Development and Environment Institute did not understand fully master planning, and as a result, the output of environmental protection master plan will not be combined with the master plan closely. Besides, the output of environmental protection master plan only provides evaluation of environmental status quo of Xiamen, and there are no proposals to solve the problems. Its actual effect is not ideal.

3 Optimization Proposal

Because of the above-mentioned disadvantages, the compiling of urban planning is neither able to get a professional and reasonable guidance in advance nor able to get specific assessment and correction afterward. As a result, the high quality of urban climate and environment is out of the question. A series of proposals should be adopted to optimize the planning approaches.

3.1 Planning Process

In order to increase the impact of PEIA on urban planning in Xiamen, on one hand, PEIA should be included in the process of compiling the master plan as early as possible to guide the draft of master plan. On the other hand, the problems and results mentioned in PEIA need to be expressed with diagrams which can make it more intuitive and more comprehensible and able to guide the urban spatial layout and form creation.

Introducing EIA into process of compiling regulatory plan is an important measure to test and optimize the impact on environment exerted by projects. Firstly, reasonable EIA of regulatory plan can be an effective link between EIA of master plan and EIA for projects. Secondly, EIA of regulatory plan can reduce burden of EIA of master plan to a certain extent, which can optimize the process of compiling the master plan by making EIA of master plan to be finished early. Thirdly, according to 'Environmental Impact Assessment Law of the PRC,' EIA is an indispensable element of spatial planning at all levels.

3.2 Procedure

Subject researches on urban heat island, wind environment, air pollution, and other aspects should be increased, such as urban ventilation channel planning, urban climatic maps, and other research projects, which can make optimizing suggestions for urban planning and design, and provide specific and practical evaluation indicators for EIA as well. According to the experience of developed countries, research on urban climate map is the only way of connecting local urban climate issues and urban planning. In practice, researches on urban climate issues are difficult to be applied to guide urban planning because of the lack of detailed, timely, and accurate climate information.

Urban climate analysis tools such as Airpak, PHOENICS, and Ecotect are effective means to improve efficiency and reduce cost. At present, there are a wide range of these tools, and the technology used has become increasingly advanced. For example, by using software based on computational fluid dynamics (CFD), we can accurately simulate the flow, quality, heat, pollution, thermal comfort, and other issues of urban atmosphere. It is worth being extended.

Statutory maps can directly control the construction of project. Including reasonable urban climatic constraints in statutory maps can make the output of subject researches effectively implemented. Furthermore, relevant departments should appropriately enhance the binding force of statutory maps, which can ensure that urban climatic constraints can be obeyed, and it also can ease the contradictions that caused due to many times of revisions of statutory maps and the contradictions between statutory maps and plans at all levels above it.

3.3 Organization Structure

By introducing appropriate trade-off mechanism into the organization structure, and increasing impact on the planning and design decisions of Meteorological Bureau and Environmental Protection Bureau, and increasing the proportion of climatic factors in urban planning, various factors in the process of urban planning can be considered fully, and the plans can be fully optimized as well. In this regard, we can refer to trade-off mechanism in Germany. According to the second paragraph of Article 1a of 'Building Codes' of Germany, environmental protection requirements mentioned in paragraph 7 of Article 1 must get attention while taking trade-off decision. Target system will be reasonably built by coordination and restriction on behalf of the institutions of public interest.

Spatial planning system should be set, which can strengthen the correlation and binding among the planning tools at all levels. Docking sectors should be set among the departments that involved in urban planning and make them participate in each other's work to a certain extent, which can make the purpose of mutual adaptation and mutual coordination more accessible. For instance, in Germany, 'Raumordnungsgesetz' and 'State Planning Law' have made special provisions for requirements, goals, special requirements, and many other concepts of regulations of urban space, which can avoid inconsistent concepts caused by the differences among kinds of plan. The information of goals and intentions of spatial regulation can accurately convey to all levels of spatial planning, which make urban planning and special planning can have indirect impact on specific urban constructive activities (Li 2009).

Building a collaboration platform among the urban planning agencies can make them to share resources, technology, and information, and by doing this, the convergence among plans from different agencies will be improved and good cooperation among the urban planning agencies would be formed. In that regard, informal planning tool 'Framework Plan' in the planning system in Germany is an effective collaboration platform. Framework plan can harmonize construction requirements and determine the potential of regional development, showing the future of land-use status rough. It will be often undertaken at the design competition stage to propose clear and specific development goals or to protect specific development elements. Framework plan has no standard procedures and no legal force, but it can make planning departments to collaborate more effectively.

The establishment of a collaborative approach with certain force (such as Public Institutions Planning Participation System) is recommended to provide the basis for close cooperation among the institutions. In Germany, the objects, procedures, depth, duties, rights, obligations, and supervision mechanism of planning participation of public institutions should be clarified by local laws and regulations at least.

Establishing a long-term relationship between departments engaging in compiling urban planning and other departments such as universities and other institutions for scientific research is a feasible method to equip departments engaging in compiling urban planning with the ability of technology and to get long-term technical support for environmental protection. Universities and other institutions for scientific research are more effective in the areas of studies on the urban climate issues than departments engaging in compiling urban planning because of their advantages of human resources and technology. But planning departments can provide a lot of practical platforms, information of city, and other resources, which are the shortcomings of institutions for scientific research. Establishing a long-term relationship between them can not only provide technical support for urban planning and design but also solve the problem of technical personnel reserves to a certain extent.

3.4 Resource

Increase the financial support for studies on urban climate and its implementation, which can ensure that subject researches can be done in a timely manner and updated constantly. In fact, promotion of scientific research ability can promote local economic growth. According to the precedents in Germany, the USA, and other countries, environmental protection industry can create a great deal of economic output and a large number of employment opportunities (Economic and Commercial Section of the Embassy of the People's Republic of China in the Federal Republic of Germany).

Table 2 Scope of climate information gathering for urban ventilation channel planning in Stuttgart

Climatic elements	Scope of climate information gathering
Air temperature	Distribution of surface temperature in the evening, morning, and day, ratings of temperature distribution, distribution of annual maximum and minimum temperature, temperature fluctuation, frequency of temperature inversion, the number of days that daily maximum temperatures exceeded 25 and 30 °C, the number of days during which temperatures are always below 0 °C
Precipitation	Distribution of annual precipitation, distribution of the number of days that daily precipitation exceeded 10 mm, distribution of the number of snow-covered days
Wind	Overview of the wind farm at 6:00, 15:00, and 24:00 when daily maximum temperatures exceeded 25 °C, wind rose in all regions, distribution of average wind speed, ventilation condition
Air pollutants	Distribution of quantity of radiation of CO_2, SO_2, PM10, distribution of SO_2 emissions from transport facilities, distribution of residential districts with serious air pollution, distribution of capability of summery, hibernal, and annual renewal of air
Noise	Traffic load in days that traffic exceeded 1,000 and 10,000, loads of traffic noise in the daytime and nighttime, areas free from noise, acres of space free from traffic disturbance
Integrated climatic index	Distribution of hot bioclimatic days and cold bioclimatic days, distribution of population suffering from bioclimatic load

Source Liu and Shen (2010). *Note* Humidity change is seriously subjected to temperature change, and distribution of radiation can be fully reflected in the distribution of air temperature and wind conditions. So the humidity and the radiation are not considered as climatic elements

There are a lot of works to be done in regard to information support. In order to promote launching and development of studies on urban climate, relevant departments should publish urban climatic and environmental data for scientific research institutions and the public. Complex urban data collection is basis for urban climate studies. In Stuttgart, the necessary data includes temperature, precipitation, wind, air pollutants, noise, integrated indicators of climate and other elements, as well as topography and land use information (Table 2).

4 Conclusion

The investigation shows that the structural flaws of Xiamen urban planning system in process, procedure, organization structure, resource, and other aspects make it hard to effectively integrate studies on urban climate to guide urban planning at all levels. Once again, the decisive impact of work approach on product quality has been proved in urban planning. Because the quality of cities is based on each individual land-use plan, the systematic, standardized, and targeted optimizing of work approaches in urban planning is logically necessary.

On the one hand, there are significant differences between China and developed countries in terms of social, economic, and environmental conditions, and planning ‘benchmarking’ seems unreasonable. On the other hand, comparative study on multisystems made the existing defects of our system and the comparative advantages of advanced systems more clear and specific, which formed the foundation of optimizing working approaches. Discussions on the integration of solution to urban climate issues in urban planning should be an opportunity for China urban planning system toward scientification nowadays.

Acknowledgments Supported by the National Natural Science Foundation of China (Grant No.51408516); Soft science research project in Fujian Province (Grant No.2013R0097); the Fundamental Research Funds for the Central Universities (Grant No.2013121027); Xiamen Construction Science and Technology Plan Projects (Grant No.XJK2013-1-7).

References

Dai X, Xiao R, Liu Y (2010) Integrating environmental evaluation with urban-rural planning. Planners 08:82–86

Edwards Deming W (1994) The new economics: for industry, government, education. Massachusetts Institute of Technology, Cambridge, p 33

He Z (2009) The ‘hard bundling’ and ‘soft bundling’ between regulatory detailed plans—comment on ‘bidding, auction and listing’ planning consultation in Xiamen. Planners 11:76–81

HJ/T 130-2003 (2003) Technical guide for plan environmental impact assessment (Trial)

Li Y (2009) Synergy mechanism between comprehensive spatial planning and urban planning in Germany: a comparative study. J Wuhan Univ Sci Technol 2:54 (Social Science Edition)

Liu S, Shen J (2010) Urban ventilation channel planning method based on local circulation: a case study of Stuttgart, Germany. J Zhejiang Univ 10:1985–1991

Ministry of Housing and Urban-Rural Development (2005) Urban planning-making method

Shi X (2010) Environmental impact assessment of urban master plan in Xiamen (2010–2020). Environmental Impact Assessment in Xiamen University, Xiamen

Standing Committee of the National People’s Congress (2002) Environmental impact assessment law of the PRC

Xiamen Planning Bureau (2010) Technical regulation for urban planning management in Xiamen

Xiamen Planning Bureau (2014a) Kaiyuan industrial district B-14 Lot Statutory Map [EB/OL]. (7 Aug 2014). http://www.xmgh.gov.cn/zfxxgk/zfxxgkml/ghcg/kzxxxgh/201408/t20140807_45677.htm

Xiamen Planning Bureau (2014b) Weibo paper lot statutory map [EB/OL]. (7 Aug 2014). http://www.xmgh.gov.cn/zfxxgk/zfxxgkml/ghcg/kzxxxgh/201408/t20140807_45678.htm

Optimization of a Combined Heat and Power Generation System with Ice Thermal Storage

Cheng-gang Cui, Xiao-fei Yang, Feng Tian, Ting-yu Gao and Zhi-cong Zhu

Abstract This paper explores the optimization of a combined heat and power generation system (CCHP) with ice thermal storage air-conditioning system in consideration of minimal cost. A CCHP requires the simultaneity of the electrical and thermal demands. However, the thermal demand is strongly time-varying or even discontinuous. The ice thermal storage provides an attractive solution to the CCHP by shifting the electrical power while more profitably exploiting the thermal energy. The optimization of the simultaneous use of a gas turbine, an absorption chiller, a brine chiller, and an ice-storage tank to satisfy given electricity, heat, and cooling demands is considered. An MINLP algorithm is used to develop optimal operating strategies for the cogeneration system. Case study is based on a building belonging to Shanghai Advanced Research Institute. Finally, the results of economic analysis providing some guidelines of operation strategies to a CCHP with ice thermal storage.

Keywords Combined heat and power generation system · Ice thermal storage · Optimization · Case study

1 Introduction

Combined heat and power generation system (CCHP) is widely acknowledged as a key alternative for thermal and electric energy generation with respect to the separate production (SP) of cooling, heat, and electricity. It consists of the simultaneous production of electricity and heat and yields a high energy-saving effect, especially if compared to the traditional energy supply configuration. This leads to lower fuel consumption generated at a lower cost and in a more environmentally friendly way

C. Cui (✉) · X. Yang · F. Tian · T. Gao · Z. Zhu
Shanghai Advanced Research Institute, Chinese Academy of Sciences,
Shanghai 201210, People's Republic of China
e-mail: cuicg@sari.ac.cn

S. Feng et al. (eds.), *Low-carbon City and New-type Urbanization*,
Environmental Science and Engineering, DOI 10.1007/978-3-662-45969-0_2

(Wu and Wang 2006). Due to the interaction between hot and cold electric system load, which may lead generation system can only bear part of the load electrical load and thermal load coincide. This makes the system to heat, and power load matching problem has become an important research topic (Gu et al. 2014).

Energy storage technologies have become an increasingly popular technology (Basecq et al. 2013). They can make the energy been stored during off-peak times and dispatched during peak times. In order for energy storage to make a significant impact, inexpensive storage technologies, thermal energy storage (TES), the storage of heat or cooling has been incorporated into the CCHP systems (Barbieri et al. 2012). TES can dramatically reduce payback periods in addition to improve the project's return on investment since it stores energy by using relatively simple and inexpensive technologies. However, CCHP systems that require storage exhibit transient behavior and the storage processes themselves are transient (Raine et al. 2014). Therefore, it is critical to develop effective operating strategies for CCHP systems with TES technologies (Ameri et al. 2005).

This work considers how a CCHP system with ice thermal storage can be operated in order to provide economic benefits. A MINLP model for tackling the short-term operation problem which aims at minimizing the operating cost of a CCHP system with ice thermal storage. The proposed model can take into account the situation that cooling is provided by directly using the chiller or by discharging the ice bank when electricity prices are high. The model has been tested on a CCHP system test case in south China. The computational results are discussed in terms of the quality of the solutions and the influence of ice thermal storage.

2 System Description

The CCHP system with ice thermal storage consists of a power generation unit (PGU), a waste recovery system, a cooling system, and an ice-storage air condition system as shown in Fig. 1 (Wang et al. 2010). In this paper, a cooling storage system of ice-storage air-conditioning system is selected, which is composed of double-duty chiller units and ice-storage tank (Al-Abidi et al. 2012). The system load consists of cooling and electrical load. The solid and dash lines represent the cooling load flows and the electric power, respectively. PGU which consumes natural gas produces electricity and heat simultaneously to meet electrical and heat load demand. If there is excess energy, it will be stored in ice-storage air-conditioning system. If the thermal load is larger than that of the amount PGU provides, the ice-storage air-conditioning system provides additional energy to the system.

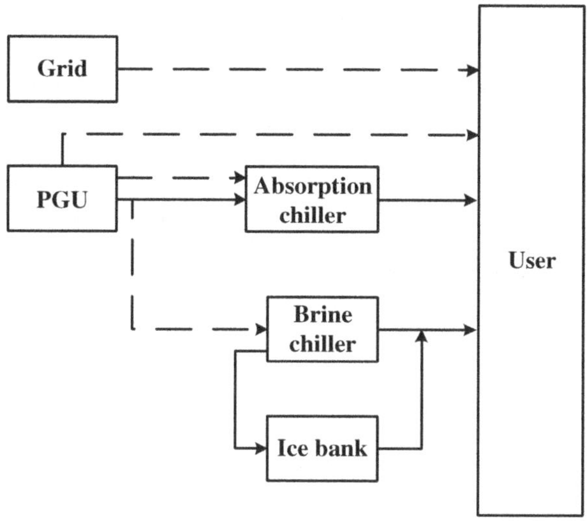

Fig. 1 Energy flow diagram of CCHP system

3 Model Formulation

3.1 Problem Description

The aim of this paper was to develop a method to generate the optimal operation of a CCHP system installed with ice-storage system in which the electric grid and a local thermal network have been considered. The total net present cost is selected as the evaluating criteria, which is used as the optimal objective to be minimized. An ice TES unit is used to avoid the discontinuous operation of the CCHP system and to increase the number of operating hours.

The optimal operations of CCHP problem we want to tackle can be stated as follows:

- Generation units with fixed size, performance curves, and start-up/warm-up times;
- Ice-storage tank with fixed capacity and constant loss rate;
- Time-dependent price of electricity;
- Determine the following for each time period of a time T horizon $[t_0, t_f]$:
  - The set of units to be switched on
  - The value of the operative variables of each unit
  - The storage tank level

which minimize the operating cost while satisfying the demands of thermal and electric.

3.2 Model Description

A mixed integer nonlinear programming (MINLP) model concerning the aforementioned scenario problems has been developed. The mathematical model is reported in the following by a set of constraints and an objective.

1. Set the energy optimization scheduling period as $[t_0, t_f]$, the energy scheduling interval is T, the scheduling period includes N_t time intervals T, and $t_i = [t_0 + (i-1)T, t_0 + iT]$, where $i = 1, \ldots, N_t$.
2. Set energy type as $j = 1, \ldots, N_j$, where N_j is the total number of energy types. For example, set energy type $j = 1$ for power, energy type $j = 2$ for heating, and energy type $j = 3$ for cooling in CCHP system.
3. Set device number as $k = 1, \ldots, N_k$. Set N_k represents all the devices, including N_s represents the number of types of energy conversion devices, N_c represents the number of complementary multi-energy devices, N_m represents the number dynamic energy devices, and N_n represents the number of the energy storage devices.
4. Set U as the devices, and Q as the energy generating or using by the devices. For a device, it can be expressed as U_k. The generating or using energy can be represented as $Q_{k,j}$. For example, the internal combustion engine can be expressed as U_1; when generating electric power, it is represented as $Q_{1,1}$, when generating heat, it is represented as $Q_{1,2}$; and when generating cooling, it is represented as $Q_{1,3}$.

3.2.1 Parameters and Variables

1. Time-varying parameter
   Time-varying parameters are the parameters that may change over time periods as demand fluctuations or the ambient temperature. Therefore, they have different values for each time period t:
   - $p_j(i)$: the energy demands at time t;
   - $\mathrm{pb}_j(i)$: the energy purchase price at time t;
   - $\mathrm{ps}_j(i)$: the energy sale price at time t;
2. Constant
   - $\mathrm{cm}_{k,j}$: device maintenance costs, which is proportional to the amount of energy produced;
   - cs_k: start-up costs, which is proportional to the number of starts;
   - $\mathrm{ch}_{k,j}$: the loss of energy storage devices, which is proportional to the length of time;

3. Continuously variable
   - $Q_{k,j}(i)$: the energy j supply or consumption of device k at time I;
   - $\Delta Q_{k,j}(i)$ represents the amount of energy j stored in the energy storage device variation between time i and $i-1$;
   - $S_{k,j}(i)$: the energy j stored in energy storage devices k at time i;
4. Binary variable
   - $z_k(i)$: the state variables of device k at time i;
   - $m_j(i)$: state variables represent whether energy j is purchased;

3.2.2 Objective

The objective is shown as follows:

$$\min \sum_{i=1}^{N_t} C_b(i) - \sum_{i=1}^{N_t} C_s(i) + \sum_{i=1}^{N_t} C_{\text{on}}(i) + \sum_{i=1}^{N_t} C_l(i).$$

$C_b(i)$ is the energy cost of all the devices during time i. It is given by the sum of the amount of fuel each device is utilizing multiplied by its specific cost.

$$C_s(i) = \sum_{j=1}^{N_j} \sum_{k=1}^{N_k} p_s(i) Q_{k,j}(i)$$

$C_{\text{on}}(i)$ is the start-up cost of all the devices during time i. It is given by the extra cost associated to the start-up procedure.

$$C_{\text{on}}(i) = \sum_{k=1}^{N_k} \text{cs}_k(i) Z_k(i)$$

$C_s(i)$ is calculated by multiplying the amount of electric energy hourly purchased or sold to the grid.

$$C_s(i) = \sum_{j=1}^{N_j} \sum_{k=1}^{N_k} p_s(i) Q_{k,j}(i)$$

3.2.3 Constraints

1. **Energy load balance constraint**
   The balance constraint ensures that customer requirements of energy are always fulfilled by devices generating this energy j during time i as follows:

$$\sum_{k=1}^{N_s} Q_{j,k}(i) + \sum_{k=1}^{N_c} Q_{j,k}(i) + \sum_{k=1}^{N_m} Q_{j,k}(i) + \sum_{k=1}^{N_n} \Delta Q_{j,k}(i) = P_j(i),$$

   where $Q_{k,j}(i)$ is the energy j generated or consumed by device k and $\Delta Q_{j,k}(i)$ is the variation of energy j stored in device k during time i.
   To make it easy to satisfy the constraints, a slack variable sl_j is introduced to solve the constraints.

$$\sum_{k=1}^{N_s} Q_{j,k}(i) + \sum_{k=1}^{N_c} Q_{j,k}(i) + \sum_{k=1}^{N_m} Q_{j,k}(i) + \sum_{k=1}^{N_n} \Delta Q_{j,k}(i) + \mathrm{sl}_j = P_j(i)$$

2. **Device start-up constraint**
   The "start-up" constraints are used in order to set a maximum number of start-up procedures that can be tolerated by each device in order to avoid damages.

$$\sum_{i=1}^{N_T} \Delta T_{k,i} \leq N_k$$

   To ensure that the variable has value 1 at time i if, and only if, device k was off at time i, we introduce the following constraints

$$\Delta T_{k,i} \geq Z_k(i) - Z_k(i-1)$$
$$\Delta T_{k,i} \leq 1 - Z_k(i-1)$$
$$\Delta T_{k,i} \leq Z_k(i)$$

3. **Performance constraint**
   The performance constraint ensures the performance of the device. For each device, the input variables (consumed fuel and electricity) are related to the output variables (generated heat, electricity, cooling) by a performance function.

   (a) Gas turbine
   As the prime mover, the performance of gas turbine will directly affect the performance of CCHP systems. The main factors that affect the performance of the gas turbine are temperature, absolute humidity, altitude, air inlet pressure loss, and exhaust port back part load rate. For the sake of simplicity of the model, the gas turbine power output, fuel consumption,

and the corresponding heat exhaust of recyclable are considered in this paper. The model coefficients are fitted by gas turbine operating data.
Fuel consumption constraints:

$$Q_{k,0}(i) = \sum_{j=1}^{m} a_n^j Q_{k,j}^n(i) + \cdots + a_2^j Q_{k,j}^2 + a_1^j Q_{k,j}(i) + b_j.$$

Heat recycling constraints:

$$Q_{k,j'}(i) = c_n^j Q_{k,j}^n(i) + \cdots + c_2^j Q_{k,j}^2(i) + c_1^j Q_{k,j}(i) + b_j.$$

(b) Chiller
In the CCHP system, chiller system is operated at part load at most time. Different device load will case a different performance coefficient. The dual-mode chillers are modeled by the electricity consumed, where the model is divided into a cooling condition and an ice conditions model. The regression coefficients of each model are different from each other.

$$Q_{k,j'}(i) = a_n Q_{k,j}^n(i) + \cdots + a_2 Q_{k,j}^2(i) + a_1 Q_{k,j}(i) + b$$

(c) Ice-storage system
The ice-storage air-conditioning device is different from the conventional air-conditioning device. The ice bank is not a power device, but the performance of ice melting has a great impact on the operation of the ice-storage system. For simplicity, the maximum ice-melting efficiency is a constant in this paper.
Ice-storage constraint:

$$Q_{\min,j} \le Q_{k,j}(i) \le Q_{\max,j}$$

where $Q_{k,j}(i)$ is the ice stored in the device during time i, $Q_{\min,j}$ and $Q_{\max,j}$ are the minimum and maximum value of ice stored in the device.
Melting rate constraint:

$$Q_{k,j}(i) - Q_{k,j}(i-1) \le \Delta Q_{\max,j}$$

where $\Delta Q_{\max,k,j}$ is the max melting rate between time i and time $i - 1$;
Variation constraints:

$$Q_{k,j}(i) = Q_{k,j}(i-1) + \Delta Q_{k,j}(i)$$

where $\Delta Q_{k,j}(i)$ is the variation of ice between time i and time $i - 1$;
All the performance functions are fitted by a polynomial curve fitting method.

3.2.4 Solving Method

The model is written in the YALMIP algebraic modeling language (Lofberg 2004) and the resulting MINLP is solved with Bonmin optimizer.

4 Case Study

4.1 Problem Description

In order to validate the proposed mathematical model-based approach, a CCHP system with ice thermal storage located in the south of China has been taken into account. A proper dataset including both thermal and electric energy consumptions in the year 2013 was considered. The system is operated during a day by 1 h. The hourly power load demand is [300, 300, 300, 300, 320, 320, 662, 852, 852, 980, 972, 972, 972, 972, 862, 760, 665, 654, 495, 423, 300, 300, 300] and the cooling load demand is [0, 0, 0, 0, 0, 0, 262, 372, 470, 471, 472, 672, 672, 672, 462, 360, 161, 154, 95, 0, 0, 0, 0, 0] ton. The gas price is 2.4 yuan, the electricity prices are 1.287 yuan in peak period (8:00–11:00 and 18:00–21:00), 0.817 yuan in normal period (6:00–8:00, 11:00–18:00, and 21:00–22:00), and 0.305 yuan in valley period (22:00–6:00 the next day).

4.2 Device Model

In the case the cooling load is provided by adsorption chillers or ice-storage systems and electrical load is supplied by gas turbine generator or the grid, set energy $j = 1{:}3$ is electric, cooling, and gas; device $k = 1{:}4$ is the gas turbine, dual chiller in cooling, dual chiller in ice, and ice bank. All the device models are shown as follows:

1. **Gas turbine Model**
   Gas consume:

$$Q_{1,4}(i) = 0.333 * Q_{1,1}(i) + 239.2$$

   Heat recycling:

$$Q_{1,2}(i) = 2.56039 * Q_{1,1}(i) + 1{,}546.024$$

2. **Duplex-mode chillers**
   Cooling mode:

$$Q_{2,3}(i) = \left(10\left(Q_{2,1}(i)\right)^3 - 280\left(Q_{2,1}(i)\right)^2 + 23.64Q_{2,1}(i) + 0.8392\right)/1.15\; Q_{2,3}(i) \leq 100$$

Ice mode:

$$Q_{3,3}(i) = 0.565 * (10 * (Q_{3,1}(i))^3 - 280 * (Q_{3,1}(i))^2 + 23.64 Q_{3,1}(i) + 0.8392)\, Q_{3,3}(i) \le 100$$

3. **Ice bank**
   Maximum ice-storage constraint:

$$Q_{4,3}(i) \le 500$$

Ice-storage constraint:

$$Q_{4,3}(i) = Q_{4,3}(i-1) + \Delta Q_{4,3}(i)$$

Maximum melt constraint:

$$\Delta Q_{4,3}(i) \le 100$$

4.3 Results and Analysis

The numerical results obtained by the MINLP-based optimization tool of the case are shown in Fig. 2. The reader can notice that the operation is greatly affected by the electric price and the demand load in peak and valley price policy. The duplex system works at ice mode when there is no cooling demand during electric valley

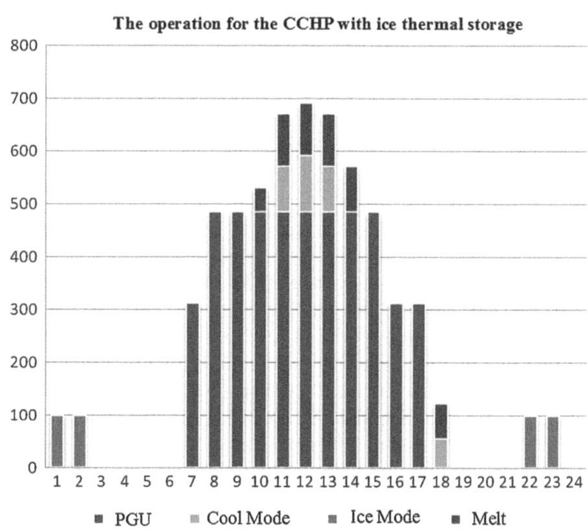

Fig. 2 The operation for the CCHP with ice thermal storage

price period. The prime mover works under the variable operating conditions if the cooling load is greater than the economic cooling load during the electricity price peak and flat period. The prime mover works under the variable operating conditions when the cooling demand is larger than the nominal load of prime mover. The ice bank system works at melt mode, and it is operating as a supplement to the prime mover where the cooling load is greater than the nominal load of prime mover and less than the sum of the nominal load of prime mover and melt load of ice thermal storage. The ice bank system works at melt mode, and the duplex chiller works at cooling mode where the cooling load is greater than the sum of the nominal load of prime mover and melt load of ice thermal storage. They are both operated as a supplement to the prime mover. They also work at the same mode when the cooling load is less than the nominal load of prime mover. From the figure, we can notice that the ice-storage system can reduce the cost of the CCHP system by a suitable operation.

5 Conclusion

To improve the economics of the energy system, an ice thermal storage system is introduced to the CCHP system. We also developed a MINLP optimization model for the systems operation. The MINLP model allows to deal with the system involving ice thermal storage. The model can handle units with nonlinear performance and solved by a MINLP optimizer. The case study shows that suggest that, under TOU conditions, the ice thermal storage system can effectively improve the economy of the CCHP system, and make the operation of the system more flexible.

Acknowledgments This work was supported in part by the Chinese Ministry of Housing and Urban-Rural Development under Grant 2013-K8-25, the Chinese Academy of Sciences Innovation Engineering under Grant KGCX2-EW-321 and the Strategic Priority Research Program of the Chinese Academy of Sciences, Grant No. XDA06000000.

References

Al-Abidi AA, Bin Mat S, Sopian K et al (2012) Review of thermal energy storage for air conditioning systems. Renew Sustain Energy Rev 16(8):5802–5819

Ameri M, Hejazi SH, Montaser K (2005) Performance and economic of the thermal energy storage systems to enhance the peaking capacity of the gas turbines. Appl Therm Eng 25(2):241–251

Barbieri ES, Melino F, Morini M (2012) Influence of the thermal energy storage on the profitability of micro-CHP systems for residential building applications. Appl Energy 97:714–722

Basecq V, Michaux G, Inard C et al (2013) Short-term storage systems of thermal energy for buildings: a review. Adv Build Energy Res 7(1):66–119

Gu W, Wu Z, Bo R et al (2014) Modeling, planning and optimal energy management of combined cooling, heating and power microgrid: a review. Int J Electr Power Energy Syst 54:26–37

Lofberg J (2004) YALMIP: a toolbox for modeling and optimization in MATLAB. In: IEEE international symposium on computer aided control systems design. IEEE, pp 284–289

Raine RD, Sharifi VN, Swithenbank J (2014) Optimisation of combined heat and power production for buildings using heat storage. Energy Convers Manag 87:164–174

Wang JJ, Jing YY, Zhang CF (2010) Optimization of capacity and operation for CCHP system by genetic algorithm. Appl Energy 87(4):1325–1335

Wu DW, Wang RZ (2006) Combined cooling, heating and power: a review. Prog Energy Combust Sci 32(5):459–495

Energy-Related Carbon Emissions in Shanghai: Driving Forces and Reducing Strategies

Chun-zeng Fan, Hai-ying Gu and Hong Jiang

Abstract This paper calculated the energy-related carbon emissions from production, household and energy transformation sectors in Shanghai and decomposed the effects of their changes in carbon emissions resulting from 11 causal factors of reflecting the changes in socioeconomic activity, intensity of energy and the structure by logarithmic mean divisia index. The results show that the changes of economic activity (EA), population size (PS), total energy consumption in transformation and energy consumption per capita (ECPC) increase CO_2 emissions obviously. The changes of energy intensity (EI), urban and rural population distribution structure, energy mix of household and mix of energy in transformation drive the decrease of CO_2 emissions. The changes of economic structure (ES), energy mix of production, and energy transformation structure (ETS) can't increase or decreased CO_2 emissions continuously in 3 periods respectively. Therefore, adjusting ES, ETS, energy mix of transformation and decreasing the EI of each production sector will be the main routes to reduce CO_2 emissions. Developing clean energy to substitute fossil energy and enforcement of carbon capture will be necessary in the future.

Keywords Energy · Carbon emissions · Factor decomposition · Shanghai

C. Fan (✉) · H. Gu
Antai College of Economics and Management, Shanghai Jiaotong University, Shanghai 200052, People's Republic of China
e-mail: cuzfan@sjtu.edu.cn

C. Fan
Economics Department, University of Western Ontario, London, ON N6A 5C2, Canada

H. Jiang
College of Social Science, Shanghai University, Shanghai 200444
People's Republic of China

S. Feng et al. (eds.), *Low-carbon City and New-type Urbanization*,
Environmental Science and Engineering, DOI 10.1007/978-3-662-45969-0_3

1 Introduction

Because the energy-related carbon emissions account for the main portion of global anthropogenic carbon emissions. The growth of energy consumption is the prime factor responsible for the increasing CO_2 emissions (Intergovernmental Panel on Climate Change (IPCC) 2007). As the world biggest CO_2 emitter, China emitted 8.9 billion metric tons CO_2, accounting for 27.09 % of global CO_2 emission in 2010 (Olivier et al. 2011). Reducing CO_2 emission to prohibit global warming is an important responsibility for every country including China. Therefore, Chinese government proposed reducing carbon emissions per unit GDP by 40–45 % in 2020 compared with 2005 levels in 2009. China's 12th FYP (2011–2015) proposed that energy consumption should be cut by 16 % per unit of GDP and cut 17 % CO_2 emissions by 2015. Cities are the most important energy consumer and CO_2 emitter, it is also CO_2 reducer supported by its abundance capital and technology.

As China's biggest metropolis, energy consumption and CO_2 emission in Shanghai has increased greatly as a result of rapid economic growth. In 2008, Shanghai was chose to be one of the two (Shanghai and Baoding) pilot cities of constructing low carbon city. In the implemented 12th FYP (2011–2015), one of the important goal of Shanghai was to reduce energy consumption and CO_2 emission. Therefore, studying the energy-related carbon emissions in Shanghai will both help the government make good policies and construct a successful low carbon model city for other regions to formulate.

So far, a lot of researchers have created a substantial amount of literature in this field. For example, Peters and Hertwich (2008) and Chen and Chen (2011a, b), Chen (2011) estimated the global energy-related carbon emissions. Paul et al. (2004), Wang et al. (2005), Tunc et al. (2009) studied national energy-related carbon emissions. Dhakal (2009), Zhao et al. (2009) and Zhou et al. (2010) studied this subject on different cities. However, these studies calculated the energy-related carbon emissions mainly according to 3 production sectors (primary industry, secondary industry, and tertiary industry) and 4 kinds of energy (coal, oil, natural gas and electricity).

Studies about Shanghai, such as Shao et al. (2011), Chen et al. (2013), Zhao and Tan (2010), Li et al. (2011), Liu et al. (2012), Steffen (2013) and so on, also considered a relatively small number (four or fewer) of types of energy. This decreases the accuracy of accounting, and only creates a coarse resolution.

Carefully studying of carbon emissions and the proportion of total emissions accounted for by each sector will do good to policy making and solve more practical problems.

There are a lot of methods of factor decomposition. LMDI is commonly used because of its ability to provide perfect decomposition without residual results and robust theoretical foundations, strong adaptability to a range of situations, e.g., Ang and Zhang (2000), Ang and Liu (2001), Ang (2004, 2005). A lot of Studies of factor decomposition use this method, e.g., Zhang et al. (2009), Oh et al. (2010), Liu (2007).

In this current research, the production sectors have generally been divided into three or four sectors, and the factors that influence carbon emissions have usually been divided into EA, ES, EI, and the energy mix consumed by each sector (Zhang et al. 2009; Oh et al. 2010). The household sector, composed by urban and rural sectors, influences carbon emissions by the changing of PS, of URPDS, of household ECPC, and of household energy mix Zhang et al. (2013). The energy transformation sector can be divided into 8 sub-sectors (Thermal Power, Heating Supply, Coal Washing, Cokeing, Petroleum Refineries, Gas Works, Natural Gas Liquefaction and Briquettes) (NBS DES, National Bureau of Statistics Department of Energy Statistics). Shanghai's energy-related CO_2 emission in the course of energy transformation mainly comes from the first 2 sectors. Therefore the CO_2 emission of energy transformation can be divided into 3 sectors: thermal power, heating supply and the other sector. In recent studies, the production sectors have been divided at a coarse resolution or at a detailed resolution on secondary industry (Zhao and Tan 2010) and haven't consider the energy transformation, while the data is relative older, and haven't given the main reasons of CO_2 emission in the energy transformation. So the reasons why carbon emissions continue to increase is unclear. In general, researchers have not provided data that can be directly used to guide the management of carbon emissions in Shanghai. With the rapid growth of population and economy, Shanghai's energy consumption and CO_2 emissions increase quickly. In order to decrease the CO_2 emissions, it is urgently necessary to conduct a full calculating of Shanghai's energy-related carbon emissions and find its main driving factors.

In this paper, we collected detailed raw data for a larger number of sectors and energy types than were used in previous research. The goal of this paper is to use LMDI to analyze: (1) Shanghai's total energy-related carbon emissions and its structure. (2) the main forces of driving the recent changes in Shanghai's energy-related carbon emissions. (3) the countermeasures to control Shanghai's energy-related carbon emissions. This will help the government design feasible and effective policies to control and reduce carbon emissions.

2 Data and Methodology

2.1 Data Collection

We contributed Shanghai's energy-related carbon emissions in energy transformation, energy consumption of the production and the household sectors. We defined 6 production sectors: primary industry (farming, forestry, animal husbandry, fisheries and water conservation; secondary industry; construction; TSPS (transportation, storage, and postal systems); WRTHR (wholesale, retail trade, hotel, and restaurant sector) and other production), 2 household sectors (urban households and rural households) and 3 transformation sectors (thermal power, heat supply and other energy transformation). Shanghai's main source of CO_2 emission in energy transformation are electricity, heat, coking, petroleum refineries, briquettes and gas works sectors. This paper divides the energy transformation into 3 sectors: electricity, heat and the other energy transformation sectors. We obtained data on the output values for the production sectors based on constant prices (with 2000 as the base year) and population data of urban and rural households from Shanghai Statistical Yearbooks from 1996 to 2010 (BMBS and NBS). To improve the resolution of the energy analysis, we defined 19 kinds of energy: raw coal, cleaned coal, other washed coal, charcoal briquettes, coke, coke oven gas, other gas, crude oil, gasoline, kerosene, diesel oil, fuel oil, liquefied petroleum gas (LPG), refinery gas, natural gas, other petroleum products, other coking products, heat and electricity. Energy consumption data are obtained from China Energy Statistical Yearbooks that contained data from 1991 to 2010 (NBS DES, National Bureau of Statistics Department of Energy Statistics). We adopted the final energy consumption from the energy-balance tables in these yearbooks as the energy consumption of the first 8 sectors. We adopted the net input (input–output) of energy transformation from the energy-balance tables in these yearbooks as the energy consumption of the other 3 sectors.

Energy consumption data are converted into coal equivalents based on the standard coal-equivalent coefficients obtained from China Energy Statistical Yearbook (NBS DES, National Bureau of Statistics Department of Energy Statistics). In this paper, we assumed complete combustion of energy sources to generate carbon emissions. Energy-related carbon emissions (C) can be calculated based on the following formula:

$$C = \sum_{i} E_i \times Y_i \tag{1}$$

where i denotes the different kinds of energy (i = 1, 2, 3, …, 19), E_i denotes the consumption of energy type i, and Y_i denotes the carbon-emission coefficient for energy type i. The carbon-emission coefficients for the 19 kinds of energy were obtained from IPCC (2006). Since the carbon emissions from the processes of electricity and heat generation from fossil energy were already calculated, and consumption of electricity and heat does not cause carbon emissions. This paper assumed the carbon emission coefficients for electricity and heat were 0. The carbon emission coefficients for the output fossil energy by input in the transformation sectors were also 0, because they hadn't been burned and emitted CO_2. Non-energy use of fuels (i.e., as industrial materials) was subtracted from the final energy consumption of secondary industry.

2.2 LMDI Model

The aim of this study is to calculate related-energy carbon emissions of the 11 key sectors and decomposed the changes of CO_2 emissions into several factors. Considering the different characteristics of carbon emissions in the course of production, household consumption and energy transformation, we analyzed them separately.

EI was mainly affected by the technology level of the energy utilization processes the energy consumption structure, the energy consumption quantity and the mixture of energy.

Energy transformation is also an important source of CO_2 emission, including total energy transformation (TET), energy transformation structure (ETS), and energy mix of transformation (EMT).

Using LMDI model, we decomposed energy-related carbon emissions from production, households and energy transformation into the following 11 driving factors: (1) economic activity (EA) (i.e., GDP), (2) economic structure (ES), (3) energy intensity (EI) (i.e., energy consumption per unit of output), (4) the energy mix of the production sectors (EMP) (i.e., the proportion of total energy provided by each of the 19 fuels), (5) population size (PS), (6) the urban and rural population distribution structure (URPDS) (i.e., the proportions of total population accounted for by urban and rural households), (7) energy consumption per capita (ECPC), (8) the energy mix consumed by the two household sectors (EMH), (9) TET, (10) ETS and (11) EMT. According to the model developed by Ang (2005), Shanghai's energy-related carbon emissions (C) can be formulated as follows:

$$
\begin{aligned}
C &= \sum_{ij} \mathrm{GDP} \times \frac{Q_i}{\mathrm{GDP}} \times \frac{E_i}{Q_i} \times \frac{E_{ij}}{E_i} \times \frac{C_{ij}}{E_{ij}} + \sum_{ki} P \times \frac{P_k}{P} \\
&\quad \times \frac{E_k}{P_k} \times \frac{E_{ki}}{E_k} \times \frac{C_{ki}}{E_{ki}} + \sum_{ni} E_{\mathrm{tr}} \times \frac{E_n}{E_{\mathrm{tr}}} \times \frac{E_{ni}}{E_n} \times \frac{C_{ni}}{E_{ni}} \\
&= \sum_{ij} \mathrm{GDP} \times S_i \times \mathrm{EI}_i \times \mathrm{ES}_{ij} \times Y_{ij} + \sum_{ij} P \times S_k \\
&\quad \times \mathrm{EI}_k \times \mathrm{ES}_{ki} \times Y_{ii} + \sum_{ni} E_{\mathrm{tr}} \times S_n \times \mathrm{ES}_{ni} \times Y_{ni}
\end{aligned}
\qquad (2)
$$

In the formula (2), C_{ij} denotes the CO_2 emissions caused by consumption of energy type i in production sector j; C_{ik} denotes the CO_2 emissions caused by consumption of energy type i by urban or rural households (k); E_j denotes the energy consumption by production sector j; E_{ij} denotes the consumption of energy type i by production sector j; E_k denotes the energy consumption by urban or rural households (k); E_{ik} denotes the consumption of energy type i by urban or rural households (k); EI_j denotes the EI of production sector j; E_k denotes ECPC (k); ES_{ij} denotes the proportion of TEC by production sector j accounted for by consumption of energy type, that is, energy mix of production; ES_{ik} denotes the proportion of total energy consumption by urban or rural domestic households (k) accounted for by consumption of energy type i, that is, energy mix of household consumption; E_{tr} denotes TET, $E_{\mathrm{tr}n}$ denotes the energy transformation of n sector, $E_{\mathrm{tr}ni}$ denotes the transformation of sector n accounted for by consumption of energy type i; GDP denotes the gross domestic product in constant (2000) prices; i denotes the different types of energy (i = 1, 2, 3, …, 19); j denotes the production sectors (j = 1, 2, 3, …, 6); k denotes the household sectors (k = 1, 2); n denotes energy transformation senators (n = 1, 2, 3); P denotes the total population; P_k denotes the urban or rural population (k); Q_j denotes the output value of production sector j; Y_{ij} denotes the carbon-emission coefficient for energy type i in production sector j; Y_{ik} denotes the carbon-emission coefficient for energy type i of urban or rural households; Y_{trin} denotes the carbon-emission coefficient for energy type i in transformation sector j; S_j denotes the proportion of GDP accounted for by the output value for production sector j, that is, ES; S_k denotes the proportion of the total population accounted for by urban and rural populations (k); E_{tr} represents the total net fossil input by energy transformation; $ES_{\mathrm{tr}ni}$ denotes the ETS; $ES_{\mathrm{tr}ni}$ denotes EMT. We assumed that the carbon-emission coefficients for each energy source was constant, thus the carbon-emission intensify of a certain energy did not change. The carbon-emission change (ΔC) from year 0 to year T can be expressed as follows:

$$\begin{aligned}\Delta C = C_T - C_0 = \Delta C_{\mathrm{GDP}} + \Delta C_{S_j} + \Delta C_{\mathrm{EI}_j} + \Delta C_{\mathrm{ES}_{ij}} + \Delta C_P \\ + \Delta C_{S_K} + \Delta C_{\mathrm{EI}_k} + \Delta C_{\mathrm{ES}_k} + \Delta C_{E_{\mathrm{tr}}} + \Delta C_{\mathrm{EI}_{\mathrm{trn}}} + \Delta C_{\mathrm{ES}_{\mathrm{trn}}}\end{aligned} \tag{3}$$

where ΔC denotes the change in CO_2 emissions from year 0 to year T; C_T denotes the CO_2 emissions in year T; C_0 denotes the CO_2 emissions in year 0; ΔC_{GDP} denotes the impact of changes in EA, ΔC_{S_j} denotes the impact of changes in ES, ΔC_{EI_j} denotes the impact of changes in EI, $\Delta C_{\mathrm{ES}_{ij}}$ denotes the impact of changes in the EMP, ΔC_P denotes the impact of changes in PS, ΔC_{S_K} denotes the impact of changes in the URPDS, ΔC_{EI_k} denotes the impact of changes in ECPC, ΔC_{ES_k} denotes the impact of EMH, $\Delta C_{E_{\mathrm{tr}}}$ Denotes the impact of changes in the TEC, $\Delta C_{\mathrm{EI}_{\mathrm{trn}}}$ denotes the impact of changes of the ES; $\Delta C_{\mathrm{ES}_{\mathrm{tr}ni}}$ denotes the impact of changes in the EMT. According to the additive decomposition of LMDI, formula (3) can be reformatted as follows:

$$\Delta C_{\mathrm{GDP}} = \sum_{ij} \frac{C_{ij}^T - C_{ij}^0}{\ln C_{ij}^T - \ln C_{ij}^0} \ln\left(\frac{\mathrm{GDP}^T}{\mathrm{GDP}^0}\right) \tag{4}$$

$$\Delta C_{S_j} = \sum_{ij} \frac{C_{ij}^T - C_{ij}^0}{\ln C_{ij}^T - \ln C_{ij}^0} \ln\left(\frac{S_i^T}{S_i^0}\right) \tag{5}$$

$$\Delta C_{\mathrm{EI}_j} = \sum_{ij} \frac{C_{ij}^T - C_{ij}^0}{\ln C_{ij}^T - \ln C_{ij}^0} \ln\left(\frac{\mathrm{EI}_i^T}{\mathrm{EI}_i^0}\right) \tag{6}$$

$$\Delta C_{\mathrm{ES}_{ij}} = \sum_{ij} \frac{C_{ij}^T - C_{ij}^0}{\ln C_{ij}^T - \ln C_{ij}^0} \ln\left(\frac{\mathrm{ES}_{ij}^T}{\mathrm{ES}_{ij}^0}\right) \quad (7)$$

$$\Delta C_P = \sum_{ki} \frac{C_{ik}^T - C_{ik}^0}{\ln C_{ik}^T - \ln C_{ik}^0} \ln\left(\frac{P^T}{P^0}\right) \quad (8)$$

$$\Delta C_{S_k} = \sum_{ki} \frac{C_{ik}^T - C_{ik}^0}{\ln C_{ik}^T - \ln C_{ik}^0} \ln\left(\frac{S_K^T}{S_K^0}\right) \quad (9)$$

$$\Delta C_{\mathrm{ES}_{ik}} = \sum_{ki} \frac{C_{ik}^T - C_{ik}^0}{\ln C_{ik}^T - \ln C_{ik}^0} \ln\left(\frac{\mathrm{ES}_{Ki}^T}{\mathrm{ES}_{Ki}^0}\right) \quad (10)$$

$$\Delta C_{\mathrm{EI}_k} = \sum_{ki} \frac{C_{ik}^T - C_{ik}^0}{\ln C_{ik}^T - \ln C_{ik}^0} \ln\left(\frac{\mathrm{EI}_K^T}{\mathrm{EI}_K^0}\right) \quad (11)$$

$$\Delta C_{E_{\mathrm{tr}}} = \sum_{ni} \frac{C_{ni}^T - C_{ni}^0}{\ln C_{ni}^T - \ln C_{ni}^0} \ln\left(\frac{E^T}{E^0}\right) \quad (12)$$

$$\Delta C_{EI_{\mathrm{tr}}} = \sum_{ni} \frac{C_{ni}^T - C_{ni}^0}{\ln C_{ni}^T - \ln C_{ni}^0} \ln\left(\frac{E_n^t}{E_n^0}\right) \quad (13)$$

$$\Delta C_{ES_{\mathrm{tr}}} = \sum_{ni} \frac{C_{ni}^T - C_{ni}^0}{\ln C_{ni}^T - \ln C_{ni}^0} \ln\left(\frac{E_{ni}^T}{E_{ni}^0}\right) \quad (14)$$

In these formulas, superscripts 0 and T denote the values for year 0 and year T respectively.

3 Results

3.1 Figures and Tables

Table 1 indicates that energy-related carbon emissions in Shanghai have continued to increase. From 1995 to 2000, carbon emissions increased by 646.02 × 10^4 t, an average annual growth rate of 3.91 %. From 2000 to 2005, emissions increased by 1054.15 × 10^4 t, an average annual growth rate of 5.13 %. From 2005 to 2010, emissions increased by 1475.18 × 10^4 t, an average annual growth rate of 5.857 %. Therefore, Shanghai's energy-related carbon emissions increases quickly. In the different sectors, the secondary industry, the TSPS and thermal power sector emitted over 80 % CO_2 in the year 1995, 2000, 2005 or 2010. The emissions of the

Table 1 Trends of energy-related carbon emissions in Shanghai ($\times 10^4$ t)

	1995	2000	2005	2010
Primary industry	21.94	44.89	47	26.05
Secondary industry	1,320.77	1,554.97	1,404.33	1,612.62
Construction	11.24	40.95	83.41	83.95
TSPS	163.33	336.36	799.32	1,169.38
WRTHR	16.82	28.77	91.05	161.64
Other production sectors	35.99	78.48	106.34	232.81
Urban households	57.84	87.45	108.41	196.27
Rural households	72.09	61.91	59.9	69.54
Thermal power	1,010.8	1,312.63	1,782.95	2,250.85
Heat supply	144.15	128.79	185.99	209.95
Other energy transformation sectors	204.86	30.84	91.58	221.9
Total	3,059.85	3,706.02	4,760.28	6,234.96

secondary industry show a decreasing trend, while the emissions of TSPS and thermal power show increasing trends.

During the 3 periods, the average annual growth rates of the TSPS's carbon emissions were 15.54, 18.90, and 7.91 % respectively. The average growth rates of construction's carbon emissions were 29.51, 15.29, and 0.13 %, respectively. And the average growth rates of the WRTHR were 11.33, 25.91, and 12.16 %, respectively.

These 3 sectors had the fastest-growing carbon emissions among the production sectors. The composition of Shanghai's energy-related carbon emissions (Fig. 1) shows that the proportion of total carbon emissions accounted for by the emissions from secondary industry fell sharply. The proportion of total carbon emissions from the power sector accounted for over 30 %. The proportion of emissions by rural households sector and heat supply sector decreased.

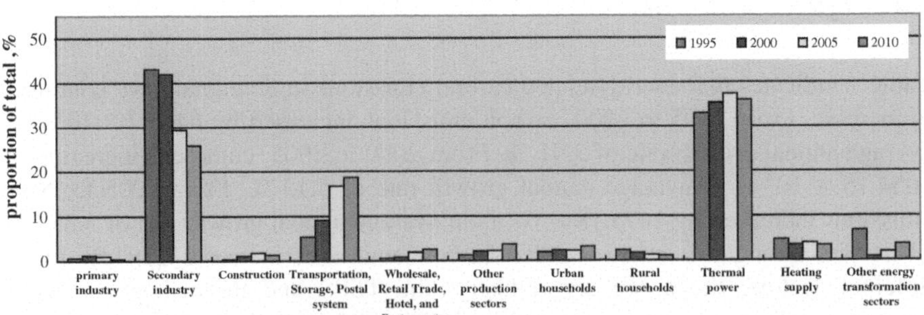

Fig. 1 Trends of Shanghai's total energy-related carbon emissions

The proportion of emissions by the TSPS and WRTHR increased. While, the proportion of emissions by the other production sectors, urban household sector and coking sector increased.

The proportions of emissions by primary industry and construction initially increased and then decreased.

The proportion of emissions by thermal power sector fluctuates repeatedly. TSPS gradually became one of the major sources of carbon emissions, increasing from 5.34 % in 1995 to 18.76 % of the total in 2010. The Secondary industry emitted 43.18 % of total CO_2 emissions in 1995, decreased to 25.86 % in 2010. The thermal power sector emitted 33.04 % of total CO_2 emissions in 1995, decreased to 36.10 % in 2010.

In addition, the proportion of emissions from the other production increased from 1.187 % in 1995 to 3.73 % in 2010. Although the proportions of emissions by urban and rural households did not exceed 5 % in any year, the proportions of emissions by urban households increased from 1.89 % in 1995 to 3.15 % in 2010. The growth of emissions by urban households should be closely monitored.

The heat supply sector and the other energy transformation sectors emitted 4.17 and 6.70 % of the total CO_2 emissions in 1995, and decreased to 3.37 and 3.56 % in 2010 respectively.

3.2 *Driving Forces of Increasing Energy-Related Carbon Emission*

In the 3 study periods, EA, PS, ECPC and the total energy consumption in transformation were stimulatory factors that increased emissions (Fig. 2). Industry structure, EI, the population of urban and rural distribution, and the mix of energy consumption in transformation obviously decreased CO_2 emission. The energy mix of household sector was an inhibitory factor during the first and second periods and a stimulatory factor during the third period. The energy consumption structure in transformation was inhibitory factor during the first and third periods and a stimulatory factor during the second period (Fig. 2).

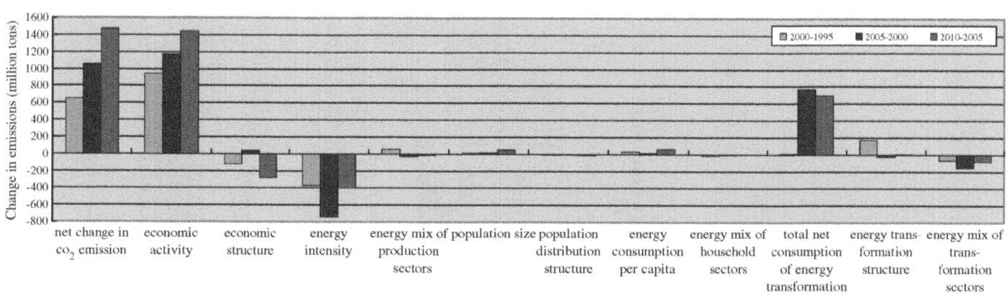

Fig. 2 Emissions changes of different sectors

3.2.1 The Main Driving Forces of Increasing CO_2 Emission

During each of the 3 periods, Shanghai's carbon emissions increased as a result of economic growth, energy transformation enlargement, population increase, and ECPC growth. However, the stimulatory effects of these factors on carbon emissions were different. EA was the dominant stimulatory factor, but the effects of PS and ECPC is smaller. The EA factor contributed to 946.06×10^4, 1172.39×10^4, and 1448.41×10^4 t carbon emissions, respectively, during the three study periods (Fig. 2). Shanghai's GDP increased from 265.83×10^9 RMB in 1995 to 1315.35×10^9 RMB in 2010 (an increase of 394.80 %), with annual growth rates during the three periods of 11.25 %. Over 10 % annual economic growth rate inevitably let the EA become the dominant factor in stimulating carbon emissions.

The growth of energy transformation increased 7.58×10^4, 767.65×10^4, and 697.23×10^4 t carbon emissions respectively in the 3 periods. The contributions of PS to changes in carbon emissions during the three periods were 15.16×10^4, 17.41×10^4, and 49.69×10^4 t respectively (Fig. 2).

The population of Shanghai increased from 14.14×10^6 persons in 1995 to 23.03×10^6 persons in 2010, with average annual growth rates during the three periods of 2.61, 2.38, and 4.94 %, respectively. Because the population growth rate increased throughout the whole study period, PS played an increasingly important role in stimulating carbon emissions. The contributions of ECPC to changes in carbon emissions during the three periods were 29.17×10^4, 15.60×10^4, and 60.74×10^4 t, respectively (Fig. 2). Urban and rural household ECPC increased throughout the study period (Fig. 3). Urban ECPC increased from 0.0784 tce/person in 1995 to 0.1824 tce/person in 2010, an increase of 1.4 times, while rural ECPC increased from 0.2463 tce/person in 1995 to 0.4178 tce/person in 2010, an increase of 69.63 % times. Obviously urban and rural ECPC increased, and stimulated the carbon emissions obviously during the studying periods.

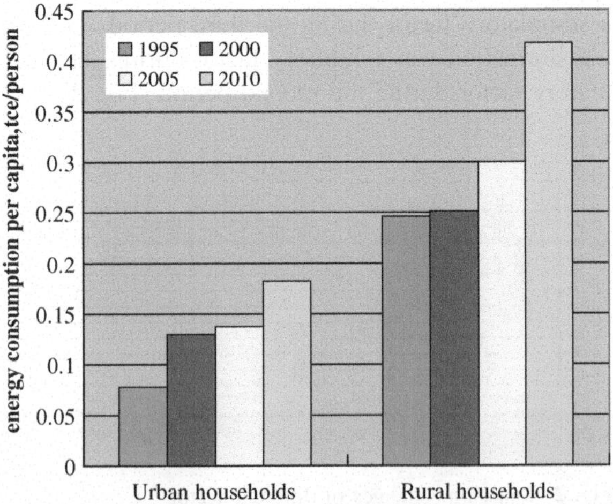

Fig. 3 Trends in energy consumption per capita of Shanghai's households

In the energy transformation, the main material is still raw coal, and its consumption increased continuously. This increased CO_2 emission greatly.

3.2.2 The Main Driving Forces of Reducing CO_2 Emission

Figure 2 shows that EI, URPDS, EMT and EMH are the main driving forces of reducing CO_2 emission. EI is the dominant inhibitory factor. In the 3 study periods, changes of EI bring to carbon emissions reduction of 373.49×10^4, 742.49×10^4 and 398.32×10^4 t respectively (Fig. 2).

Figure 4 demonstrates the trends for the EI of production sectors in Shanghai. The EI of secondary industry decreased significantly, from 1.56 tce/10^4 RMB in 1995 to 0.38 tce/10^4 RMB in 2010 (a decrease of 76.92 %).

Figure 4 also shows that the energy intensities of primary industry and construction increased first and then decreased slightly, whereas the energy intensities of TSPS and WRTHR increased quickly. The energy intensities of the other sectors show an increase-decrease-increase trend.Although the proportion of total energy consumption accounted for by secondary industry fell from 1995 to 2010. It still consumed about a quarter of the total energy. Thermal power sector consumes over 30 % total energy, and it showed an increasing trend (Fig. 5). Such that, these decreased EI of secondary industry played a key role in decreasing carbon emissions. Among these production sectors, the secondary industry, TSPS, and primary industry have higher energy intensities. While the other 3 production sectors have lower energy intensities.

Therefore, we should pay special attention to the sectors with higher energy intensities: secondary industry, TSPS, and primary industry. Thermal power sector emitted over 30 % of the total CO_2 emissions. It also needs to be paid more attention.

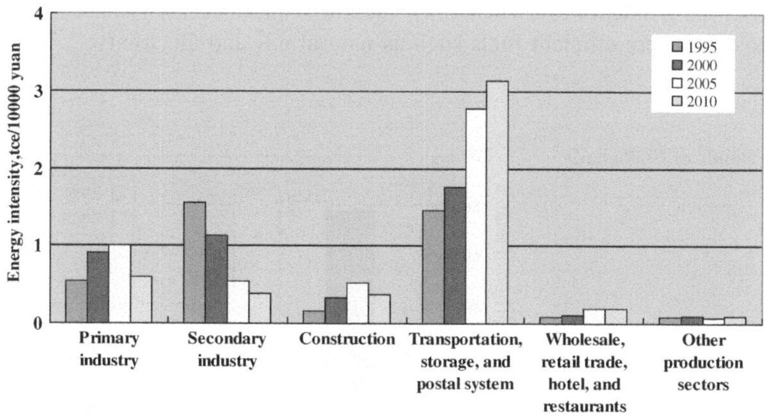

Fig. 4 Trends of energy intensify for Shanghai's production sectors

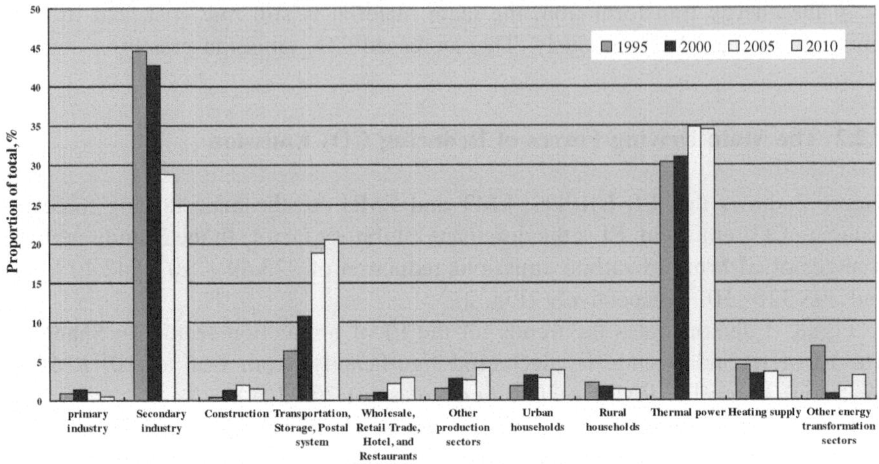

Fig. 5 Trends of total energy consumption by Shanghai's 11 sectors

During the 3 study periods, the changes of URPDS reduced 9.30×10^4, 6.10×10^4, and 11.74×10^4 t carbon emissions respectively (Fig. 2).

Total population in Shanghai increased continuously, the ECPC of rural households was higher than that of urban households. The proportion of rural households decreased (Fig. 6). Due to the need to provide part of agriculture products and retain some rural land as an ecological buffer for the city, it isn't practical to reduce carbon emissions by further decreasing the rural proportion of Shanghai's total population. Emission reduction should rely on reducing the ECPC and the clean energy alternatives both for urban and rural households.

In the household sectors, the proportion of urban household energy consumption accounted for by raw coal decreased and the proportions accounted for by gasoline and natural gas increased. This change is the result of government efforts to provide incentives for residents to switch from fuels that produce high carbon emissions (e.g., coal) to more efficient fuels such as natural gas and electricity.

Fig. 6 Trends of Shanghai's URPDS

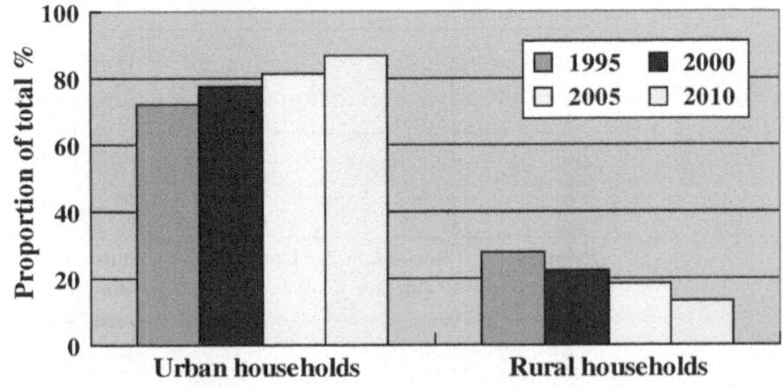

The energy consumption of households changed from raw coal, natural gas and electricity into natural gas, gasoline diesel oil and electricity. The net result is that the energy mix consumed by the 2 household sectors became less carbon-intensive. During the 3 study periods, the changes of household energy mix reduces 15.04×10^4, -8.08×10^4, and -1.12×10^4 t CO_2 emission respectively (Fig. 2). The inhibitory effect of the energy mix consumed by the household sectors was becoming small in the 3 study periods.

Due to the energy mix improving, the changes of EMT brings 66.00, 155.57 and 80.74 carbon reduction, while the changes of EMH brings 15.04, 8.08 and 1.12 t carbon reduction respectively during the 3 study periods. EMT and EMH improvement act as 2 of the important factors to reduce CO_2 emission in Shanghai.

3.2.3 The Fluctuating Driving Forces of CO_2 Emission

During the 3 study periods, the contribution of changes in ES to the changes in carbon emissions fluctuated, from reduction of 117.12×10^4 t during the first period to increase of 42.13×10^4 t during the second period and finally to a decrease of 275.93×10^4 t during the last period (Fig. 2). The proportion of GDP accounted for by the output value of primary industry increased and then decreased, whereas the proportion of GDP accounted for by secondary industry, construction, transportation, storage and postal system sector, WRTHR and other production sectors increased continuously (Fig. 7).

The results also show that the energy intensities of secondary industry, of TSPS, and of the primary industry were greater than those of the construction, WRTHR, and the other production sectors (Fig. 4).

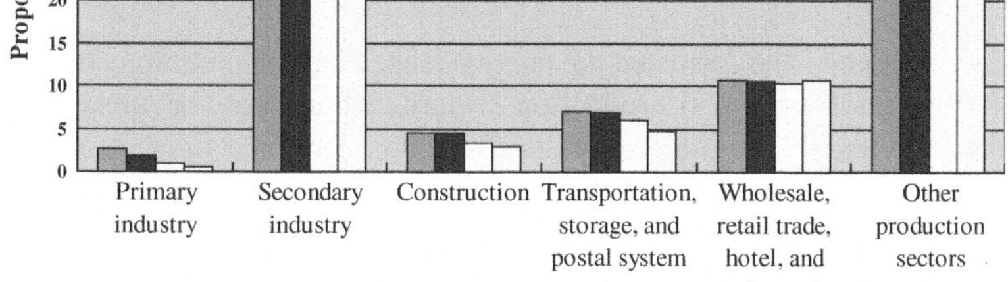

Fig. 7 Trends of Shanghai's economic structure

The ES of Shanghai has been adjusted since 1995, the proportions of GDP accounted for by the primary sector, by the transportation, storage and postal sector, and by construction decreased. While the proportions accounted for by secondary industry, by WRTHR, and by the other production sectors showed fluctuation. Because the adjustments of the ES sometimes increased and sometimes decreased carbon emissions, the proportion accounted for by secondary industry fluctuated, which causes the synchronous fluctuations in sector's effect on carbon emissions.

During the 3 study periods, the contributions of changes in EMP to the changes in carbon emissions were 58.51×10^4, -25.60×10^4, and -19.12×10^4 t, respectively (Fig. 2).

The main changes in EMP are that the proportion of secondary industry's energy consumption accounted for by raw coal firstly increased quickly and then slowly. The proportion of energy consumption by the other production sectors accounted for by raw coal decreased quickly, while the proportions accounted for by natural gas and electricity surged. And the proportion of energy consumption by WRTHR accounted for by LPG and heat increased.

The carbon-emission coefficients are highest for coal, followed by oil and various forms of gas. The coefficients for electricity and heat were assumed to be 0 in this study. In the energy transformation sectors, if the coal or oil has not combusted and hasn't given off CO_2, then transformation is just from one coal and oil product into another coal or oil products, the coefficients for the transformed coal products and oil products are 0. Because the proportions accounted for by energies with higher carbon-emission coefficients decreased and the proportions accounted for by energies with lower coefficients increased, this led to the less carbon-intensive EMP.

The adjustment of the energy mix consumed by the production sectors strengthened throughout the study period, so the inhibitory effect of the energy mix continued to increase (Fig. 7).

Due to the unstable energy demands, the changes of ETS bring to 170.50 and 6.08 t carbon emission increase in the first and third periods, but bring to 23.39 t carbon emission reduction in the second period. Therefore, ETS is one of the fluctuating driving forces of CO_2 emission.

4 Conclusions

This paper estimated and analyzed the energy-related carbon emissions by using the data of 19 type of energy, 6 production sectors, 2 household sectors, 3 transformation sectors and 11 driving factors in Shanghai. That is a finer resolution than that of the previous studies. It will provide a scientific basis of designing emission-reduction measures for policy-makers.

From 1995 to 2010, Shanghai's energy-related carbon emissions increased continuously, which is mainly caused by production, household and energy transformation sectors. Secondary industry, thermal power and TSPS accounted for a large proportion of total carbon emissions. Changes in EA, PS, ECPC and TET stimulated emissions throughout the study period. Whereas changes in EI, the URPDS, EMT and EMH decreased emissions. In addition, changes in the ES, EMP and ETS had fluctuating effects on emissions. Changes in EA belong to the dominant stimulatory factor. However changes in EI belong to the dominant inhibitory factor. The EI of secondary industries decreased continuously, which strongly reduced the CO_2 emissions. Changes in the ES played a strong role in reducing carbon emissions from 2005 to 2010. Changes in EMH decreased emissions slightly. Changes in the energy mix consumed in the energy transformation obviously decreased the CO_2 emission. The effect of changes in the EMP show a weak increasing trend.

The results show that the decrease of CO_2 emission caused by the changes of EI, URPDS, and EMT can not offset the CO_2 emission increase caused by EA, PS, ECPC and TET. The total CO_2 emission increased quickly in the studying periods.

Therefore, the main approaches to reduce Shanghai's carbon emissions in the future will be continuing to reducing the EI of the production sectors, to adjust the ES to favor less energy-intensive sectors, to improve energy transformation efficiency and to improve the energy mix to favor energy sources with lower carbon emissions per unit of energy consumed and enforcement of carbon capture. Certainly, enhancing clean energy alternative to fossil energy and carbon capture will be the fundamental way of reducing carbon dioxide emissions.

This study can provide a more scientific basis for policy-makers to develop policies to reduce carbon emissions due to the analysis with finer resolution. Certainly, there is still room for improvement in the study. For example, the production and energy transformation sectors were divided more coarsely than is desirable because of no enough available data for higher resolution. This study assumed that all combustion processes are 100 % efficient, this was unrealistic. Among the carbon emissions from energy transformation processes, we only considered emissions from electricity, heat generation and the other that was divided more coarsely too.

In the future research, we plan to divide the production and energy transformation sectors more finely as data with higher resolution becomes available, incorporate a coefficient to account for incomplete combustion, and add carbon emissions from other energy transformation processes (e.g., petroleum refineries, gas work). Moreover, carbon emissions result from many processes (e.g., iron and steel smelting, non-ferrous metal smelting, cement production, et al.), and other indirect emission from electricity and heat should be included in the analysis.

Acknowledgments This research is supported by National Science Foundation of China (No. 71333010, 70973076) and National Social Science Foundation of China (No. 11AZD080).

References

Ang BW (2004) Decomposition analysis for policymaking in energy: which is the preferred method? Energy Policy 32(9):1131–1139

Ang BW (2005) The LMDI approach to decomposition analysis: a practical guide. Energy Policy 33(7):867–871

Ang BW, Liu FL (2001) A new energy decomposition method: perfect in decomposition and consistent in aggregation. Energy 26(6):537–548

Ang BW, Zhang FQ (2000) A survey of index decomposition analysis in energy and environmental studies. Energy 25(12):1149–1176

BMBS and NBS (1996–2010) Shanghai Statistical Yearbook. China Statistics Press, Beijing

Chen GQ (2011) An overview of energy consumption of the globalized world economy. Energy Policy 39(10):5920–5928

Chen GQ, Chen ZM (2011a) Greenhouse gas emissions and natural resources use by the world economy: ecological input–output modeling. Ecol Model 222(14):2362–2376

Chen ZM, Chen GQ (2011b) Embodied carbon dioxide emission at supra-national scale: a coalition analysis for G_7, BRIC, and the rest of the world. Energy Policy 39(5):2899–2909

Chen F et al (2013) Theoretical research on low-carbon city and empirical study of Shanghai. Habitat Inter 37:33–42

Dhakal S (2009) Urban energy use and carbon emissions from cities in China and policy implications. Energy Policy 37(11):4208–4219

Intergovernmental Panel on Climate Change (IPCC 2006) Guidelines for national greenhouse gas inventories: vol. 2 Energy. Available at http://www.ipcc-nggip.iges.or.jp/public/006gl/ol2.tml

IPCC (Intergovernmental Panel on Climate Change). Climate Change 2007: Synthesis Report. Geneva, Switzerland

Li L, Chen C et al (2011) Energy demand and carbon emissions under different development scenarios for Shanghai, China. Energy Policy 38:4797–4807

Liu CC (2007) An extended method for key factors in reducing CO_2 emissions. Appl Math Comput 89(1):440–451

Liu Z, Liang S, et al (2012) Features, trajectories and driving forces for energy-related GHG emissions from Chinese mega cites: the case of Beijing, Tianjin, Shanghai and Chongqing, Energy 37:245–254

National Bureau of Statistics Department of Energy Statistics (NBS DES) (1991–2010) China Energy Statistical Yearbook. China Statistics Press, Beijing

Oh I, Wehrmeyer W, Mulugetta Y (2010) Decomposition analysis and mitigation strategies of CO_2 emissions from energy consumption in South Korea. Energy Policy 38(1):364–377

Olivier JGJ, Janssens MG, Peters JAHW, Julian W (2011) Long-term trend in global CO_2 missions 2011 report, The Hague: PBL/JRC

Paul et al (2004) Decomposition of energy and CO_2 intensities of Thai industry between 1981 and 2000. Energy Econ 26(5):765–781

Peters GP, Hertwich EG (2008) CO_2 embodied in international trade with implications for global climate policy. Environ Sci Technol 42(5):1401–1407

Shao S et al (2011) Estimation, characteristics, and determinants of energy-related industrial CO2 emissions in Shanghai, 1994–2009. Energy Policy 39:6476–6494

Steffen L (2013) Low-to-no carbon city: lessons from western urban projects for the rapid transformation of Shanghai. Habitat Int 37:61–69

Tunc GI, Türüt-Asik S, Akbostancı E (2009) A decomposition analysis of CO_2 emissions from energy use: Turkish case. Energy Policy 37(11):4689–4699

Wang C, Chen JN, Zou J (2005) Decomposition of energy-related CO2 emission in China: 1957–2000. Energy 30(1):73–83

Zhang JY et al (2013) Estimation of energy-related carbon emission in Beijing and factor decomposition analysis. Ecol Model 252:258–265

Zhang M, Mu HL, Ning YD, Song YC (2009) Decomposition of energy-related CO_2 emission over 1991–2006 in China. Ecol Econ 68(7):2122–2128

Zhao M, Tan L (2010) Decomposing the influencing factors of industrial carbon emissions in Shanghai using the LMDI method. Energy 35:2505–2510

Zhao M, Zhang WG, Yu LZ (2009) Carbon emissions from energy consumption in Shanghai City. Res Environ Sci 22(8):984–989 (in Chinese)

Zhou SY, Chen H, Li SC (2010) Resources use and greenhouse gas emissions in urban economy: ecological input–output modeling for Beijing 2002. Commun Nonlinear Sci Numer Simul 15 (10):3201–3231

Renewable Energy Investment Project Evaluation Model Based on Improved Real Option

Chun-yan Dai, Yi-xian Wang, Dong Li and Yan-ling Zhou

Abstract This study constructed a real option valuation model that the risk could be adjustable based on the characteristics of renewable energy projects and the idea of the Capital Asset Pricing Model and used this model to valuate the renewable energy projects considering selling green electricity certificate revenue. And then, it verified the effectiveness and rationality of this model through the investment projects of the wind power.

Keywords Capital Asset Pricing Model · Real option · Renewable energy

1 Introduction

As the main body of the social economy, enterprise plays a vital role in the renewable energy popularization and the industry development. For the renewable energy, power generation industry in China is in the early stages of the industry, and there are still many uncertainties. All of those factors, such as policy, economy, technology, energy, and resources, will affect the evaluation of the value of renewable energy power generation projects. The scientific and prudential evaluation of the project value will be related to the interests and development of the enterprises and the whole society. Traditional investment decision methods are

---

C. Dai (✉)
Energy Management and Low Carbon Development Research Center, Chongqing Technology and Business University, Chongqing 400067, People's Republic of China
e-mail: daichunyan@tsinghua.org.cn

C. Dai · D. Li · Y. Zhou
Management Department, Chongqing Technology and Business University, Chongqing 400067, People's Republic of China

Y. Wang
Engineering Center, Chongqing Technology and Business University, Chongqing 400067, People's Republic of China

S. Feng et al. (eds.), *Low-carbon City and New-type Urbanization*,
Environmental Science and Engineering, DOI 10.1007/978-3-662-45969-0_4

difficult to accurately assess the value of the project due to the uniqueness of energy and its industry development. Real option theory which is a quantitative tool of the option value of investment project makes an effective mining to the value of the management flexibility and the strategic value of the investment. At present, there have been some researches which make an application of real option concept in making an evaluation of renewable energy investment project in foreign. Lee and Shih (2010) made a point that "the policy effectiveness evaluation model" integrated the cost-effective curve which was brought by renewable energy technology innovation with the real options model. The model quantified the effectiveness of the policy through the analysis of the uncertainty in fossil fuel prices and policy-related factors and studied the current renewable energy policies in Taiwan through empirical analysis (Lee and Shih 2010). Through a study of ethanol-expanded reproduction investment, Pederson and Zou (2009) established a model which embedded the risk factors of market price in investment decision and made use of the available historical market price data, using real options method and Monte Carlo simulation method to evaluate investment decisions of expanding ethanol reproduction problems (Pederson and Zou 2009). Cheng et al. (2011) revealed the changes about clean energy investment strategic choices and strategic value which is due to the lead time of power generation equipment investment and uncertainty of future electricity demand with the binomial real options model which is based on continuous compound option. However, the current studies had a lack of analysis about systemic risks which cannot be distributed outside the enterprise. In general, we always directly divide the value into two parts (the discounted value of future cash flows and option value of investment projects); when we use the real options to valuate the value of investment projects, we mainly use the risk-free rate to discount future earnings without taking the market risk factors that cannot be dispersed into account, and to some extent, resulting in partial distortion about the option value of the project. This paper which is based on analyzing the investment characteristics of renewable energy projects and drawing the idea about Capital Asset Pricing Model brings the non-dispersed systemic risks into the assessment of the project value and builds an improved real option valuation model with the risk adjusted. Then, this paper will make an evaluation of renewable energy investment projects which are under the quota system; we can verify the effectiveness and feasibility of the model through analyzing a typical case of wind power industry and then make some proposals on the investment and operation of renewable energy projects.

2 Model Construction

2.1 Investment Characteristics of Renewable Energy Project

According to the provisions on the type of renewable energy under the "Renewable Energy Law" and "renewable energy quota management approach (draft)," renewable energy generation projects mainly refers to wind power, solar power, biomass power, geothermal, ocean energy power generation, and other non-hydro-renewable energy generation projects. There are obvious differences between renewable energy generation project investments and traditional thermal power project investments. Firstly, renewable energy generation project has a characteristic of investment irreversibility. Because of the higher degree of specialization of renewable energy generation equipment, the majority or all of the initial investment may become a sunk cost as long as making a decision to invest in a renewable energy project that makes the investment irreversible. Secondly, renewable energy return on investment is more uncertain. Many factors will have an enormous impact on the financial income of renewable energy power generation projects, such as policy changes, international and domestic market environment, and technological innovation. Huge differences may exist between the actual yield and the expected yield of investments in renewable energy projects since there are many uncertainties. On the one hand, the investment may generate spillover benefits in a good external environment and internal management; on the other hand, if the risk brings an assault on the project, and the manager could not manage risk well, it will certainly bring losses to the project; third, renewable energy generation projects are more strategical. The exhaustion conventional fossil fuels and increasing awareness of social environment will make renewable energy get a huge development although all kinds of uncertainties exist in renewable energy industry. The current renewable energy investment may not bring corporate large amount of cash incomes, but as the situation develops, the rise of renewable energy sources will make the value of the project gradually reflected, and early investments also lay the foundation for future competition of companies in related fields. Meanwhile, it can promote local employment and local investment, establish good relations with government departments and social image, and can also bring more potential revenue for enterprises, if the enterprise actively participates in renewable energy project investments. So, companies should analyze the impact of renewable energy generation projects for the future development of enterprises, from a strategic point of view.

We could see the option property of renewable energy projects after we analyzed the investment characteristics of renewable energy projects; the option value can be expressed as a function $V = f(S, X, \delta, T, r)$; here, S is the present value of expected future cash flows of a renewable energy project, X is the exercise price which is the initial investment to get this option, δ is the uncertainty about the value of renewable energy projects, T is the validity of the option, and r is the risk-free rate. According to the investment decision criteria, investors will decide to carry out the execution when the present value of cash flow of the project and the option value are greater than the value of the initial investment. In general, the value assessment of the investment project is directly divided into two parts: the discounted value of future cash flows and the option value of the investment projects in the current value assessment research based on the real options. Among that, it uses the risk-free rate for discounting future earnings when calculating the net present value, and the option value is calculated based on a point of project risk management, using the risk rate, so risk caliber is inconsistent in the calculation of different parts of the same project value. The reasons for analyzing this contradiction are that the non-fragmented market risk factors are not taken into account in the calculation of the net present value (NPV) of the project and the option value of the project depends on the project NPV calculations; this also causes partial distortion of the project option value to a certain extent. Therefore, this thesis draws on Lenos' (2007) analysis of market risk theory to make a risk adjustment for the expected return on investment project. A real investment market can be obtained through an analogy with financial market, if, making analysis of the real assets from a financial market perspective, we bring related investment data into model when calculating financial assets with CAPM, and the obtained expected returns on investment should be the basis of the discounting of investment projects; if not, company can choose other more competitive projects to invest. This also avoids the situation with market segmentation among projects in the market, and yet, contradiction that calculated caliber of different values appears unified in the same project. We need to identify types of options that the project contains and select the appropriate real option pricing method when evaluating the option value of energy power generation projects. Considering that renewable energy power generation investment projects have growth value and strategic value and management flexibility value, the option type that project contains can be defined as growth option. According to the optional time characteristic and the convenience of data acquisition of the renewable energy power generation project, this thesis adopts the B–S option model for renewable energy power generation projects to evaluate the option value. In February 2012, the National Development and Reform Commission issued a "renewable energy quota management approach (draft)"; the quota system would affect renewable energy generation projects valuation; this study considered the benefits the relevant provisions of green electricity certificate could produce for the projects. Consequently, this study holds that the total value of renewable power generation project consists of three parts including the risk-adjusted net present value, the option value, and the income of selling green electricity certificate. What must be clear is that, there has not been a corresponding market which can carry out the relevant certificate transaction and the "draft" only gives a rule on the corresponding quota proportion of 2015. This study brings the sale of green power certificate into the evaluation of the total value of the project in order to have a better reflect on a good growth in the future of the renewable power generation project.

2.2 Model Construction

1. Risk-adjusted NPV
   Firstly, consider the project's NPV of expected cash flows in the case that items are not adjusted for risks. According to NPV, risk-adjusted NPV formula is (1):

$$\mathrm{NPV}_r = \sum_{t=1}^{n} \mathrm{CI}_t/(1+r)^t - \sum_{t=1}^{n} \mathrm{CO}_t/(1+r)^t \tag{1}$$

where n is the investment period, CI_t is the cash inflows in t year, CO_t is the cash outflows in t year, and r is the discount rate. Further in reference of the idea of CAPM, achieving the desired rate of return on investment projects through the risk analysis of the renewable energy industry, and making risk adjustment, we get (2):

$$k = E(R_i) = r + \beta_i[E(R_m) - r] \tag{2}$$

Here, k is the risk-adjusted discount rate, which means the discount coefficient of the return on investment at the market average level; $E(R_i)$ is the expected equilibrium returns on asset i; $E(R_m)$ represents the expected return on the market portfolio of all assets; β_i indicates the system risk of assets i that is the sensitivity of asset i when faced with the changes of market m; and r is the risk-free rate. If we brought the risk-adjusted discount rate into the project NPV calculations, we can get formula (3):

$$\mathrm{NPV}_k = \sum_{t=1}^{n} \mathrm{CI}_t/(1+k)^t - \sum_{t=1}^{n} \mathrm{CO}_t/(1+k)^t \tag{3}$$

Looking at the meaning of the coefficients in formula (3), in addition to NPV_k and k, the other coefficients are the same meanings as formula (1). NPV_k represents the risk-adjusted NPV; k is the risk-adjusted discount rate.

2. Risk-adjusted option value of the project
   Through risk-adjusted NPV, we will post the data on behalf of the related adjustments into B–S option pricing model, calculating the option value of the projects included. The simplified model formula which is obtained by partial differential equations is:

$$C = SN(d_1) - Xe^{-r(T-t)}N(d_2) \tag{4}$$

   Here, $d_1 = \frac{\ln(S/X)+(r+\delta^2/2)(T-t)}{\delta\sqrt{T-t}}$, $d_2 = d_1 - \delta\sqrt{T-t}$.
   The representative meanings of coefficients in these three formulas are as follows: C is the option value of the project; S is the market value of the underlying asset; X is the option exercise price; δ is the price volatility of the underlying asset; r is the risk-free rate; T is the expiration date of the option, t is the current time, $T–t$ is maturity of the option; and $N(\bullet)$ is the cumulative normal distribution density function. Calculating the option value depends on the NPV calculations; here,

$$S = \sum_{t=1}^{n} \mathrm{CI}_t/(1+k)^t, \quad X = \sum_{t=1}^{n} \mathrm{CO}_t/(1+k)^t$$

3. Assessing incomes from selling green electricity certificates of the project.
   The income from sale of green electricity certificates W_c is affected by some variables, such as the annual generation capacity Q, the proportion of green electricity quota ρ, the certificates price P_c, and so on; it can be expressed as formula (5):

$$W_c = Q^*(1 - \rho) * P_c / 1000 \tag{5}$$

   Thus, the NPV of income from selling green electricity certificates during the full life cycle project period of the renewable energy generation can be expressed as formula (6):

$$\text{NPV}_{W_c} = \sum_{t=1}^{n} W_c/(1+k)^t \tag{6}$$

4. The total value of the project
   Finally, adding the project's risk-adjusted net present value, option value, and the present value of the project to sell green electricity certificate together, incorporating into the calculation of the total value of the project, we can get the total value assessment model of renewable energy project based on improved real options, as formula (7):

$$\begin{aligned} W &= C + \text{NPV}_k + \text{NPV}_{W_c} \\ &= SN(d_1) - Xe^{-r(T-t)}N(d_2) + \sum_{t=1}^{n} CI_t/(1+k)^t \\ &\quad - \sum_{t=1}^{n} CO_t/(1+k)^t + \sum_{t=1}^{n} W_c/(1+k)^t \end{aligned} \tag{7}$$

3 Case Analysis

Company A is a large listed enterprise which specializes in new energy development, design, construction investment, and the development, promotion, and application of low-carbon technologies; it was successfully listed in Hong Kong in late 2010. It is ready to invest a wind power project W which is located in province H and plans an installed capacity of 49.5 MW. The wind power project is in line with our strategic planning for renewable energy power development; it plays a great, significant, and positive role in the optimization of the structure of H provincial power source, increasing the reliability of supply and servicing for local economic development, achieving sustainable development; therefore, the selected case here is a representative and typical one. The whole cycle of wind power project

W will last 21 years, the construction period is 1 year, the completed full productive capacity is 49.5 MW, the production period is 20 years, and the annual full-load running time is 2,500 h. The initial investment is 5.00 million yuan. All of the funds are invested in the first year, and 4.33 million yuan is the fixed capital, and the rest is the circulating capital and others. According to state regulations, the acquisition-subsidized electricity price is 0.62 yuan/KWh when power generation equipment operates within 30,000 h, exceeding 30,000 h; the acquisition may refer the local average price 0.41 yuan/KWh; the project runs with higher cash inflow in the first 12 years. The residual rate of fixed assets is 5 % in the end of project operations, approximately converting into 21.65 million yuan, withdrawing liquidity is 1.49 million yuan, and the annual generating capacity of the wind farm is expected to be 1.2375 billion kWh.

1. Risk-adjusted NPV and NPV calculations are not risk adjusted
   If we use risk-free rate for the expected future net cash flows discounted items without considering environmental risk factors. Through the formula (1), the NPV of the project is 132.53 million yuan. Now, the NPV is greater than zero; we should invest in this project in accordance with judgment criterion of investment decision.
   Considering the impact of systemic risk factors that the project faces, we should use a risk-adjusted discount rate k on risk-adjusted NPV. As the long-running cycle of the project, and the entire operating period is 20 years, we refer to the Ministry of Finance-issued 20-year bonds in the recent years, choosing the 20-year half-year interest-bearing debt which was published in by the ministry of finance, national debt code 010713, taking its 4.52 % annual interest rate as the risk-free interest rate r of this case. We can obtain the coefficient $\beta_i = 1.155$ while using the method in reference Xia et al. (2004). Then, selecting the difference between the closing price of Hang Seng Index 17,668.83 in September 23, 2011, and the closing price of Hang Seng Index 19,411.46 in June 29, 2012, as a return on investment, on this basis, we can obtain the return on investment in this period $R_m = 9.86$ %. Taking the above parameter values into formula (2), the expected return on investment assets is the risk-adjusted discount rate:

$$k = E(R_i) = r + \beta_i[E(R_m) - r] = 10.69\,\%$$

   The risk-adjusted NPV is −82.93 million yuan when we put k into formula (3) to calculate. At this point, the risk-adjusted NPV is less than zero; the investor should invest in the project according to the investment decision criterion. Through the comparison between NPV_r and NPV_k, you can see that the NPV of the project is reducing when we consider the systemic risk of the project; this is because of taking no account of systemic risk, and investors expect the same investment returns; lower risk will increase the expected returns of investors, and relatively higher risk will lead to lower expected returns. When the systemic risk was brought into reward system, according to the thought of discounted cash flow, investors would reduce a certain level expectations of expected cash flows because of the increased risk. Seen from the difference between the results, we need to take environment and systemic risks that the investment faces into account when calculating the NPV of the project; otherwise, that poses a potential danger to project operation without considering the risks to make the investment value artificially high.

2. Option Value of the Project
Here, we use B–S option pricing model to evaluate option value of the project; this model needs definite volatility of the underlying asset. This study selects the stock's closing price of company A between February 15, 2012, and July 10, 2012; then, using historical volatility method to forecast the volatility of revenue growth of this investing enterprise, the day volatility of assets δD is obtained as 2.15 %, so the annual volatility δY is 34.11 %. According to the formula of B–S option model, and bringing the above parameters into formula (4), we obtain $d_1 = 1.2728$, $N(d_1) = 0.8985$; $d_2 = 0.2903$; $N(d_2) = 0.3853$. The option value of project $C = 300{,}050{,}000$ yuan.
3. Green electricity certificate revenue
At present, although there are no specific detailed rules for the implementation of renewable energy quota index, the practice of renewable energy quota system has been set, so the evaluation of the renewable energy power generation project investment needs to consider the increased revenue that the quota brings for the enterprise. "Renewable energy power quota management approach (draft)" provides the quota proportion of large-scale renewable energy generation enterprises that accounted for 6.5 % for their own, while a prescribed amount of green electricity certificates is 1,000 K/Wh. The price of the certificate must be defined if you would have revenue from energy renewable green electricity certificates inserted into the value evaluation of the project. Ideally, the price of green electricity certificates should be the difference between the marginal cost of renewable and conventional electricity power. Selecting the average of the lower limit 0.15 yuan/KWH and the upper 0.5 yuan/KWH of wind power green electricity certificate estimates in literature Li et al. (2012) to valuate the price for green power certificate (Xia et al. 2004), the price of a green electricity certificate is obtained as 325 yuan. According to the basic data of wind farms, when the annual generating capacity is 1.2375 trillion KWH, the number of green power certificates that companies can sell is 115,706 besides the need for companies to meet their quota for the proportion of renewable energy. Due to the differences between the marginal cost of different renewable energy relative to conventional electricity, you can use multiple certificates to be adjusted for different green electricity certificate designing different weights; for instance, a certificate of solar green electricity can be used for five copies and wind power is one, and the multiple choice of this paper for wind power project certificate is double which conforms to cost expectation of current wind power electricity. Taking the data into calculation formula (6) of green power certificate revenue, we can get the project-sale income of green electricity certificates as 37.6

million yuan a year; if the project runs for 20 years, the total NPV is RMB 343.19 million.

4. The total value of the project
   Adding C, NPV_k, and $NPVw_c$, the project's final value is 560,310,000 yuan. According to the investment decision criterion, we should invest in the project. Based on risk adjustment of interest rate and the evaluation of the option value, you can dig deeper into the value of the project and avoid overestimation of risk returns, and the impact of systemic risk is also brought into the project evaluation.
5. Sensitivity analysis
   Further, find out the sensitivity factors of project financial data to make reference for investment decisions. This study selected annual full-load generation time, initial investment, and feed-in tariff; these three indicators affect prime operating revenue of the project; then, we use a single sensitivity analysis to separately calculate the sensitive degree of the whole project value when index increases or decreases by 5 % or 10 %. The specific calculation results are shown in Table (1).
   As shown in the table, the uncertainly factor which affects the project value most is annual full-load generation time, and its sensitive coefficient is around 4.5, followed by a feed-in tariff with sensitivity coefficient of about 4.35; the last is the initial investment, sensitivity coefficient of around −2.25. Through break-even analysis of the project, we can get the break-even annual full-load generation time of the project as 1,937.8 h, and feed-in tariff level of profit and loss balance as the current 76.73 % of feed-in tariff level. In the operation of the project, therefore, people need to strengthen the maintenance and repair of wind turbine in order to ensure the benefits and the value of the project, to ensure that

Table 1 Sensitivity analysis

Uncertain factors	Rate of change (%)	Change rate of the project value (%)	Sensitivity coefficient (%)
Original scheme	0	0	0
Annual full-load generation time	−10	−44.78	4.48
	−5	−22.44	4.49
	5	22.53	4.51
	10	45.15	4.51
Initial investment	−10	22.66	−2.27
	−5	11.30	−2.26
	5	−11.26	−2.25
	10	−22.47	−2.25
Feed-in tariff	−10	−43.30	4.33
	−5	−21.70	4.34
	5	21.80	4.36
	10	43.67	4.37

the unit can carry on the production capacity according to the plan without loss due to the equipment failure; at the same time, it also shows that projects face a greater uncertainty when it is the insufficient run caused by the natural environment. The sensitivity coefficient of the initial investment is relatively smaller; enterprises need reasonable budget and strict control to ensure that the project will not appear larger as initial investment increases, and yet minimize the amount of investment. Although the feed-in tariff is not controllable, it can reflect the sensitivity of the project value when the project faces price level fluctuations, so the influence of electricity price fluctuation on the project value cannot be ignored due to the primary business of project electricity sales.

4 Summary

The investment value of renewable energy projects is affected by multiple internal and external factors, considering the non-diversifiable systemic risk and selling green electricity certificate revenue in the quota system; this study constructed a real option valuation model in which the risk could be adjustable and improvements can be made to evaluate the value of renewable energy projects. This model brings potential losses which can be created by external systemic risk of the firm into value assessment system, providing a new idea for value assessment of that type of investment projects to effectively reduce investment risks. The mode still has some limits. First, the market risks cannot be measured correctly; the project risk coefficients of the model are only adjusted by financial markets data with the help of the relation of the financial markets and the real investment. This method will impact the calculation results. Secondly, due to the long operating cycle of renewable energy power generation projects, when we use option valuation model to calculate the value of the project, the method of historical volatility may not accurately reflect the project's asset volatility of future changes in a longer period of time, so how to better determine the volatility of project assets worth further investigation.

Acknowledgments Supported by the humanity society, and scientific research funds of the Ministry of Education for the western and border areas. (Project No: 13XJC630003).

References

Cheng C-T, Loa S-L, Lin TT (2011) Applying real options analysis to assess cleaner energy development strategies. Energy Policy 39:5929–5938

Lee S-C, Shih L-H (2010) Renewable energy policy evaluation using real option model—the case of Taiwan. Energy Econ 9(1):67–78

Lenos T (2007) Real option: managerial flexibility and strategy in resource allocation. Tsinghua University Press, Beijing, pp 49–54

Li F, Liu Z, Jia X (2012) The evaluation model of wind power project investment risk based on the theory of whole life cycle. East China Electr Power 40(4):0531–0535

Pederson G, Zou T-y (2009) Using real options to evaluate ethanol plant expansion decisions. Agric Finan Rev 6:23–35

Xia H, Zeng Y, Tang X (2004) The review of the real option method for technology innovation strategy investment. J Manag 2:88–96

Multi-connected Air-Conditioning System Standards Analysis and Annual Operating Performance Evaluation

Dong-liang Zhang, Xu Zhang and Ning Cai

Abstract Problems and limitations of existing multi-connected air-conditioning unit standards are analyzed in this chapter. Based on irrationality of the existing evaluation criteria, the evaluation index and method for evaluating whole year energy consumption and efficiency of multi-connected air-conditioning system are studied. The evaluation index, calculation formula, weights, and testing condition parameters are proposed. The correctness of the evaluation method is verified. The results show that eight region average classification method based on dynamic load is feasible to evaluate seasonal energy efficiency ratio of multi-connected air-conditioning system. The evaluation method is applicable to evaluate DVM or variable frequency-controlled multi-connected air-conditioning system. The evaluation method can be applied to evaluate annual energy consumption and annual integrated energy use efficiency of multi-connected air-conditioning system in the building, as well as to compare energy efficiency among different air-conditioning systems.

Keywords Multi-connected air-conditioning system · Performance evaluation · Standards analysis

1 Introduction

Multi-connected air-conditioning (heat pump) unit has been widely used and rapidly developed due to its several advantages, such as easy maintenance, wide range capacity output, precise capacity control, and high seasonal energy efficiency ratio (SEER). It has become one of the most commonly used central air-conditioning

D. Zhang (✉) · N. Cai
College of Energy and Power Engineering, Nanjing Institute of Technology, Hongjing Road, Nanjing 211167, Jiangsu, People's Republic of China
e-mail: zhangdongliang@njit.edu.cn

X. Zhang
College of Mechanical Engineering, Tongji University, Shanghai 200092 People's Republic of China

S. Feng et al. (eds.), *Low-carbon City and New-type Urbanization*, Environmental Science and Engineering, DOI 10.1007/978-3-662-45969-0_5

systems in small commercial buildings and civil buildings. Compared with centralized or semi-centralized air-conditioning system, multi-connected air-conditioning system is simpler in structure and easier to install. Outdoor unit is smaller and occupies less space than that of centralized or semi-centralized air-conditioning system. Each indoor unit of multi-connected air-conditioning system is independently controlled. And its energy loss is lower than that of the other two systems due to its only once heat exchange in heat transfer process.

At present, there have been several studies relevant to multi-connected air-conditioning system. Goetzler et al. (2004) analyzed energy saving potential and market factors of refrigerant systems that vary flow and volume. Zhang et al. (2009, 2011) and Zhao (2009) addressed experimental study on performance of digital variable multiple air-conditioning system under part load conditions. Huang et al. (2008) conducted an experimental study on operating characteristics of ducted air-conditioning unit with digital scroll compressor and conventional scroll compressor under refrigerating and heating condition. Hu and Yang (2005) discussed the relationship between the opening degree of electronic expansion valves and the compressor output ratio. Zhou et al. (2008) developed a simulation module for variable refrigerant volume (VRV) air-conditioning system on the basis of EnergyPlus. Several papers addressed control strategy of VRV air-conditioning system by Shi (2000), Shi et al. (2003a, b, c). Kim et al. (1999) applied fuzzy logic to control refrigerant distribution for the multi-type air conditioner. Wu et al. (2005) and Xia (2005) studied control scheme and optimization control of VRV air-conditioning system. And evaluation standard for multi-connected air-conditioning (heat pump) unit (GB/T 1883; JRA4048-2006; Office of Energy Efficiency and Renewable Energy Department of Energy 2006; ANSI/AHRI 2010) has been formulated.

The studies above are mainly focused on operating characteristics, control strategies, and evaluation method of multi-connected air-conditioning system. However, up to now, seldom researches have been conducted on standards analysis and annual operating performance evaluation of multi-connected air-conditioning system. In this chapter, problems and limitations of existing multi-connected air-conditioning unit standards are analyzed and the existing evaluation criteria, the evaluation index, and the method for evaluating multi-connected air-conditioning system annual energy consumption and integrated efficiency are studied.

2 Multi-connected Air-Conditioning Unit Standards Analysis

2.1 Chinese Multi-connected Air-Conditioning Unit Standard

The first multi-connected air-conditioning unit standard GB/T18837-2002 was born in China in 2002. In this standard, IPLV(C/H) (Refrigerating or Heating Integrated Part Load Value) is used to evaluate refrigerating or heating performance of multi-connected air-conditioning unit.

Take refrigerating condition as example, IPLV(C) is calculated by Eq. (1).

$$\begin{aligned} \text{IPLV(C)} = & \frac{(\text{PLF}_1 - \text{PLF}_2)(\text{EER}_1 + \text{EER}_2)}{2} \\ & + \frac{(\text{PLF}_2 - \text{PLF}_3)(\text{EER}_2 + \text{EER}_3)}{2} \\ & + \frac{(\text{PLF}_3 - \text{PLF}_4)(\text{EER}_3 + \text{EER}_4)}{2} \\ & + (\text{PLF}_4)(\text{EER}_4) \end{aligned} \quad (1)$$

The above algorithm considers part load performance and on-off loss of compressor under nominal condition. IPLV(C) represents average cycle energy efficiency ratio under nominal condition. It is more reasonable to evaluate performance of multi-connected air-conditioning unit than EER. However, there exist the following two problems: (1) ±10 % part load ratio change range of test condition may affect the performance evaluation result when performance of different multi-connected air-conditioning units is compared. (2) The testing condition parameters should reflect actual operating performance and operating time of multi-connected air-conditioning system when evaluating its actual operating performance. Therefore, the evaluation index in GB/T18837-2002 cannot be used to evaluate actual operating performance of multi-connected air-conditioning system.

2.2 Japanese Multi-connected Air-Conditioning Unit Standard

In JRA4048 standard, cooling seasonal performance factor (CSPF), heating seasonal performance factor (HSPF), and annual performance factor (APF) are used as evaluation index. The testing condition parameters are the same as that of variable frequency-controlled air conditioner unit. When nominal refrigerating capacity of multi-connected air-conditioning unit is less than 28 kW, two indoor units with the same capacity are experimented, and nominal refrigerating capacity and energy consumption of one indoor unit and two indoor units are measured.

Take refrigerating condition as example, as shown in Fig. 1, calibration points are solid dots with 35 °C outdoor dry bulb temperature, 27 °C indoor dry bulb temperature, and 19 °C indoor wet bulb temperature.

Using variational condition characteristic modified coefficient gained from statistical average of Japanese products, refrigerating capacity and energy consumption under 29 °C outdoor dry bulb temperature may be calculated, as shown in hollow dots in Fig. 1. Considering on-off loss of compressor under low part load ratio, the model of refrigerating capacity and energy consumption was developed and combined with air-conditioning load curve of typical building and operating

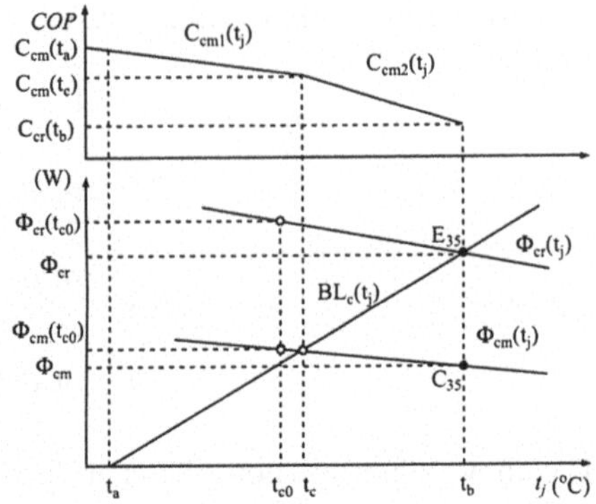

Fig. 1 Typical building load line and multi-connected air-conditioning unit performance model for CSPF evaluation

time distribution with outdoor temperature, the cooling seasonal performance factor CSPF or SEER is calculated as shown in Eq. (2).

$$\text{CSPF} = \frac{\text{CSTL}}{\text{CSTE}} \tag{2}$$

where CSTL is total refrigerating capacity, and CSTE is total energy consumption during cooling season.

In JRA4048 standard, variable working condition characteristics were considered, while variable air-conditioning load performance was not included. In addition, using indirect calculation method to gain performance parameter under 29 °C outdoor dry bulb temperature condition cannot reveal differences among different products. And two points cannot reflect operating performance of product under large range operating condition. What is more, the relationship between air-conditioning load of the building and outdoor temperature is simplified as a linear in JRA4048, which also does not reflect actual air-conditioning load characteristics of the building. Therefore, CSPF or SEER in JRA4048 standard cannot reflect actual performance of multi-connected air-conditioning system.

2.3 American Multi-connected Air-Condition Unit Standard

ANSI/AHRI1230-2010 was proposed in the USA in 2010. In this standard, CSPF or HSPF is used as evaluation index when refrigerating capacity is less than 19 kW, which exists the same problem as that of JRA4048 standard. While refrigerating capacity is more than 19 kW, IEER is used as evaluation index, which is shown in Eq. (3).

$$\text{IEER} = a \times A + b \times B + c \times C + d \times D \tag{3}$$

where A, B, C, and D are the energy efficiency ratio under 100, 75, 50, and 25 % part load condition, respectively; and a, b, c, and d are the weight coefficients of the percentage of operating time corresponding to each part load ratio, respectively.

ANSI/AHRI1230-2010 considers building type, meteorological parameters, and operating time, comprehensively. However, CSPF/HSPF or IEER is not available to evaluate annual energy consumption of multi-connected air-conditioning system. On the one hand, testing condition parameters cannot reveal typical condition of annual operating, which are determined by dynamic air-conditioning load characteristic and operating time distribution. On the other hand, weighted coefficient cannot reveal energy consumption characteristic of actual operating condition of multi-connected air-conditioning system. Using operating time percentage as weighting coefficient in ANSI/AHRI1230-2010 may only reflect the operating time proportion of each region, but not reflect air-conditioning load of the building. So, weighting coefficient in ANSI/AHRI1230-2010 cannot reflect seasonal energy consumption characteristic of multi-connected air-conditioning system.

3 Annual Operating Performance Evaluation Method of Multi-connected Air-Conditioning System

Algorithm of seasonal energy efficiency of multi-connected air-conditioning system is similar to IPLV index of chiller, which includes three factors, air-conditioning load characteristic of the building, operating time distribution, and part load performance. Typical operating parameters are obtained by the statistic of seasonal operating parameters of multi-connected air-conditioning system. And seasonal energy consumption performance indexes of multi-connected air-conditioning system will be expressed by weighted average value of energy efficiency ratios corresponding to several typical operating points.

Part load performance is effected by outdoor air temperature and air-conditioning load of each region. Therefore, outdoor air temperature and part load ratio are used as testing condition parameters. The specific calculation steps are as follows:

1. Calculate air-conditioning load characteristic of air-conditioning area and determine typical testing condition parameters and weighting coefficients.
2. Measure operating performance parameters of multi-connected air-conditioning system under typical testing condition and calculate corresponding EER.
3. Calculate seasonal energy efficiency and seasonal energy consumption of multi-connected air-conditioning system.
4. Accumulate seasonal cooling capacity, seasonal heating capacity, and seasonal energy consumption of each multi-connected air-conditioning system, and annual energy efficiency and energy consumption of the building are calculated.

Divide air-conditioning region into several regions, calculate weight coefficient of each region, and seasonal energy efficiency is calculated by weighting EER or

COP on weight coefficient of typical test condition in each region, which is named IPLEER or IPLCOP as shown in Eqs. (4) and (5).

$$\mathrm{IPLEER} = a \times \mathrm{EER}_{\mathrm{I}} + b \times \mathrm{EER}_{\mathrm{II}} + c \times \mathrm{EER}_{\mathrm{III}} + d \times \mathrm{EER}_{\mathrm{IV}} + \cdots \quad (4)$$

$$\mathrm{IPLCOP} = a \times \mathrm{COP}_{\mathrm{I}} + b \times \mathrm{COP}_{\mathrm{II}} + c \times \mathrm{COP}_{\mathrm{III}} + d \times \mathrm{COP}_{\mathrm{IV}} + \cdots \quad (5)$$

where *a*, *b*, *c*, and *d* are the weighting coefficient of each region, respectively.

On the basis of CSTL, HSTL, IPLEER, and IPLCOP, seasonal energy consumption and integrated energy efficiency are calculated, which are named OPC (operating power consumption) and IPF (integrated performance factor), respectively, as shown in Eqs. (6) and (7).

$$\mathrm{OPC} = \mathrm{CSTL}/\mathrm{IPLEER} + \mathrm{HSTL}/\mathrm{IPLCOP} \quad (6)$$

$$\mathrm{IPF} = \frac{\mathrm{CSTL} + \mathrm{HSTL}}{\mathrm{OPC}} \quad (7)$$

In this chapter, average region division method is used, and the typical testing condition parameters are determined by weighting outdoor air temperature and part load ratio on operating time. If the proportion of each region's operating time to total operating is taken as weighting coefficient, only operating time proportion is revealed, but air-conditioning load characteristic of the building is not considered. While the proportion of part load ratio and its corresponding operating time product of each region to total product of part load ratio and its corresponding operating time is taken as weighting coefficient, it may not only reflect operating time but also the air-conditioning load characteristic of the building, which is more reasonable.

4 Verification for Annual Operating Performance Evaluation of Multi-connected Air-Conditioning System

In this chapter, annual energy consumption performance of multi-connected air-conditioning system composed of one outdoor unit (14 HP) and twelve indoor units is studied. The type and rated capacity of multi-connected air-conditioning unit are shown in Table 1. Building model is presented in Fig. 2. Boundary conditions of the model are selected in accordance with limit value of Public Building Energy Saving Design Standard GB 50189-2005.

Table 1 Indoor and outdoor unit type and rated capacity/kW

	Type	Rated refrigerating capacity	Rated heating capacity
Indoor unit	AVXCMH040EF	4	4.5
Outdoor unit	RVXVHT140GF	40	45

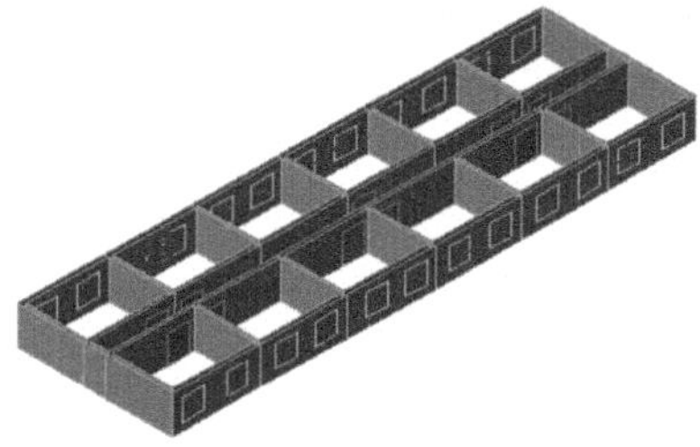

Fig. 2 Building model

Four typical hot summer and cold winter cities, Shanghai, Nanjing, Wuhan, and Chongqing, are selected as study object. Dest software is used to calculate annual air-conditioning load of the building. And annual energy consumption characteristics are gained on the basis of annual air-conditioning load of the building and energy consumption calculation model of multi-connected air-conditioning system. The above data are the baseline of verification for annual operating performance evaluation of multi-connected air-conditioning system.

The calculation results of eight region average classification method are shown in Fig. 3. The relative error between calculation results and baseline is less ±10 %, which is in accordance with the engineering permissible precision range.

The calculation results of 12 region average classification method are shown in Fig. 4. The relative error between calculation results and baseline is less ±10 %, which is in accordance with engineering permissible precision range. The calculation accuracy of 12 region average classification method is better than that of 8 region average classification method.

Therefore, 8 region average classification method not only guarantee the precision but also is simple to implement. And due to the similarity of part load performance of multi-connected air-conditioning system, the evaluation method is applicable to evaluate DVM or variable frequency-controlled multi-connected air-conditioning system. The evaluation method can be applied to evaluate annual

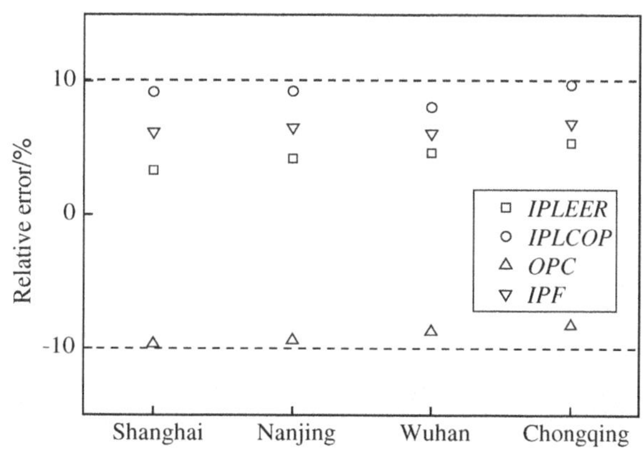

Fig. 3 Relative error distribution

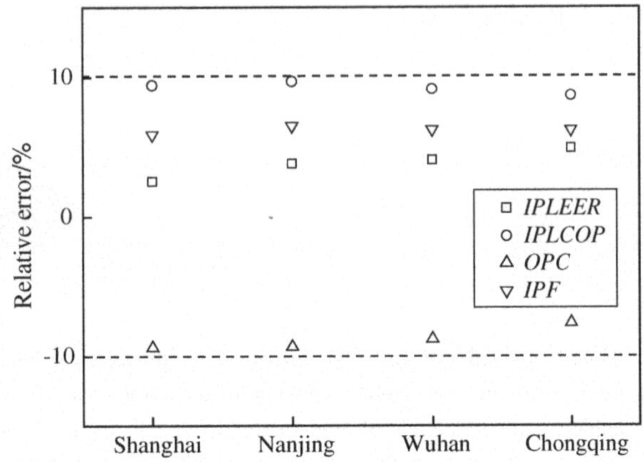

Fig. 4 Relative error distribution

energy consumption and annual integrated energy use efficiency of multi-connected air-conditioning system in the building, as well as to compare energy efficiency among different air-conditioning systems.

5 Conclusions

1. Multi-connected air-conditioning unit standards provide evaluation index and testing method for the evaluation of the performance of the product, but they are not available to evaluate the actual operating performance.
2. Eight region average classification method based on dynamic load is feasible to evaluate seasonal energy efficiency ratio of multi-connected air-conditioning system, which not only guarantee the precision but also is simple to implement.
3. The evaluation method proposed in this chapter can be applied to evaluate annual energy consumption and annual integrated energy use efficiency of multi-connected air-conditioning system in the building, as well as to compare energy efficiency among different air-conditioning systems.

Acknowledgments This research was supported by the National Natural Science Foundation of China under grant No. 51408302, Science and Technology Project of Housing and Urban-rural Construction Ministry under grant No. 2013-K1-6.

References

ANSI/AHRI1230-2010 (2010) Performance rating of variable refrigerant flow (VRF) multi-split air-conditioning and heat pump equipment

GB/T 18837 (2002) Multi-connected air-condition (heat pump) unit [S] (in Chinese)

Goetzler W, Roth KW, Brodrick J (2004) Variable flow and volume refrigerant system. ASHRAE J 46:164–165
Hu SC, Yang RH (2005) Development and testing of a multi-type air conditioner without using AC inverters. Energy Convers Manag 46:373–383
Huang H, Li QH, Yuan DX et al (2008) An experimental study on variable air volume operation of ducted air-conditioning with digital scroll compressor and conventional scroll compressor. Appl Therm Eng 28:761–766
JRA4048-2006 Annual performances factor of package air conditioners. The Japan Refrigeration and Air Conditioning Industry Association (in Japanese)
Kim JH, Kim KS, Sim MS et al (1999) An application of fuzzy logic to control the refrigerant distribution for the multi type air conditioner. In: Proceedings of IEEE international fuzzy systems conference, Seoul, Korea, 22–25 Aug 1999, pp 1350–1354
Office of Energy Efficiency and Renewable Energy Department of Energy (2006) 10 CFR Part 430, Part II, Energy Conservation Program for Consumer Products: Test Procedure for Residential Central Air Conditioners and Heat Pumps (Proposed Rule)
Shi WX (2000) Study on variable refrigerant flow air conditioning system performance and control strategy: (dissertation). Tsinghua University, Beijing (in Chinese)
Shi WX et al (2003a) Simulation study on control strategy of MVRV air conditioning system with heat recovery. Heat Ventilating Air Conditioning 33:12–17 (in Chinese)
Shi WX, Yan QS, Peng XF (2003b) Experimental study on control strategy of an MVRV air conditioning system with heat recovery. Heat Ventilating Air Conditioning 33:46–50 (in Chinese)
Shi WX, Yan QS, Li XT et al (2003c) A network model to simulate performance of variable refrigerant volume refrigeration systems. ASHRAE Trans 109:61–68
Wu DX, Fan GD, Shi M (2005) Control scheme for multi-connected air conditioning(heat pump) units. Heat Ventilating Air Conditioning 35:103–105 (in Chinese)
Xia JJ (2005) Research on optimization control of variable refrigerant flow (VRF) air-conditioning system (dissertation). Tsinghua University, Beijing (in Chinese)
Zhang DL, Zhang X (2009) An experimental study on operating characteristics of DVM air conditioning system under part load condition. In: 6th international symposium on heating, ventilating and air conditioning, ISHVAC, Nanjing, China, 6–9 Nov 2009, pp 452–459
Zhang DL, Zhang X, Liu J (2011) Experimental study of performance of digital variable multiple air conditioning system under part load conditions. Energy Build 43:1175–1178
Zhao W (2009) Study on performance under part load condition of variable refrigerant flow system (MS thesis). Tsinghua University, Beijing (in Chinese)
Zhou YP, Wu JY, Wang RZ et al (2008) Simulation and experimental validation of the variable-refrigerant-volume (VRV) air-conditioning system in EnergyPlus. Energy Build 40:1041–1047

Research on Transport Subsidies for Public Transit and Cars

Fei Shi

Abstract Creating a transit-oriented built environment and implementing the important economic policy of transport subsidy are effective means to promote the use and accelerate the development of public transit. The importance of the transport economic policy must not be ignored because a policy preference will possibly affect travel behavior to a great extent. Taking subsidy as an angle and the city Nanjing as an example, this article focused on analyzing and comparing the operating subsidy and disguised external cost subsidy for public transit and cars use. The result showed that the unit subsidy for cars is 11.8 times of that for public transit. This is extremely unfair and it goes against the established transit-oriented development direction. Furthermore, this article, based on stated preference survey data, analyzed the transfer of travel mode on the condition that free parking is cancelled. The result showed that more than 1/3 customers would turn to public transit and slow-moving traffic under that condition. In conclusion, this article brought forward policy suggestions, such as eliminating the subsidy for parking prices, setting maximum index of appertaining parking facilities instead of minimum index, subsidizing public transit, and increasing car-using tax, in order to achieve society equity and transit-oriented development.

Keywords Transport subsidy · Public transit · Car · Selection of travel mode · Development orientation

1 Introduction

Transit Metropolis is an urban strategy adopted to cope with the high-speed growth of cars, traffic congestion, and lack of transport and environmental resources. The connotation of Transit Metropolis is certainly not only limited to the "public transit

F. Shi (✉)
School of Architecture and Urban Planning, Nanjing University,
No. 22, Hankou Street, Nanjing 210093, Jiangsu, People's Republic of China
e-mail: shifei@nju.edu.cn

S. Feng et al. (eds.), *Low-carbon City and New-type Urbanization*,
Environmental Science and Engineering, DOI 10.1007/978-3-662-45969-0_6

priority plan" which belongs to the government system, but also it is about creating an urban environment beneficial to public transit from the perspectives of urban development and physical construction, such as urban spatial structure, land development, and new town design. Copenhagen, Curitiba, and Singapore have become world-famous Transit Metropolises for the harmonious symbiosis between their public transit service and urban physical environment (Susuki et al. 2013).

However, this article will not discuss further about those mentioned above. Because other than physical environment construction, the economic factor that cannot be ignored for promoting the use and development of public transit is another extremely important factor. Here is a simple example: The subsidy for public transit (including surface public transit and metro) in Beijing has reached up to 11 % of the whole city's financial expenditure, higher than the 7.2 % for social security and employment, but the embarrassing fact of surface traffic congestion has not changed yet. Therefore, this article attempts to explore more about the economic means, specifically subsidy.

As an important infrastructure of cities, urban public transit has an important influence on the overall benefit of the whole society. However, for public transit operators, it is difficult to ensure both social benefit and economic benefit, which determines the necessity of financial subsidy. At present, subsidy is provided for public transit in Chinese cities, especially megalopolises. But meanwhile, we feel that the subsidy generally considered for public transit also apply to car using, such as invisible subsidy mentioned later in the article—marketplace provides free parking in order to attract customers. Obviously, this goes against the policy of developing public transit preferentially, and brings about the transfer of travel mode and social injustice (Small and Verhoef 2007). Based on this cognition, this article emphasis on operating subsidies, using Nanjing as an example, to calculate and compare the travel subsidies for public transit and cars, and reveal the influence of subsidies on travel modes based on preference survey results. Through comparisons of subsides and studies of subsidy behavior, researchers hope to partly explain the preference of travel modes in Chinese cities and put forward policies and recommendations accordingly.

2 Review

2.1 Transport Subsidy Provided by the Government

Public transit is usually regarded as a public welfare undertaking. The majority of cities in the world are not making ends meet in public transit operations. The tickets income cannot cover the operating costs, and such loss is usually taken up by the government in order to ensure the normal operation of public transit.

For example, the Bay Area Rapid Transit system (BART) in the metropolitan area of San Francisco, USA, realized a fare income of only USD66 million, but had operating cost of up to USD134 million from June 1983 to June 1984. Among the USD68 million subsidies, 86 % comes from the consumption tax collected by the local government, and the balance comes from property tax and others (Wachs 1993). The proportion of subsidy for public transit is also very high in Western European countries, and has reached up to 50–70 % in cities such as Amsterdam, Barcelona, Paris, Roman, and Glasgow (Abou-Zeid et al. 2012; Litman 2005). In China, the local government pays increasing attention to public transit undertaking. Taking Beijing, the capital city, as an example, the amount of subsidy for surface public transit and metro has increased from around RMB1.0 billion at the beginning of this century to RMB18 billion in 2013.

2.2 Transport Subsidy Provided by Employers

In the West, especially in the USA, it has become a custom for employers to provide commuter subsidy, or fringe benefit, such as employer-paid free parking in urban central areas. However, some states have broken this convention. For example, in the Parking Cash out Law promulgated in 1992, California required employers to provide another option for employees, and more employees selected cashing out (Willson and Shoup 1990). In 2008, the Bicycle Commuting Subsidy Bill signed by the American former president George W. Bush enabled bicycle commuters to obtain employer subsidy together with driving commuters and bus commuters (Heinen et al. 2010).

Things in China are different from those in the USA. Transport subsidy was included in the payroll even in the era of planned economy. But for a long time, the transport subsidy has not aimed at any certain traffic mode, and has been mixed together with other incomes. So it does not affect the selection of travel mode. Along with the social and economic development, some enterprises have begun to provide market-based subsidy aiming at customers' specific travel mode for certain purposes, and this behavior has affected the selection of travel mode, and so aroused the attention of this article.

2.3 Brief Summary

As a whole, the West seems to be more mature on this matter since it guarantees the implementation of transport subsidy policy by legislation. The subsidy for commuter travel has experienced the evolution from single subsidy for car commuters to the subsidy for multiple commuting modes including cycling, and this process is also accompanied by the gradual ripening of the society's recognition of the transportation, environmental, and social problems. The calculation of demand elasticity proves that public transit subsidy is an effective but not the best method for raising ridership. The West has developed many researches about the influence of parking subsidy on travel behavior, and obtained convincing and credible conclusions. The subsidy for the use of cars is mostly explicit in the West.

At present, China pays increasing attention to public transit pricing mechanism and subsidy policy (Yang et al. 2010), but lacks studies on subsidy for nonpublic transit travel and contrastive studies on subsidy for various travel modes, as well as researches on the influence of subsidy on the selection of travel mode. In addition, the subsidy for use of cars in China is implicit, and is not perceived by most people.

Actually, at home and abroad, especially in the West, many researches have been conducted on the pricing of public transit subsidy and free parking. But surprisingly, there are few contrastive studies on the city subsidy for travels by public transit and by cars. This will be the topic mainly discussed in this article, and will reflect the guidance quality of urban transport development and the fairness of subsidy at present.

3 Method

In this article, the model used is not very complicated, and the research method is also relatively simple. This article focuses on the operating subsidy behavior in travel by surface public transit and cars. Such subsidy is possibly explicit, such as public transit subsidy, or it could be implicit, such as marketplace's policy of free parking based on shopping reaching a certain amount in order to attract customers. In addition, subsidy behavior occurs in the dynamic process of travel, such as public transit operation, and also in the static process of travel, such as cars parking. External cost is also a type of subsidy and will be discussed below. The calculation of subsidy amount for travels by surface public transit and cars will be detailed to every person time, in order to enhance the comparability.

The calculation of transport subsidy below takes the city of Nanjing as an example. Nanjing is the capital city of Jiangsu province with a population of 7 million. Owing to its superior geographic location, the city has remained the transportation center of the Yangzi Delta region. Nanjing is one of the historical and cultural cities in China. Nanjing bears the reputation of the Capital of Ten Dynasties, since the year 229 AD, it served as the capital of Wu, Eastern Jin, Song, Qi, Liang, Chen, Southern Tang, Ming, Taiping Kingdom, and Republic of China subsequently. Nowadays, Nanjing has developed into a multiple-producing industrial base in eastern China, an important hub of transportation and communication center, one of the China's four major scientific research and educational central cities. The gross national products of the whole city in 2013 amounted to more than 800 billion yuan, which has a 12 % increase compared to the previous year.

4 Calculation and Comparison

4.1 Calculation of Subsidy for Travels by Public Transit

Developing public transit is an inevitable choice for constructing a sustainable urban transport system, but public transit operation enterprises' policy loss is an indisputable fact. With Nanjing as an example, the government's subsidy for surface public transit was more than RMB500 million in 2010. Therefore, if ridership is clear as given, the amount of subsidy for each person time can be calculated (Table 1).

It may be known from the above table that the cost for travels by public transit is around 1.97 Yuan/person time, wherein, 1.40 Yuan is the amount actually consumed by passengers, and the remaining 0.57 Yuan/time is the governmental subsidy. This is equivalent to a subsidy of 0.562 Yuan/time for passengers (excluding senior citizens, under-aged passengers, and passengers living by government relief) taking surface public transit in Beijing in 2005 (Hao et al. 2009). However, if the utilization factor of subsidy fund is considered, then the actual subsidy for single travel by public transit will probably be less than 0.5 Yuan for management losses.

In such a case, in comparison with the subsidy for travels by cars, does such a subsidy appear high or low?

Table 1 Calculation table of the operating cost of and operating subsidy for surface public transit of Nanjing in 2010

	Parameters	Data	Calculation method
a	Ridership of surface public transit (billion person times)	10.08[a]	
b	Total cost (billion Yuan)	19.87[a]	
c	Cost per person time (Yuan/person time)	1.97	=b/a
d	Total tickets income (billion Yuan)	14.16	
e	Actual ticket consumption per person time (Yuan/person time)	1.40	=d/a
f	Average subsidy for single travel by public transit (Yuan/person time)	0.57	=(e − c)
g	Average time in bus per person time (min)	25.0[a]	
h	Average speed of surface public transit (km/h)	15.0[a]	
i	Average riding distance in bus (km)	6.25	=g × h
j	Subsidy per person time kilometer (Yuan/person time kilometer)	0.091	=f/i

Note [a] Nanjing Urban Planning Bureau 2011

4.2 Calculation of Subsidy for Free Parking

The full-process expense for travel by cars generally includes the following 3 parts: gas fee, parking fee, and cars depreciation. Currently, the domestic gasoline price in China mainly fluctuates according to the international gasoline price. And, since there is no governmental subsidy for cars depreciation, this article does not consider the subsidy for these two sectors, but only focuses on the subsidy of parking fee.

In China, most employers do not provide car parking space for employees. But in recent years, some large-scale commercial institutions have successively issued the policy of free parking for consumption reaching a certain amount in order to attract more customers, and this behavior is worthy of our consideration. For example, the Golden Eagle Emporium at Nanjing Xinjiekou Business Circle states in its free parking policy that customers may enjoy 2 h/4 h/8 h (upper limit) free parking for shopping of accumulatively over 400 Yuan/800 Yuan/1,500 Yuan on the same day, and the excessive part needs payment according to the standard of parking charge in Nanjing. As a matter of fact, two metro lines and dozens of bus lines pass by the Nanjing Xinjiekou Business Circle, and it is not really necessary to travel there by car.

On November 2, 2013, we conducted an investigation at the parking lot of Golden Eagle Emporium, and obtained 147 effective questionnaires in total. Data showed that these cars enjoyed 3.35 h free parking on average. According to the parking charging standard of Nanjing, they obtained a parking subsidy of averaged 19.5 Yuan in total actually. Each car carried 2.25 persons on average, so the unit subsidy amount was 19.5/2.25 = 8.67 Yuan/person time, about 15.2 times of the subsidy for surface public transit, as shown in Fig. 1. The calculation process is as shown in the following table (Table 2).

Obviously, free parking has brought about a bigger customer flow to the emporium, but also it has given rise to a new problem: The parking subsidy arising from free parking was by far higher than the governmental subsidy for public transit operation. According to the result of interviews, the respondents welcomed free parking; but when asked what if free parking were cancelled, over 30 % of them

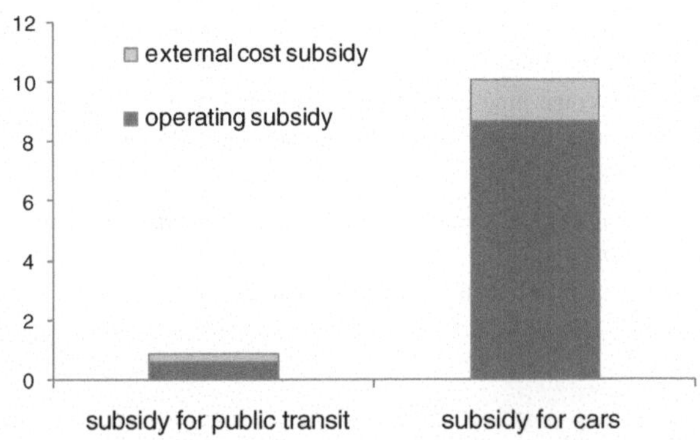

Fig. 1 Comparison chart of subsidy for public transit and cars

Table 2 Calculation table of subsidy amount under the free parking policy of the parking lot constructed by the Nanjing Golden Eagle Emporium

	Parameters	Data	Calculation method
a	Average duration of free parking enjoyed (hour)	3.35	
b	Parking subsidy actually obtained (Yuan)	19.5[a]	
c	Average passengers in each car (person time)	2.25[b]	
d	Unit subsidy amount (Yuan/person time)	8.67	=b/c
e	Unit subsidy for car/unit subsidy for public transit	15.2	=d/0.57

Note [a] According to the parking charging standard newly implemented in Nanjing in 2012, the parking fee/15 min charged by the off-road parking in first-tier areas was 1.5 Yuan, and the parking for the first 15 min was free

[b] *Data source* investigation data of November 2, 2013

expressed that they would consider changing the travel mode. Therefore, free parking has actually induced social injustice and the ambiguous orientation of transportation development (Shoup 1997). The practice of Golden Eagle Emporium is not a single case. Nowadays, this policy is also implemented by more than a dozen of commercial retail giants in the Nanjing City Center (Xinjiekou Business Circle, and Gulou Business Circle) and large-scale warehousing shopping malls such as IKEA and METRO in the outskirts of the city. Nanjing Jinrunfa Supermarket provides 2 h free parking on condition of shopping of only 50 Yuan, and becomes the merchant presenting the most lenient free parking conditions. In addition, large-scale merchants in multiple medium- and large-scale cities such as Beijing, Shanghai, and Guangzhou, also have the policy of free parking for shopping of over a certain amount, and undoubtedly, this attracts citizens' patronage by driving, and becomes one of the important factors inducing urban traffic congestion and air pollution.

4.3 Calculation of Subsidy for External Cost

Currently, China does not charge fees related to external cost (including air pollution, noise pollution, climatic warming, and traffic accident) during the driving, so it may be regarded as an implicit subsidy. The estimated total external cost of the four factors combined is 0.275 Yuan/VKT in 2005. Similarly, the bus has external cost, which is around 0.864 Yuan/VKT (Wang 2011). In case the external cost of buses is on average amortized to individuals, and the average passenger capacity of a bus is 35 passengers, then the per capita value of public transit subsidy for only external cost will be 0.864/35 = 0.025 Yuan/VKT, only 1/11 of that for car.

According to investigation, 147 drivers' average round-trip mileage for shopping is 11.56 km, so the external cost of travel by car is 11.56 × 0.275 = 3.18 Yuan, and the per capita value is 3.18/2.25 = 1.41 Yuan/person time; if there are 35 passengers in a bus, then the external cost arising from the same-distance driving by bus will be

Table 3 Calculation table of external cost of travel

	Parameters	Data	Calculation method
a	External cost of unit car VTK (Yuan/VTK[a])	0.275[b]	
b	External cost of unit bus VTK (Yuan/VTK)	0.864[b]	
c	External cost of unit bus passenger VKT (Yuan/person time VKT)	0.025	=b/35
d	Average round-trip mileage of cars (km)	11.56[c]	
e	Average number of people carried by car (person time)	2.25[c]	
f	Per capita external cost of each travel for shopping by car (Yuan/person time)	1.41	=a × d/e
g	Per capita cost of each travel for a same mileage by bus (Yuan/person time)	0.285	=c × d

Note [a] VKT is short for vehicle kilometers traveled
[b] Wang 2011
[c] *Data source* data of investigation on November 2, 2013

0.864 × 11.56/35 = 0.285 Yuan/person time, which is by far lower than that of travel by car (Table 3).

The comparison of subsidy for travels by public transit and car is as shown in Fig. 1. Obviously, no matter whether in terms of operating subsidy or the implicit subsidy for external cost, the whole society's subsidy for travels by public transit is much less than that for travels by car, and the total subsidy for car (10.08 Yuan/person time) is 11.8 times of that for public transit (0.855 Yuan/person time), which is extremely unreasonable. This creates a social polarization and an unfair policy for the dominant urban transportation mode. For travelers and consumers, they are encouraged to use cars more by broad Downs Law (Downs 1992).

5 Conclusion and Discussion

According to the analysis, the subsidy arising from both the operation section and external cost for public transit is by far lower than that for car, and after the two items are lumped together, the difference in unit subsidy is discovered to be even more than 10 times. This actually embodies "car priority," instead of "public transit priority." What is more, compared with the commuter subsidy in North America, these subsidies seem to be more implicit, and have aroused few people's attention. The investigation on residents sufficiently shows the stimulative function of free parking on travel by car. Obviously, transport subsidy deeply affects travel behavior, which in turn affects the determined urban transportation development orientation.

Concerning subsidies for various travel modes, we shall first eliminate the welfare of parking prices, namely, de-welfarization. Under the circumstances that urban land resources are extremely scarce, the behavior of executing low price not corresponding to the high cost of parking space resources, and even providing free parking to the public is distinctly featured by the maximization of suppliers' welfare, instead of social welfare, and is actually partial to the high-income group. The substance of subsidy rests with wealth redistribution, and can only flow from the high-income group to the low-income group. Therefore, free parking toward high-income group is naturally neither reasonable nor fair.

Another problem shall be emphasized, namely, the appertaining parking facilities. The government shall prudentially consider the demand-driven element, that is, the scale of appertaining parking spaces, especially in urban central areas, and shall not calculate the number of parking spaces to be constructed with trip generation rate index widely applied to Traffic Impact Analysis (TIA), but shall lower the standard of appertaining parking spaces constructed, and make the highest, instead of the lowest, construction standard. Under such circumstances, developers even do not have to construct appertaining parking spaces for the reason of cost control, and so, it may be called "cancelling the construction of appertaining parking facilities." Considering relatively low supply level, the market will only tighten the supply of parking spaces, and raise the parking fee (Shoup 1999; Willson 1995, 2013). Since the 1990s, Singapore has started to tighten the quantity of parking spaces in downtown area, and allowed developers to reduce the construction of appertaining parking spaces by 20 % at the most. A highlighted case is that the Market Street Auto Park located in CBD and initially constructed in the 1960s provided 704 parking spaces, but these parking spaces were closed in 2011 (Land Transport Authority (LTA) 2013).

It is an effective way to promote the establishment of a sustainable and transit-oriented urban transportation development mode through so-called push-pull effect, i.e., the subsidy for travels by public transit and the taxation during the use of the car (Bly and Oldfield 1986; Sakai and Shoji 2010; Savage and Schupp 1997). Obviously, starting with the construction of transit-oriented urban transportation development mode, we shall provide subsidy for public transit, eliminate all subsidies for the car, and raise various taxes from owning to using cars. The taxation aiming at cars and the subsidy aiming at public transit shall be corresponding, that is, the various over-levied taxes shall be applied to the development of public transit. Specifically, we shall raise fuel tax and innovatively charge parking adjustment fee, and a tax reform aiming at both driving and parking of cars is undoubtedly advisable, and will be effective eventually.

Finally, the author considers that the built environment and public transit facilities should play an important role in promoting public transit development and building Transit Metropolis, but reemphasizes that the economic factor should never be ignored, since various improper subsidies will surely lead to consequences not in accordance with the original intentions of urban and transportation planning. When built environment is out of order and fails in creating transit-oriented circumstance, reasonable economic means could be able to turn the passive situation around.

Acknowledgments This research work was supported by National Natural Science Foundation of China (No. 51308281) and Jiangsu Provincial Natural Science Foundation of China (No. BK2012728).

References

Abou-Zeid M, Witter R, Bierlaire M, Kaufmann V, Ben-Akiva M (2012) Happiness and travel mode switching: findings from a Swiss public transportation experiment. Transp Policy 19 (1):93–104

Bly PH, Oldfield RH (1986) The effects of public transport subsidies on demand and supply. Transp Res Part A Gen 20(6):415–427

Downs A (1992) Stuck in traffic: coping with peak-hour traffic congestion. Brookings Institution Press, Washington, DC

Hao J, Zhou W, Huang H, Guan H (2009) Calculating model of urban public transit subsidy. J Transp Syst Eng Inf Technol 9(2):11–16

Heinen E, van Wee B, Maat K (2010) Commuting by bicycle: an overview of the literature. Transp Rev 30(1):59–96

Land Transport Authority (LTA) (2013) Land transport masterplan. LTA, Singapore

Litman T (2005) Part 2: rail transit and commuter rail: impacts of rail transit on the performance of a transportation system. Transp Res Rec J Transp Res Board 1930(1):21–29

Nanjing Urban Planning Bureau (2011) Nanjing Municipal Commission of Housing and Urban-Rural Development, Nanjing Transportation Bureau, Nanjing Institute of City & Transport Planning Co., Ltd., Annual Report on Transport Development in 2010 of Nanjing (in Chinese)

Sakai H, Shoji K (2010) The effect of governmental subsidies and the contractual model on the publicly-owned bus sector in Japan. Res Transp Econ 29(1):60–71

Savage I, Schupp A (1997) Evaluating transit subsidies in Chicago. J Public Transp 1:93–117

Shoup DC (1997) The high cost of free parking. J Plan Educ Res 17(1):3–20

Shoup DC (1999) The trouble with minimum parking requirements. Transp Res Part A Policy Pract 33(7):549–574

Small KA, Verhoef ET (2007) The economics of urban transportation. Routledge, London and New York

Susuki H, Cervero R, Iuchi K (2013) Transforming cities with transit: transit and land-use integration for sustainable urban development. The World Bank, Washington, DC

Wachs M (1993) Learning from Los Angeles: urban form, and air quality. Graduate School of Architecture and Urban Planning, University of California, Los Angeles

Wang R (2011) Autos, transit and bicycles: comparing the costs in large Chinese cities. Transp Policy 18(1):139–146

Willson RW (1995) Suburban parking requirements: a tacit policy for automobile use and sprawl. J Am Plann Assoc 61(1):29–42

Willson RW (2013) Parking reform made easy. Island Press, Washington, DC

Willson RW, Shoup DC (1990) Parking subsidies and travel choices: assessing the evidence. Transportation 17(2):141–157

Yang Y, Qi K, Qian K, Xu Q, Yang L (2010) Public transport subsidies based on passenger volume. J Transp Syst Eng Inf Technol 10(3):69–74

Supporting Industry of Xinjiang Urbanization Based on Low Carbon Economy

Jie Zhang

Abstract The aim of this paper is to analyze the supporting industry of urbanization in Xinjiang, with particular reference to low-carbon city development, and to choose eco-friendly industry which gradually transits to low-carbon urbanization. Industrial development and urbanization are two important aspects in the process of regional economic development, they promote each other, restrict each other. The precondition of realizing urbanization, which is based on the industrial development and support, the development of the new type of urbanization, is more based on the industry with support role. Using WT model to choose new-type urbanization supporting industry in Xinjiang, we choose 15 supporting industries among environment-oriented perspective, employment-oriented perspective, and tax-oriented perspective, respectively. The paper concludes that a move toward low-carbon urbanization is contingent on changing supporting industry with evolving regional industry for urban development and regeneration.

Keywords Low carbon economy · Supporting industry · Urbanization · Xinjiang

1 Introduction

Cities are tended to create a lot of employment opportunities, but per capital carbon emissions are huge. During the urbanization, environmental pollution and greenhouse gas emissions are becoming the most significant environmental issues in Xinjiang; thus, the sustainable development and revival of the region is impossible to use the conventional path of encouraging urbanization at the expense of the environment. How to provide enough employment opportunities and reduce per capita carbon emissions during the process of urbanization in Xinjiang is a vital problem.

J. Zhang (✉)
School of Economics and Management, Shihezi University, Beisi Road, Shihezi 832003, Xinjiang, People's Republic of China

S. Feng et al. (eds.), *Low-carbon City and New-type Urbanization*, Environmental Science and Engineering, DOI 10.1007/978-3-662-45969-0_7

1.1 Urbanization in Xinjiang

Urbanization is a feature and consequence of Xinjiang's economic development process. According to statistical yearbook of Xinjiang (2013), 9.82 million people belong to urban population in 2012. In 21 cities, the average population of each city is only 330,000 except Urumqi. During the past 12 years, urbanization level of Xinjiang is higher than before, but the development still relatively lags behind, and the development speed is slow. From Fig. 1, the urbanization rate of Xinjiang continues to rise, as well as the urban population increases, but compared with other regions of China, the rising level is still relatively backward. The urbanization rate of Xinjiang is 33.75 % in 2000, while that of China is 36.22 %, and the gap of them is just 2.47 %. But the gap is gradually increasing to 8.57 % in 2012. For the last twelve-year urbanization rate which rose from 33.75 to 44 % in Xinjiang, average annual growth rate is just 0.85 %, compared with the national urbanization rate that increased from 36.22 to 52.57 %, and average annual growth rate is 1.36 %, and the fact is that Xinxiang's urbanization development speed is obviously slow.

Non-agricultural industrial employment population is growing steadily. During the period 2004–2005, the ratio improved greatly, but more than half of the proportion is agricultural population, which is still bigger than the normal level. On the other hand, with urban agglomeration and industrial development rapidly moving forward in Xinjiang, the cities are guided by "a new round of urban and rural construction planning," and the construction level of urban infrastructure has improved significantly; meanwhile, the government has been building urban parks and green space, pushing forward the construction of the urban living environment, so that the urban environmental quality improves steadily.

In addition, urban primacy ratio is 4.04 in Xinxiang, 2 times higher than normal. The low proportion of big cities in Xinjiang, the imbalance between medium and large cities, transition less node city, and urban scale structure are unreasonable and lack big cities with population between 0.5 and 1 million. At the same time, the differences of urbanization level between the south, the north, and the east of

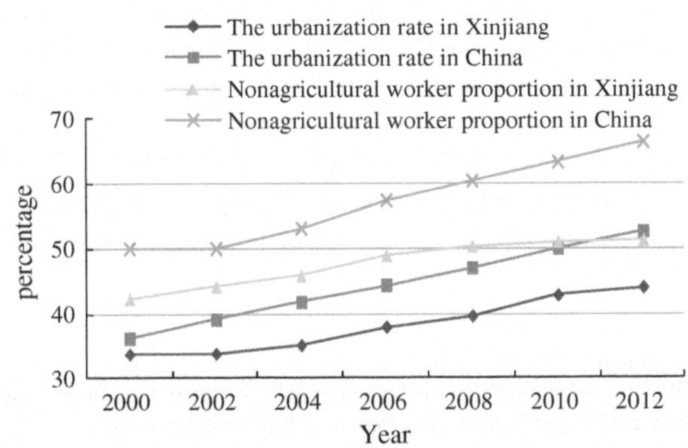

Fig. 1 Comparison of urbanization between Xinjiang and China from 2000 to 2012

Xinjiang, and the imbalance of the urban population distribution restrict the transfer mechanism of city function, influence the radiation, and play leading role between cities. It is hard to produce synergies.

1.2 Choice of Low-carbon Industry to Drive Urbanization

As a common practice in China, the industrial structure evolution refers to the change in proportionate relationships involving three major industries. That is, the structure in which primary industry takes the first place, the secondary industry the second, and the tertiary industry the third initially develops into the inverse structure ultimately (Li and Li 2003). Lin and Chen (2011) argue that heavy-industry-oriented development strategy will result in lower urbanization rate and higher urban-rural inequality. The greater the degree of heavy-industry-oriented development strategy is, the lower the urbanization rate is, and the higher the urban–rural inequality is (Lin and Chen 2011). In this process, urbanization develops in pace with the industrial structure evolution since agriculture development is the initial driving force, and second and tertiary industry development is the main motivation of urbanization (Wei and Li 2009).

It is hard to predict the impact of urbanization which affects the energy intensity, because urbanization increases economic activity through a higher concentration of consumption and production, but urbanization also leads to economies of scale and provides the opportunity for increases in energy efficiency (Madlener 2011; Madlener and Sunak 2011). Sadorsky's (2013) study shows the impact of urbanization on energy intensity is mixed. A positive and statistically significant coefficient on urbanization implies that the net effect of these two impacts is to increase energy intensity (Sadorsky 2013). So it is important to choose low-carbon industry driving urbanization, it requires the stimulation of markets for new, emerging "green industries," leading to the creation of more and more "green jobs." Local governments could be more effective in greening production by developing one-stop business services to help the businesses to reaching the conservation goals, supporting green industry start-ups, facilitating the development of eco-industrial parks, and supporting training tailored to local labor market needs (Piacentini 2012).

Emissions from industrials represent a large proportion of Xinjiang's total; a substantial expansion path which depends on energy-intensive industry to drive urbanization has been drawing to a close. Changing urbanization pattern is the inevitable choice to achieve sustainable development (Smulders et al. 2011). Therefore, Xinjiang must change the mode of urbanization, relying on low-carbon industry to drive urbanization. The purpose of this paper is to choose low-carbon industry based on the relationship between urbanization and industrialization using WT mode of Xinjiang. The following sections of the paper set out the empirical model, data, empirical results, implications, and conclusions.

2 Methodology

Weaver–Thomas (WT) model, put forward by Weaver and approved by Thomas, is an effective method used to select strategic industry. The principle of WT model is to establish a close approximation distribution comparing with observation distribution (physical distribution) and the hypothesis distribution, and then, according to the regional index system of the industry in all kinds of competition, to optimize core competitiveness of strategic industries of the region. This also shows that the model is adapted to the regional leading support industry choice, so this article uses this model to study the supporting industry.

2.1 Computational Method

Using WT model to analyze the choice of leading support industry, steps are as follows:

First step: According to certain indicators, put all industries in order, from big to small, and then adopt the index which has a supporting industry; keeping the first one as the index of the supporting industry, estimate its value with WT model, put all values together and find out the value of the comparison, and finally, the minimum value is determined. The minimum number of the industry is the supporting industry under the index number.

Second step: Calculate various indicators to determine the number of supporting industry average of the values obtained by using the model for all indicators to determine the number of supporting industries.

Third step, form a comprehensive sort matrix with the order value of industry relative to the index. For each weighting, calculate the order value of each industry, pick up the industries of the top places to be the support industries of Xinjiang.

The model calculation is as follows:

$$\mathrm{WT}_{nj} = \sum_{i=1}^{m} \left(C_i^n - 100\mathrm{EN}_{ij} / \sum_{i=1}^{m} \mathrm{EN}_{ij} \right)^2 \tag{1}$$

Among them, $C_i^n = \begin{cases} 100/n, & i \leq n \\ 0, & i > n \end{cases}$, n denotes the n industry after the reordering all industry by indicator j, and WT_{nj} denotes the combination of n industry's j indicator.

$$nq_j = k, \quad nq = \left(\sum_{j=1}^{m} nq_j \right) / n \tag{2}$$

And k denotes the minimum combination number of the WT position number, when meeting the conditions of $\mathrm{WT}_{1j} = \min\ \mathrm{WT}_{kj}(K = 1,\ 2,\ \ldots m)$.

$$D = \begin{bmatrix} D_{11} & D_{12} & \ldots & D_{1n} \\ D_{21} & D_{22} & \ldots & D_{2n} \\ \ldots & \ldots & \ldots & \ldots \\ D_{m1} & D_{m2} & \ldots & D_{mn} \end{bmatrix} = \{D_{ij}\}_{m\times n} \tag{3}$$

$$E_i = \sum_{j=1}^{N} e_j D_{ij} \tag{4}$$

EN_{ij} is the index value of the i industry's j indicator $(i = 1,\ 2,\ \ldots m)$, m denotes total industry number $(j = 1,\ 2,\ \ldots N)$, N is the total number of indicator, nq_j representing the number of supporting industries corresponds to indicator j, nq is the total number of supporting industry corresponding to all indicators, D is the comprehensive ranking matrix of supporting industry, D_{ij} denotes the index value of i industry compared to j indicator, its value can be positive and negative, e_j denotes the index weighting value of indicator j, and E_i is the comprehensive ranking value of supporting industry (Ming 2012; Zhang and Luo 2011).

2.2 Indicator Selection

According to supporting industry selection principles of urbanization, combining with the characteristics of Xinjiang's economy, we selected eight indicators from different sides reflecting supporting industry system as follows.

The employment scale (JL_i): $JL_i = L_i/\sum L_i$. Among them, JL_i denotes the employment scale of the i industry, L_i is average number of worker in i industry, and $\sum L_i$ is average number of employees for all industrial.

The profit tax scale (SR_i): $SR_i = R_i/\sum R_i$. Among them, SR_i denotes the size of tax for the i industry, R_i denotes the profit tax amount of i industry, and $\sum R_i$ is the profit tax amount of all industries.

The production energy consumption coefficient (NC_i): $NC_i = NM_i/Y_i$. NC_i denotes the energy consumption coefficient of i industry, NM_i the comprehensive energy consumption of the i industry, and Y_i is the output of the industry.

The output tax rate (δ_i): $\delta_i = R_i/Y_i$. Among them, δ_i is the output tax rate for the i industry, R_i is the tax rate in i industry, and Y_i is total output value of the industry.

The added value scale of industry (GY_i): $GY_i = Z_i/\sum Z_i$. Among them, GY_i denotes the added value scale of i industrial, Z_i is the added value of i industry, and $\sum Z_i$ is total added value for all industries.

The overall labor productivity (Q_i): $Q_i = \Delta Y_i/L_i$. Among them, Q_i denotes the overall labor productivity, ΔY_i is the variation of the i industry, and L_i is the labor force of the i industry.

The coefficient of elasticity of demand (ρ_i): $\rho_i = (\Delta D_i/D_i)/(\Delta W/W_0)$. Among them, ρ_i denotes the product demand elasticity coefficient of the i industry, $\Delta D_i/D_i$ is demand growth rate of the i industry's product, and $\Delta W/W_0$ is the growth rate of GDP.

The employment elasticity coefficient (E_i): $E_i = (\Delta P_i/P_i)/(\Delta W/W_0)$. Among them, E_i denotes the employment elasticity coefficient of the i industry, $\Delta P_i/P_i$ is demand growth rate of the i industry's employment, and $\Delta W/W_0$ is the growth rate of GDP.

3 Results

3.1 Calculate Each Indicator's Value of Every Industry

We find that the supporting industry in the process of new-type urbanization in Xinjiang is mainly concentrated in the non-agricultural industries. Due to the limitations of statistical data and data availability, this paper chose 37 industries to analyze; they are as follows: 1. Mining and Washing of Coal, 2. Extraction of Petroleum and Natural Gas, 3. Mining and Processing of Ferrous Metals Ores, 4. Mining and Processing of Nonferrous Metals Ores, 5. Mining and Processing of Nonmetal Ores, 6. Processing of Food from Agricultural Products, 7. Manufacture of Food, 8. Manufacture of Wine, Beverages, and Refined Tea, 9. Manufacture of Tobacco, 10. Manufacture of Textile, 11. Manufacture of Textile Wearing Apparel, Footwear, and Caps, 12. Leather, Fur, Feather and Related Products Manufacturing, 13. Processing of Timber, Wood, Bamboo, Cane, Grass Products, 14. Manufacture of Furniture, 15. Manufacture of Paper and Paper Products, 16. Printing and Copying of Medium for Record, 17. Oil Processing, Coking and Nuclear Fuel Processing, 18. Raw Chemical Material and Chemical Products, 19. Manufacture of Medicine, 20. Manufacture of Chemical Fiber, 21. Manufacture of Rubber Products, 22. Manufacture of Nonmetal Mineral Products, 23. Smelting and Pressing of Ferrous Metals, 24. Smelting and Pressing of Nonferrous Metals, 25. Manufacture of Metal Products, 26. Manufacture of General Purpose Machinery, 27. Manufacture of Special Purpose Machinery, 28. Manufacture of Automobile, 29. Manufacture of Electric Equipment and Machinery, 30. Telecom Equipment, Computer, and Other Electronic Equipment, 31. Measuring Instruments and Machinery for Cultural Activity and Office Work, 32. Manufacture and Repairing of Metal Products, 33. Electricity and Thermal Production and Supply, 34. Gas Production and Supply, 35. Water Production and Supply, 36. Construction industry and 37. Service industry. According to Xinjiang statistical yearbook 2013 and 2012 (Xinjiang Bureau of Statistics 2012, 2013), calculate each indicator's value of every industry in Table 1.

Table 1 Indicator value of each industry

Industry	Employment scale	Profit tax scale	Production energy consumption coefficient	Output tax rate	Added value scale	Overall labor productivity	Coefficient of elasticity of demand	Employment elasticity coefficient
1	0.01835	0.03331	1.41078	26.4	0.04585	24.1	1.633	−0.0917
2	0.02403	0.4493	0.59971	60.8	0.38705	153.4	−0.2412	0.5218
3	0.00382	0.01793	0.52813	27.4	0.01648	41.5	1.2008	1.877
4	0.00272	0.01629	0.61878	48.4	0.01326	46.9	0.7842	−0.3517
5	0.00112	0.00154	0.79327	18.1	0.00226	19.4	−1.2064	0.2215
6	0.00883	0.01451	0.35709	8.4	0.01581	17.2	1.3596	0.3595
7	0.00627	0.0058	0.59442	8.5	0.00901	13.4	0.8022	−0.2164
8	0.00402	0.01435	0.2279	30.4	0.01195	28.6	1.1947	1.2295
9	0.00029	0.01345	0.02969	70	0.00891	293.6	1.2785	0.5001
10	0.0125	0.00312	0.50271	4.4	0.00774	6	0.5714	0.0923
11	0.00064	0.00029	0.06061	10.4	0.00051	7.6	3.3415	3.4111
12	0.00013	0.00027	0.07177	8	0.00051	37.6	−1.2383	−2.7281
13	0.00029	0.00058	1.50618	14.7	0.00053	17.5	0.9241	−0.3479
14	0.00039	−0.00005	0.14822	−1.2	0.00033	8.2	0.9514	−13.3042
15	0.0017	0.00134	1.40978	11.8	0.0017	9.6	1.1197	0.422
16	0.00051	0.00023	0.21485	8.7	0.00053	10	0.6567	0.1231
17	0.01443	0.09501	0.78772	10.6	0.06888	56.2	0.5079	0.6148
18	0.01807	0.04564	2.38617	16.5	0.06516	25.5	0.9882	1.8135
19	0.00116	0.00191	0.30999	18.8	0.00169	14	2.4441	0.7418
20	0.00278	0.00113	0.96501	2.4	0.00413	14.3	−1.8002	−0.8955
21	0.00414	0.00484	0.17296	8.3	0.00604	14	2.9993	0.9606
22	0.01746	0.01668	1.86503	9.7	0.02639	14.6	1.243	1.3144

(continued)

Table 1 (continued)

Industry	Employment scale	Profit tax scale	Production energy consumption coefficient	Output tax rate	Added value scale	Overall labor productivity	Coefficient of elasticity of demand	Employment elasticity coefficient
23	0.01366	0.01018	1.71539	2.6	0.02902	13.6	0.6942	1.5388
24	0.00555	0.00941	0.71334	7.3	0.01822	25.3	3.857	2.7404
25	0.00258	0.00214	0.14692	5.2	0.00284	10.5	−0.6311	0.4124
26	0.00069	0.00081	0.25519	12.8	0.00094	13.6	4.1143	2.3396
27	0.00124	0.0022	0.08086	13.2	0.00239	18.6	0.7215	−0.7752
28	0.0006	−0.00037	0.03688	−2.2	0.0002	3.1	6.4405	0.478
29	0.00289	0.01841	0.03426	15.6	0.01113	37.1	0.192	0.0072
30	0.00138	0.0008	1.6422	6.7	0.00002	11.2	1.2004	0.7574
31	0.00001	0.00006	0.18282	27.3	0.00004	53.3	−11.3886	−17.1721
32	0.00016	0.00008	0	7.5	0.00017	10.5	2.2747	4.1183
33	0.01728	0.0359	1.47446	10.8	0.08765	36.3	2.17	0.7197
34	0.00118	0.00354	0.16126	18.2	0.00383	31.4	4.1572	2.6391
35	0.00085	0.00015	1.19163	4.4	0.0011	12.4	1.7432	2.3283
36	0.20347	0.04997	0.06002	5.7	0.12276	5.8	1.5708	0.8975
37	0.60482	0.17925	0.41006	12.3	0.14773	2.35	1.424	0.4439

3.2 Calculate the WT Value and the Corresponding Number of Supporting Industries

All the indicators' values of each industry are arranged in sequence from high to low, and calculate the corresponding number of supporting industry's WT value in Table 2.

According to each indicator's number of supporting industry in Table 2, its weighted average, it can determine the number of supporting industry in the process of the urbanization in Xinjiang $nq = 15$. Selected indicators of this paper have positive and inverse difference, indicator of industrial energy consumption coefficient as reverse, and the rest as positive.

3.3 Determination of Supporting Industries

According to different priorities in the new-type urbanization, the determination of supporting industry is divided into the following three aspects:

First, based on the pursuit for resource-saving, environment-friendly, and green GDP, give 30 % weight to energy consumption coefficient, distribute other indicators' weight equally, and according to the size of the final rankings, determine whether it is the basis of supporting industries.

Secondly, a new type of urbanization based on "employment" as the leading factor, the basis of the employment scale, and employment elasticity coefficient, respectively, gives the weight of 0.2, and other indicators' weights are evenly distributed.

Thirdly, to support "profits and tax" as the leading industry, the tax scale and output tax rate gives the weight of 0.2, respectively, and other indicators' weights are evenly distributed.

According to the comprehensive sequencing rank calculation, identify 15 supporting industries which dominate the urbanization process of Xinjiang in different aspects. Results in Table 3 show that, currently, the priority choice of supporting industry in Xinjiang should be as follows:

Environment-oriented perspective: 32. Manufacture and Repairing of Metal Products, Machinery, 9. Manufacture of Tobacco, 11. Manufacture of Textile Wearing Apparel, Footwear and Caps, 34. Gas Production and Supply, 24. Smelting and Pressing of Nonferrous Metals, 26. Manufacture of General Purpose Machinery, 19. Manufacture of Medicine, 28. Manufacture of Automobile, 35. Water Production and Supply, 3. Mining and Processing of Ferrous Metals Ores, 2. Extraction of Petroleum and Natural Gas, 18. Raw Chemical Material and Chemical Products, 23. Smelting and Pressing of Ferrous Metals, 8. Manufacture of Wine, Beverages and Refined Tea, and 22. Manufacture of Nonmetal Mineral Products.

Employment-oriented perspective: 32. Manufacture and Repairing of Metal Products, Machinery, 11. Manufacture of Textile Wearing Apparel, Footwear and

Table 2 WT value of each industry and number of support industry

Employment scale	Profit tax scale	Production energy consumption coefficient	Output tax rate	Added value scale	Overall labor productivity	Coefficient of elasticity of demand	Employment elasticity coefficient
2,345.06	1,103.17	1,422.38	1,198.08	836.43	1,074.32	2,061.51	844,867.41
2,643.43	1,165.99	989.32	873.85	754.73	892.82	1,698.23	843,001.76
2,811.13	1,238.93	705.44	658.64	711.47	787.65	1,486.26	841,748.75
2,945.68	1,302.36	513.42	513.08	733.03	726.32	1,378.98	840,397.83
3,053.61	1,377.6	363.33	430.13	780.51	673.44	1,304.84	839,241.98
3,139.18	1,440.63	269.86	371.18	825.41	637.41	1,255.47	838,144.01
3,210.24	1,493.49	214.39	324.53	875.93	617.19	1,238.48	837,096.94
3,274.91	1,537.37	182.07	291.34	920.05	604.91	1,229.79	836,177.09
3,333.07	1,576.74	155.45	266.21	957.78	599.07	1,225.28	835,234.35
3,383.41	1,610.34	137.94	247.6	993.25	594.56	1,227.43	834,406.08
3,428.56	1,640.35	128.57	235.32	1,025.31	592.1	1,231.69	833,615.91
3,467.89	1,670.56	121.78	225.97	1,054.13	595.67	1,238.13	832,939.07
3,502.5	1,697.84	115.77	218.36	1,081.61	599.28	1,245.04	832,315.3
3,534.14	1,725.97	113.81	212.79	1,105.98	603.61	1,252.47	831,718.58
3,562.43	1,752.02	113.5	210.16	1,128.85	608.07	1,258.82	831,170.93
3,587.75	1,776.59	117	207.63	1,150.84	613.96	1,265.12	830,637.2
3,610.7	1,799.13	122.63	206.09	1,172.38	619.84	1,272.38	830,174.07
3,632.3	1,820.35	129.52	205.7	1,192.1	625.27	1,282.46	829,730.1
3,652.24	1,839.7	138	207.18	1,210.85	630.41	1,292.06	829,340.76
3,670.56	1,857.55	146.64	208.6	1,228.3	635.2	1,301.89	828,959.93
3,687.41	1,874.21	155.05	210.13	1,244.4	639.66	1,313.38	828,602.92
3,702.91	1,889.68	163.83	211.6	1,259.6	643.82	1,324.62	828,219.86

(continued)

Table 2 (continued)

Employment scale	Profit tax scale	Production energy consumption coefficient	Output tax rate	Added value scale	Overall labor productivity	Coefficient of elasticity of demand	Employment elasticity coefficient
3,717.27	1,904.13	172.32	213.47	1,1273.65	648.44	1,335.78	827,856.31
3,730.74	1,917.72	180.47	215.71	1,287.04	653.33	1,347.47	827,480
3,743.39	1,930.35	188.49	218.16	1,299.58	658.44	1,358.34	827,133.05
3,755.19	1,942.26	195.99	221.17	1,311.51	663.18	1,369.9	826,786.66
3,766.22	1,953.55	204.83	225.1	1,322.64	667.84	1,382.47	826,409.19
3,776.61	1,964.13	213.38	229.4	1,333.07	672.5	1,398.54	826,057.23
3,786.42	1,974.07	221.66	234.27	1,342.84	677.56	1,421.69	825,695.49
3,795.7	1,983.46	229.46	238.85	1,352.12	682.65	1,449.33	825,266.92
3,804.43	1,992.33	237.39	245.02	1,360.92	688.24	1,484.28	824,834.87
3,812.74	2,000.7	244.92	251.04	1,369.23	693.63	1,517.36	824,036.66
3,820.6	2,008.67	252.14	260.25	1,377.15	700	1,557.15	821,009.9
3,828.1	2,016.38	259.64	269.91	1,384.65	706.38	1,731.05	817,343.13
Number of support industry	Number of support industry	Number of support industry	Number of support industry	Number of support industry	Number of support industry	Number of support industry	Number of support industry
2	3	18	21	6	14	12	37

Table 3 Composite value and ranking of supporting industry in Xinjiang

Environment oriented		Employment oriented		Tax oriented	
Composite value	Rank	Composite value	Rank	Composite value	Rank
11,190.38	32	21,756.23	32	10,443.93	32
10,813.54	9	21,469.06	11	10,692.85	9
10,806.55	11	21,334.48	24	10,645.42	11
10,726.05	34	21,321.69	34	10,490.6	34
10,705.45	24	21,256.91	26	10,497.47	24
10,678.15	26	21,206.44	35	10,480.34	26
10,660	19	21,165.25	3	10,459.88	2
10,657.41	28	21,138.47	9	10,605.14	19
10,641.94	35	21,113.09	19	10,829.45	35
10,620.1	3	21,101.91	18	10,476.21	3
10,609.13	2	21,094	23	10,805.35	28
10,586.63	18	21,065.59	8	10,458.79	8
10,584.83	23	21,056.22	22	10,471.04	23
10,576.38	8	21,036.81	21	10,414.82	18
10,564.82	22	21,034.26	28	10,507.02	22
10,563.22	21	21,018.91	2	10,495.44	21
10,550.72	36	21,002.98	30	10,511.57	30
10,543.46	30	20,957.61	36	10,596.56	36
10,543.13	29	20,936.44	33	10,683.66	33
10,501.06	33	20,930.8	17	10,452.54	17
10,499.54	17	20,892.9	15	10,578.07	25
10,497.95	25	20,889.44	25	10,578.61	15
10,485.2	15	20,869.2	5	10,601.9	5
10,475.43	16	20,863.58	16	10,726.07	29
10,473.44	5	20,862.14	37	10,508.17	16
10,469.54	37	20,859.69	29	10,699.39	4
10,462.45	6	20,856.48	6	10,461.98	37
10,454.95	27	20,834.42	10	10,638.58	6
10,454.6	10	20,816.82	13	10,496.9	10
10,454.42	12	20,815.25	4	10,564.02	13
10,452.53	4	20,807.47	7	10,387.59	27
10,449.49	13	20,798	27	10,995.89	7
10,439.19	7	20,785.38	1	10,513.79	12
10,429.25	20	20,780.56	20	10,741.6	20
10,426.16	1	20,779.78	12	10,665.73	1
10,399.7	14	20,696.41	14	10,531.45	14
10,362.71	31	20,618.99	31	10,484.47	31

Caps, 24. Smelting and Pressing of Nonferrous Metals, 34. Gas Production and Supply, 26. Manufacture of General Purpose Machinery, 35. Water Production and Supply, 3. Mining and Processing of Ferrous Metals Ores, 9. Manufacture of Tobacco, 19. Manufacture of Medicine, 18. Raw Chemical Material and Chemical Products, 23. Smelting and Pressing of Ferrous Metals, 8. Manufacture of Wine, Beverages and Refined Tea, 22. Manufacture of Nonmetal Mineral Products, 21. Manufacture of Rubber Products, and 28. Manufacture of Automobile.

Tax-oriented perspective: 32. Manufacture and Repairing of Metal Products, Machinery, 9. Manufacture of Tobacco, 11. Manufacture of Textile Wearing Apparel, Footwear and Caps, 34. Gas Production and Supply, 24. Smelting and Pressing of Nonferrous Metals, 26. Manufacture of General Purpose Machinery, 2. Extraction of Petroleum and Natural Gas, 19. Manufacture of Medicine, 35. Water Production and Supply, 3. Mining and Processing of Ferrous Metals Ores, 28. Manufacture of Automobile, 8. Manufacture of Wine, Beverages and Refined Tea, 23. Smelting and Pressing of Ferrous Metals, 18. Raw Chemical Material and Chemical Products, and 22. Manufacture of Nonmetal Mineral Products.

4 Conclusion

According to the analysis of internal mechanism of the urbanization in Xinjiang, using the Weaver–Thomas model, we chose 15 supporting industries, to calculate the value of indicator and WT. This paper argues that well-selected industries at the regional level can support the growth of the new green sectors and speed up the transition to low carbon economies. These industries are the main impetus of new-type urbanization process. Nonetheless, current development of the secondary industry in Xinjiang mainly depends on traditional manufacturing industry and heavy industry, while petroleum and petrochemical industry disproportionately dominates secondary industry, and technology enhances the industry to advance relatively slowly. Obviously, Xinjiang should endeavor to push forward the development of eco-friendly industries through policy guidance and encouragement for technological advances. Only through this way, the quality and competitiveness of industrial enterprises could be improved rapidly. As the quality of the secondary industry development increases, the urbanization rate will go up accordingly.

Acknowledgments The research work was supported by National Natural Social Science Foundation of China under Grant No. 12CJY039 and Ministry of Education of Humanities and Social Science Project in China under Grant No. 12XJJC790006.

References

Li PX, Li CG (2003) Regional industry structure evolution and urbanization timing sequences analysis. Econ Prob 1:4–6 (in Chinese)

Lin JY, Chen B (2011) Urbanization and Urban-Rural Inequality in China, a new perspective from the government's development strategy. Front Econ China 6(1):1–21

Madlener R (2011) The impact of urbanization on urban structures and energy demand in developing countries. In: Smart energy strategies conference 2011 Zurich, 21–23 Sept 2011

Madlener R, Sunak Y (2011) Impacts of urbanization on urban structures and energy demand: what can we learn for urban energy planning and urbanization management? Sustain Cities Soc 1:45–53

Ming G (2012) Research on strategic industry development of Sichuan province based on weaver thomas model. J Chongqing Technol Business Univ (Nat Sci Ed) 05:35–39 (in Chinese)

Piacentini M (2012) Rationale and policies for the green growth of cities and regional economies. Int Econ Econ Policy 9:129–146

Sadorsky P (2013) Do urbanization and industrialization affect energy intensity in developing countries? Energy Econ 37(4):52–59

Smulders S, Bretschger L, Egli H (2011) Economic growth and the diffusion of clean technologies: explaining environmental kuznets curves. Environ Resource Econ 49:79–99

Wei J, Li M (2009) Empirical analysis on the promotion of industry structure evolution to urbanization. Forum Sci Technol Chin 11:83–87 (in Chinese)

Xinjiang Bureau of Statistics (2013) Xinjiang Statistical Yearbook. China Statistics Press, 2012, 2013 (in Chinese)

Zhang Z, Luo Z (2011) The choice of strategy industries by the Weaver-Thomas model. J Finance Econ 01:51–56 (in Chinese)

Comparison of Both Old and New Versions of the Evaluation Standard for Green Building in China with LEED in American

Jing-jing Wang, Yu-rong Zhang and Yuan-feng Wang

Abstract At present, energy saving and emission reduction have become a central task for all countries and governments, and green building has become a research hot spot at the same time. Green building is the building which maximizes the conservation of resources, protection of the environment, and realization of coordinated unification between human and nature in the whole life cycle of the building. Green building is an important embodiment of sustainable development strategy. At the beginning of the 1990s, the concept of green building was introduced to China. In recent years, in order to promote the development of green building, China issued a series of related approaches and normative documents, such as Evaluation Standard for Green Building (GB/T50378-2006). Recently, in order to improve the green building standard, the Ministry of Housing and Urban Construction gives out the announcement, approving the Evaluation Standard for Green Building as the national standard, numbered GB/T50378-2014. The US Green Building Rating System (LEED) is now considered to be the most sophisticated and influential among all kinds of environmental protection assessments. In the revision of the Evaluation Standard for Green Building, certain aspects of LEED were consulted, so there is similarity between the two standards. But Evaluation Standard for Green Building also has its own characteristics based on China's conditions. So this chapter pays attention to the comparison of the Evaluation Standard for Green Building (the latest revision), Evaluation Standard for Green Building (GB/T50378-2006), and US LEED standards. It uses the following eight aspects: evaluation phase, evaluation objects, index categories, refined indicators, scoring points, evaluation methods, evaluation results, and weight distribution to show the similarities and differences among the three standards. And it can point out the advantages and shortcomings of Evaluation Standard for Green Building in China and makes reasonable suggestions.

Keywords Comparison · Evaluation Standard for Green Building · LEED · Standard

J. Wang · Y. Zhang · Y. Wang (✉)
School of Civil Engineering, Beijing Jiaotong University, Beijing 100044, China

S. Feng et al. (eds.), *Low-carbon City and New-type Urbanization*,
Environmental Science and Engineering, DOI 10.1007/978-3-662-45969-0_8

1 Introduction

Green building refers to the architecture which makes scientific planning, guidance, and advanced applicable technology to decrease the consumption of resources and energy, reduces waste generation and damage to the ecological environment, provides users a comfortable, healthy working and living environment, and ultimately achieves symbiosis with nature in its entire life cycle (Zuo and Zhen 2014).

At the beginning of the 1990s, the concept of green building was introduced to China. In recent years, in order to promote the development of green building, China issued a series of related approaches and normative documents. On June 1, 2006, China began to implement Evaluation Standard for Green Building (GB/T50378-2006) [hereinafter referred to as the "standard (2006)"] (Wan et al. 2009). On November 15, 2007, the Green Building Evaluation Standard issued, which marked that green building evaluation in our country has entered a new stage.

Recently, the Ministry of Housing and Urban Construction gives out an announcement, approving the Evaluation Standard for Green Building as the national standard, numbered GB/T50378-2014,[1] [abbreviated as "standard 2014"] and it will come into effect from January 1, 2015. New version of Evaluation Standard for Green Building has more requirements and more extensive content than the 2006 version. In the revision process, the standard summarizes recent practical experience and research results of green building evaluation, carried out a number of research and trial and evaluation, used the experience of the relevant foreign advanced standards, and solicited opinions from the relevant parties. For the revised criteria, the target range has been extended, the evaluation phase is clearer, evaluation method is more scientific, and evaluation index system is more perfect and innovative.

The US Green Building Rating System (LEED) is now considered to be the most sophisticated and influential among all kinds of environmental protection assessments, green building assessment, and building sustainability assessment criteria all over the world. And many countries used it as the model to build their own green building and building sustainability of the Evaluation Standard (Seinre et al. 2014).

In this chapter, the two versions of the Evaluation Standard for Green Building (GB/T50378-2006) (Evaluation Standard for Green Building 2006) (GB/T50378-2014) were analyzed with LEED (Green Building Council 2008), in order to identify the advantage of the "standard (the latest revision)" comparing to the "standard (2006)," and put forward recommendations for continued improvement of the Evaluation Standard for Green Building.

---

[1] Evaluation standard for green building (draft) (EB/OL). Beijing: Ministry of Housing and Urban-Rural Development of the People's Republic of China; Available from http://www.mohurd.gov.cn/zcfg/jabwj_0/jsbwjjskj/201209/t20120919_211434.html (in Chinese).

2 Comparing the Similarities and Differences of China's Evaluation Standard for Green Building and LEED in USA

In the revision of the Evaluation Standard for Green Building, certain aspects of LEED were consulted, so there is similarity between the two standards. But Evaluation Standard for Green Building also has its own characteristics based on China's conditions. This section gives detailed analysis and comparison of LEED and Evaluation Standard for Green Building (old and new versions), which includes the following: the assessment phase, evaluation objects, index comparison, refined indicators, scoring points, evaluation methods, evaluation results, and the weight setting (Li and Ling 2011).

2.1 Evaluation Phase

Buildings that adopted LEED evaluation criteria can be divided into three phases according to the progress of the project: design, procurement, and construction phase. Design phase mainly refers to take energy, water, and building comfort of use into consideration during the design, such as air-conditioning systems, water recycling, lighting, and other aspects; Construction phase mainly refers to site controlling, construction waste management, and indoor construction and air quality control; Procurement phase mainly refers to the procurement of recycle materials, local materials, renewable materials, and low volatile materials. Commissioning, energy-saving device measurement, and the thermal comfort survey can be taken after the construction is completed.

The standard (2006) of green building assessment focuses on the operational phase. It says the evaluation shall be conducted one year after the building put into use. The evaluation emphasizes on the actual performance and running effect of the building.

The assessment phase of the revised standard can be divided into the design evaluation and operation evaluation, and adding architecture design and four environmental performances evaluation to it.

Operation evaluation is not only to evaluate green measures, but also to evaluate the practical effect produced by these green measures. In addition, it also pays attention to green footprint produced in construction process of green architecture and scientific management during normal operation.

Although the design evaluation is not as comprehensive as operation evaluation, it also has a great meaning and value. First, these green measures involved in the design evaluation have been proved to be effective, and these effective measures greatly ensure that the effect is obvious. Second, the design evaluation is a forward evaluation and can find problems early, thus helps solve problems and improves the evaluation effect. Finally, from the perspective of implementation, the design evaluation is easier than operation evaluation in a wide range.

2.2 Evaluation Objects

According to different types of buildings and different stages of life cycle, the core products of LEED can be subdivided into eight categories: existing building, core and shell, new construction, school, neighborhood development, retail health care, homes, and commercial interiors (Ouyang 2008). The evaluation objects of standard (2006) are divided into residential buildings and public buildings which includes three types: office, shopping malls, and hotels. The new version green building standard extends the applicable scope to all types of civil construction.

Considering China's construction market situation at that time, when old version of the standard compiled, it mainly focused on evaluating the residential construction with large amount and public buildings with large resource consumption. The new green building standard extends the applicable scope to all types of civil construction, to meet the needs of the present stage of green building practices and evaluation work.

Compared with the LEED family of different types of buildings, the standard (2006) covers less evaluation objects, and the division is not detailed enough. The standard (the latest revision) tends to cover more comprehensively in terms of subject.

2.3 Index Categories

Table 1 is the comparison of indicator categories among LEED (with LEED-NC, for example) and the old and new versions of standard, and you can see that the top

Table 1 Index categories comparison of LEED and China standard

	LEED	Standard (2006)	Standard (latest revisions)
Similar indexes	Sustainable sites	Land saving and outdoor environment	Land saving and outdoor environment
	Water efficiency	Energy saving and energy utilization	Energy saving and energy utilization
	Energy and atmosphere	Water saving and water utilization	Water saving and water utilization
	Materials and resources	Material saving and material utilization	Material saving and material utilization
	Indoor environment quality	Indoor environment quality	Indoor environment quality
Dissimilar indexes	Innovation in design	Operation management	Operation management
	Regional priority		Construction management

five categories are similar indexes of the three standards, involving two aspects: energy and resources, and environment load and indoor environment quality.

Different indexes reflect in the following aspects: (1) LEED displays the whole building life cycle by combining the building design (LEED-NC) and operations management (LEED-EB), standard (the latest revision) increases construction management evaluation index to cover the whole life period and phase of building, and this does not reflect in the standard (2006); (2) LEED sets design and innovation index. The standard (2006) sets preference item in every type of indicators to reflect innovation. The standard (the latest revision) adds innovation items uniformly, innovation item can also be classified into seven indexes, respectively, but in order to separate the requirements and measures and encourage green building with the basic requirements in seven district areas, the standard (the latest revision) will put all the provisions of the innovative item together, listed as a separate chapter; (3) Besides, LEED adds geographical advantage item in view of the climate and resources in different areas of the United States, projects meeting the indicators can get extra points; While the Chinese green building standard reflects a region difference by setting some control items to decide whether the building can take part in the evaluation. In a word, it can be seen that the indexes of China standard have the advantage of dividing index categories more concise, but not as clear and detailed as LEED.

2.4 Refined Indicators

The specific indicators of the standard (2006) are divided into control, general, and preferences items. Among them, the control item is the essential term of green buildings; preferences item primarily refers to the items that are difficult to achieve and have higher requirements. The new version of green building standards would keep the original control item unchanged, cancel the general and preferential items, merge them into a score item, and set the innovative item at the same time. Similar to the new version standard, every index of LEED is made up of required item and score item.

Although the refined indicators are similar, the detailed contents are different. Each index control item number in Evaluation Standard for Green Building ranges from 2 to 10, and these indicators must be realized at the same time. This will reduce independent selection of green building and thus ignores the region differences; by contrast, the rigid index of LEED is much less, each category sets 1–3 control items. But, the LEED emphasizes integrality and comprehensive of performance; users can evaluate and design the building according to the technical and economic conditions of the region.

In addition, the operation management index in the Evaluation Standard for Green Buildings mainly aimed at late green construction management and property management. Developing appropriate management indicators in view of the status of national conditions will help strengthen the role of people in green building management, so as to improve the design and operation efficiency, promote the healthy and long-term development of green building (Zhang 2011).

2.5 Scoring Points

LEED evaluation points are divided into three types: Prerequisite, any eligible projects must meet all the conditions to assess the premise, otherwise it will not pass the LEED certification; Credits, that is, in the five aspects mentioned above, every level of LEED certification should meet the requirements of the corresponding points; innovation credits, if evaluating project adopts technology measures that were not mentioned in the LEED and achieves significant effect, it should be rewarded certain innovation points. In order to stimulate the project that addresses geographically specific environmental priorities, it also can get the corresponding points.

LEED clearly gives the purpose, requirements, technical measures recommended, as well as the proven documentation required to submit of each score point. Each score point contains a number of sub-items and each sub-item is based on the above score point. Meanwhile, LEED also made reference to ASHRAE Standards (American Society of Heating, Refrigerating and Air-Conditioning Engineers) and other department standard and made a clear definition to some evaluation concept that makes it easy to understand and operate.

The standard (2006) is not satisfying in terms of scoring points. The entries are scattered; most evaluation contents are qualitative and lack of necessary technical parameters and practical experience so that the operator do not know how to start. This is one of the biggest obstacles in the implementation process of the standard.

The entries of standard (the latest revision) vary widely with respect to the standard (2006), but are closer to the specific requirements of LEED. Such as in the aspect of energy-saving, both new versions standard and LEED require the use of renewable energy and optimization of energy efficiency. In terms of water saving, both are required to reduce water consumption, innovate wastewater technology, and reduce surface water erosion. As to material saving, both require building reuse, material recycling, and waste management. Considering the land, both pay attention to the ecosystem of the area and the heat island effect. Sound insulation, heat insulation, reducing harmful substances emissions, thermal comfort and view requirements, and other indoor environment requirements must be meet in both of the standard (2014) and LEED (Hu 2010).

2.6 Evaluation Methods

Both LEED and new version green standard use quantified scoring method which gives scores depend on the degree of the implementation and effects of the measures, but they are not exactly the same: LEED uses the total score as the final result, while the standard (the latest revision) evaluation grade is determined by the total score rate; standard (2006) adopts counting the number of provisions as evaluation methods.

The method of counting the provisions number in the standard (2006) has certain defects: Many of these measures are qualitative, so their accuracy and authority are not ideal. In terms of the evaluation results, only "pass" and "not pass," there is no intermediate state of the two, so the standard conditions are more demanding (if there is a requirement that is not met, the result is "not pass"). To quantify the score point is the basic requirement to guarantee the evaluation result as an objective result, therefore the standard (the latest revision) improves and establishes a quantifiable evaluation.

In contrast, the standard (the latest revision) uses the score counting method to determine the level, which is a major update element of the standard. The determined level keeps uniformity and consistency with LEED which is the international popular green building evaluation criteria. It should be said that those measures reflect that domestic green building designers absorb and inherit the international green building standard's essence and strengths. The biggest advantage of the score counting method is that it increases the flexible of space, providing a richer selection of space for green building design. The biggest advantage of the score counting method is that it increases the flexibility of space, providing a richer selection of space for green building design.

At the same time, the standard (the latest revision) also continued to a certain extent of the advantages of the standard (2006), namely control the lowest total score points rate and prevent building having "short board" in certain aspects of performance.

2.7 Evaluation Results

According to the final grades, LEED has four rating levels: certification, silver, gold, and platinum. In the standard (2006), according to the number of meeting the general and preferred items, evaluation results can be divided into three levels: one star, two star, and three star. Besides the control items should be fully met, all of the three levels should meet the relevant request: number of provisions and general items; standard (the latest revision) green building rating is determined by the total score rate, besides to meet all of the control items, the minimum score rate of every index is 50 % in order to avoid the buckets effect.

Table 2 is the comparison of evaluation results among the three standards. As can be seen from table, American standard (with LEED-NC for example) is more detailed than the Chinese standard overall, but the American standard certification's starting level is low; standard (the latest revision) (with residential building for example) certification level is slightly higher than standard (2006) (with residential building for example), especially one-star certification. Besides, China's standard is a little higher than LEED, one-star level certification in China is between certification and silver level in LEED, two-star level is between gold and platinum level of LEED, and three-star level slightly higher than the LEED platinum level.

Table 2 Comparison of certification level, score, and proportion in LEED and China standard

USA		China		
LEED			Standard (2006)	Standard (the latest revision)
Level	Total score (%)	Level	The total number of compliance (%)	The total scoring rate (%)
Certification	40–49 score (36.4–46.5 %)	One star	18–29(36.7–59.2 %)	50–65 %
Silver	50–59 score (45.5–53.6 %)	Two star	30–39(61.2–79.6 %)	65–80 %
Gold	60–79 score (54.5–71.8 %)	Three star	40–49(81.6–100 %)	80–100 %
Platinum	80–110 score (72.7–100 %)			

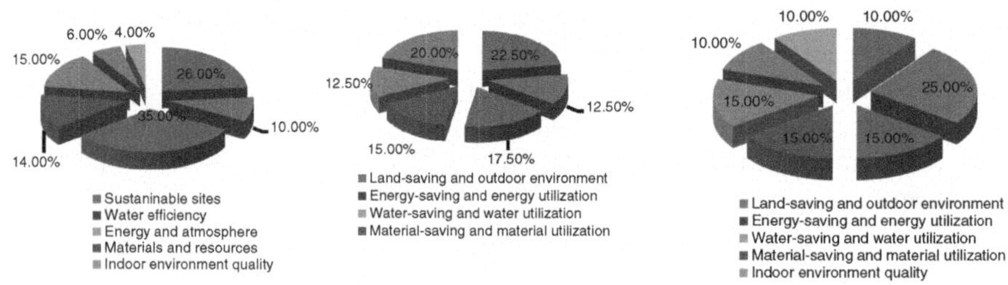

Fig. 1 Index weight distribution of USA LEED and China Standard

2.8 *The Index Weight Distribution*

Figure 1 is a comparison of index weight distribution of various categories of LEED and China standards. As shown in the Fig. 1, energy, atmosphere, and sustainable areas account a large proportion of weight distribution in LEED, reflecting the important consideration of energy reduction in LEED. China standard's weight distribution is more equal and does not reflect the focus on certain indicators.

3 Improvement of the New Evaluation Standard for Green Building

Table 3 is a general comparison of LEED and China green building standards. Through the above description and Table 3, it can be seen that the standard (the latest revision) has greatly improved in many ways, which are mainly shown in

Table 3 General comparison of USA LEED and China Standard (2006, 2014)

	The assessment phase	The evaluation objects	The index categories	The refined indicators	The scoring points	The evaluation methods	The evaluation results	The weight distribution
China Standard (2006)	Operation phase	Residential buildings and public buildings	1. Energy, resources and environment load and indoor environment quality	Control, general, and preferences items	No perfect	Counting the number of provisions	One star, two star, three star	Slightly balanced
			2. Operation management					
China Standard (the latest revision)	Design phase Operation phase	Civil construction	1. Energy, resources, and environment load and indoor environment quality	Control and score items	Not so perfect	Total score rate	One star, two star, and three star	Slightly balanced
			2. Operation management					
			3. Construction management					
LEED	Design, procurement, and construction phase	LEED family products	1. Energy, resources, and environment load and indoor environment quality	Required item and score item	Perfect	Total score	Certification, silver, gold, and platinum	Energy and atmosphere and sustainable areas account a large proportion
			2. Innovation in design					
			3. Regional priority					

following aspects: (1) Increase the assessment phase, the standard (the latest revision) are not just for the operation evaluation of a building, but also adds design evaluation. (2) The applicable scope is broader; the new national standard will extend the applicable scope to all kinds of civil construction. (3) The system structure of the standard (2014) is more compact; the standard maintains the original control item unchanged, merges the general and preferential items into a score item, adds a new construction management item, and improves the innovation item. (4) The provisions are more quantitative and qualitative; the applicability is clearer and more flexible, and the provisions adopt the method of dynamic updating. (5) Evaluation method upgrades: Standard (the latest revision) with scores counting method replaces the standard (2006) with the number counting method. (6) Modify the part of the evaluation provisions and distribute scores for all score and innovation items.

4 The Shortcomings Existing in China Standard

Compared with LEED and other more mature green building rating system in the world, China's green building standard still needs some improvements.

(1) Evaluation Standard for Green Building should increase building classification and formulate the corresponding standards.
In China, although the standard (the latest revision) will extend the applied scope to all civil building types, it does not mention new construction, expansion building, and retrofitting building. By contrast, the core products of LEED are subdivided into eight categories according to the different stages and different types of buildings and have a very comprehensive coverage. In addition, such divisions of LEED facilitate a variety of users. This is also the reason why LEED is widely accepted in the USA and around the world.
(2) Incomplete assessment phase
According to the project's process, buildings adopting the evaluation criterion of LEED can be divided into 3 phases, such as designing, procurement, and construction. Although design evaluation was included within the scope of the assessment in the standard (the latest revision) of our country and made up the lack of evaluation in the operation phase, but it does not consider ground controlling at construction stage and reusing of materials at procurement phase.
(3) Lower the threshold of evaluation system access
Reducing the threshold of evaluation system access is very important for the development of green building market and relatively easy to do. On the other hand, the mutual compensation of the indicators among energy, resources, and environment load can exist to avoid some buildings with high performance failing to participate in the evaluation.

(4) Strengthen advertisement power
The transformation of the whole society will be the most effective power to promote the concept of green building, which has been confirmed in European–American Nation. At present, we are only in "shout" stage for green building. So, we could use the network, television, newspapers, magazines, and other media to carry out rich energy saving and green building propaganda diverse forms and to improve the social awareness of the importance of the promotion of energy saving and green building.
(5) Optimize process of green building certification
LEED evaluation system has been able to achieve great success and so widely used, because of its underlying set of compact and concise reporting processes. Comparing with the relatively complicated process for green buildings in China, LEED is more mature and highly efficient. Green Building Council of America chooses their own review of the green building; that is, all of the results are from the same review team, which are more fairer and persuasive. In addition, LEED has specific provision to apply for certification for the time of each step period, and the evaluation architecture is open in the LEED information platform, which are more just and open. This is worthy of our study.

Acknowledgments The twelfth Five-Year National Science and Technology Support Program of China (Grant No. 2012BAJ12B03-02).

References

Evaluation Standard for Green Building (2006) Ministry of Housing and Urban-Rural Development of the People's Republic of China: Beijing
Green Building Council (2008) LEED 2009 New construction and major renovation rating system. U.S. Green Building Council, Washington
Hu FF (2010) The comparison of green (sustainable) building evaluation standard in China, Britain and United States. Beijing Jiaotong University, Beijing (in Chinese)
Li T, Ling CH (2011) The comparative study on the structural system of LEED and evaluation standard for green building. Architect J 3:75–78 (in Chinese)
Ouyang SHC (2008) Introduction of US green building evaluation system, leed. Build Sci 24 (8):1–3, 14 (in Chinese)
Seinre E, Kurnitski J, Voll H (2014) Building sustainability objective assessment in Estonian context and a comparative evaluation with LEED and BREEAM. Build Environ 82:110–120
Wan YM, Xue R, Huang T et al (2009) Comparison and analysis on Chinese green building evaluation standard and LEED. Build Sci 25(8):6–8 (in Chinese)
Zhang W (2011) Comparative study of green building assessment system at home and abroad. Hunan University, Hunan (in Chinese)
Zuo J, Zhen YZ (2014) Green building research-current status and future agenda: a review. Renew Sustain Energy Rev 30:271–281

Acoustic Model and Detection Method of Corona Discharge Noise

Juan Mo, Zhiyong Xu, Lei Wang and Yi Yun

Abstract A study gives rise to a new angle of the on-site detection method for the classic audible noise due to corona discharge. Based on a fundamental acoustic model, which describes major acoustic characters of the N-type wave with the consideration of the gradual distortion of the propagating sound wave, difference of Gaussians method combined with frequency comb filter is adopted for detecting corona discharge noise. The preliminary test is given to confirm the effectivity of the proposed method in more practical conditions. The research work may be helpful to the construction of new power transmission grid in China.

Keywords Corona · Discharge · Detection method · Pulsed signal · N-type wave

1 Introduction

In today's China, the problem of corona discharge on transmission line and its associated audible noise has been emerging extensively as a concern, especially in which the projects of ultra high voltage (UHV) power transmission lines are constructed. It is known that tiny protrusions such as burr, particles, water drops, snow, or ice on the conductors or connectors of high voltage overhead line facilities may

J. Mo · Z. Xu · L. Wang
State Grid Corporation of China, Beijing 100031, People's Republic of China

J. Mo
China Electric Power Research Institute, Beijing 100192, People's Republic of China

Z. Xu
Jiangsu Electric Power Company, Nanjing 210024, People's Republic of China

L. Wang
Jiangsu Electric Power Design Institute, Nanjing 210024, People's Republic of China

Y. Yun (✉)
Shanghai Advanced Research Institute, Chinese Academy of Sciences,
Shanghai 201210, People's Republic of China
e-mail: yuny@sari.ac.cn

S. Feng et al. (eds.), *Low-carbon City and New-type Urbanization*,
Environmental Science and Engineering, DOI 10.1007/978-3-662-45969-0_9

increase the electric field strength in the vicinity to reach the values where corona discharges set in. In most conditions, the sound resulted from corona discharge is perceivable as a group of crackling and hissing noise, which is situated mainly in a broad frequency range from 1 up to 20 kHz (Taylor and Chartier 1969). Moreover, corona noise generally contains an additional component of the low tones at twice the main frequency and its higher harmonics. With the rise of voltage, corona discharges may become more frequent and intensive in the new power grid, which gives rise to more problems of environmental noise, energy loss, and security implications. Hence, the monitoring, detection, and analysis of corona noise become increasingly notable for the construction of power transmission grid in China (Tang et al. 2010).

The research into the field of corona noise has never lost its topicality since the period of 1960s–1980s, when the implementation of UHV AC transmission over 1,000 kV was initially being considered (Perry 1972). As far as known, a lot of contribution to corona noise focused on the sound components in frequency domain or just in A-weighted level (Comber and Nigbor 1976; Tanabe et al. 1996; Comber et al. 1987), and yet few works concerned the details of the noise signal in time domain, which is important to detecting and monitoring corona discharge in real conditions.

2 Methodology

2.1 Fundamental Model of Corona Discharge Noise

When corona discharge occurs, the high voltage increased over a certain level, generates a high electric field that allows an efficient electron impact ionization and the production of highly reactive species, where the ions produced in the ionization-zone drift along the electrical field lines. During this progress, a significant part of the energy stored in the electronically excited gas molecules is transferred into kinetic energy of the heated gas. As a consequence, a micro-source of shock wave is initiated, of which the simplified spherical model can be described by Taylor's theory (Comber et al. 1987). During a short time in the microsecond level, a shock wave, which can be characterized by jumps in the air density, pressure, and velocity of its wavefront, propagates radially at the Mach number approximate to 1. Behind the shock wavefront, a low density region is formed with a pressure less than the atmospheric pressure. This depletion is a consequence of the air displacement due to the velocity induced by the shock wave. A classic N-type wave in the initial period is plotted in Fig. 1.

After a little time of the initial period, with rapid decrease in the pressure and velocity of wavefront, the propagating shock wave become to be an acoustic pulse temporally in a N-type waveform, of which the wave equation is given by

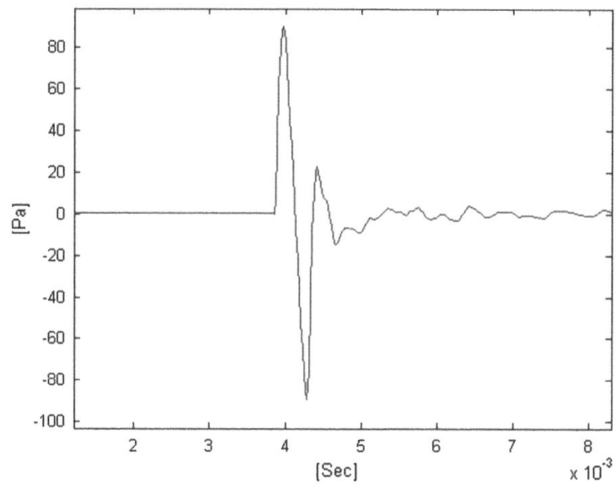

Fig. 1 Initial N-type shock wave

$$\frac{\partial v}{\partial r} + \frac{n}{2r}v - \frac{\beta}{c_0^2}v\frac{\partial v}{\partial r} = \frac{b}{2\rho_0 c_0^3}\frac{\partial^2 v}{\partial \tau^2} \tag{1}$$

where v, r, β, b are the particle velocity, the radius of wave front, the nonlinear coefficient, and acoustic absorption coefficient of air, while the transferred variable τ of time delay is expressed by

$$\tau = t - \frac{r - r_0}{c_0} \tag{2}$$

With further expansion of the wavefront due to the wave propagation, correspondingly the amplitude of the acoustic pulse becomes to be small enough so that the nonlinear wave term is negligible. In this condition, the degenerated acoustic wave equation can be solved approximately:

$$v \approx \sum_i v_0(\omega_i)\frac{r_0}{r}\exp[-\alpha(r - r_0)]\sin(2\omega\tau) \tag{3}$$

where the attenuating factor is defined by

$$\alpha \equiv \omega^2 b\left(2\rho_0 c_0^3\right)^{-1} \tag{4}$$

The equations above clearly indicate that the attenuation of higher frequency components of corona discharge noise will be exponentially increased in far field. A noise pulse initiating from an ideal N-type wave would be gradually transforming to be more smooth in the propagation process such as illustrated in Fig. 2.

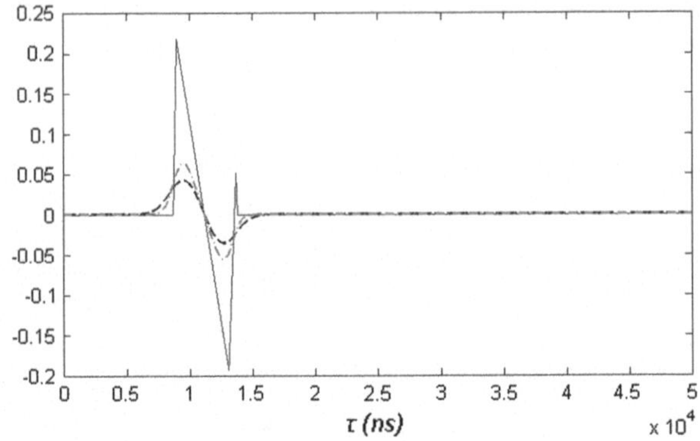

Fig. 2 Transformation of a N-type wave

2.2 Detection Method of Corona Discharge Noise

Unluckily, it is often difficult to directly extract perfect pulsed signals and detect a standard N-type wave of corona noise in real conditions. A series of unideal factors may have impacts on the on-site measurement or detection. First of all, the amplitude of $2f$ harmonic components associated with corona discharge noise is always considerable. Secondly, the sampling rate of standard acoustic measurement may be not high enough to catch the exact values of peak locations. Thirdly, different ambient disturbance extensively occurred in various environments will lead pulsed signals not as perfect as theoretical analysis.

Being motivated by the recognition method that is based on local scale-invariant features (Lowe 1999). In order to fit shock wave in real data, a developed N-type wave model is proposed as follows:

$$y(t) = \begin{cases} k_1 t, & -L_2 < t < -L_1 \ (a) \\ f_1(t), & -L_1 < t < -L_0 \ (b) \\ -k_0 t, & -L_0 < t < L_0 \ (c) \\ f_2(t), & L_0 < t < L_1 \ (d) \\ k_2 t, & L_1 < t < L_2 \ (e) \end{cases} \tag{5}$$

In Eq. (5), k_0 stands for the major descending edge of shock wave, k_1 and k_2 stand for rising edge before and after the distorted N-type wave's descending edge, and Eq. 5(b) and (d) represent N-type wave's corrupted peak and valley corresponding to the plots in Fig. 3.

As mentioned above, there may be two main kinds of unwanted sound waves contained in a period of measured signals for the detection of corona discharge noise. To get rid of the impact from the $2f$ harmonic components, a type of frequency comb filter will be applied. To minimize the impact of other stochastic audio disturbances, a type of difference of Gaussians (DoG) method is adopted, which can be utilized to distinguish them for detecting general signals of corona discharge noise. Being used by other researchers in different fields (Corrant et al. 1976), DoG method is

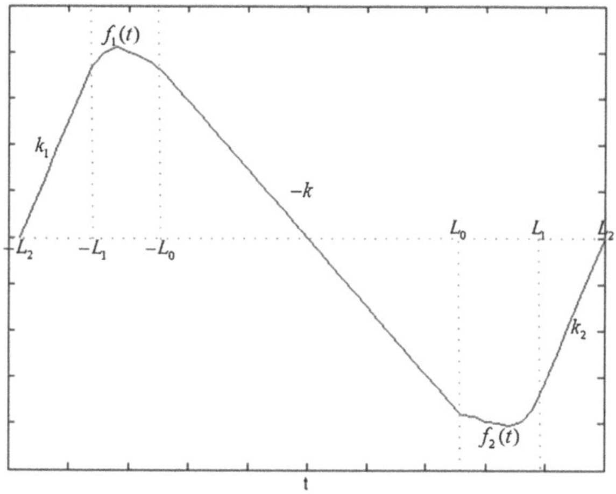

Fig. 3 Developed N-type wave model

mentioned to extract features in different scales as well as suppress noise. Motivated by their work, DoG method herein is adopted to detect shock wave and muzzle blast quickly and accurately. This method can be majorly described as follows:

$$D(t, \sigma, k) = \frac{1}{\sqrt{2\pi}\sigma} e^{-\frac{t^2}{2\sigma^2}} - \frac{1}{\sqrt{2\pi}k\sigma} e^{-\frac{t^2}{2k^2\sigma^2}} \tag{6}$$

In principle, the detection procedure includes two major steps. Firstly, measured sound signals are convoluted by $D(t, \sigma, k)$ in different scales in which wave signals of corona discharge noise have the biggest response. Then thresholds can be used to detect the discharge noise.

3 Detection Test and Results Analysis

3.1 Acoustic Separation of Corona Discharge Noise

Here, a period of practically measured sound data that contain corona discharge noise and $2f$ harmonic noise components, which is plotted in Fig. 4a, and is tested by implementing the separation algorithm, and the results are presented in Fig. 4b, c.

As shown in Fig. 4b, the additional noise mainly consisting of the $2f$ harmonic components can be filtered well so that another part of major signals consisting of the corona discharge noise is extracted as the wave plotted in Fig. 4c.

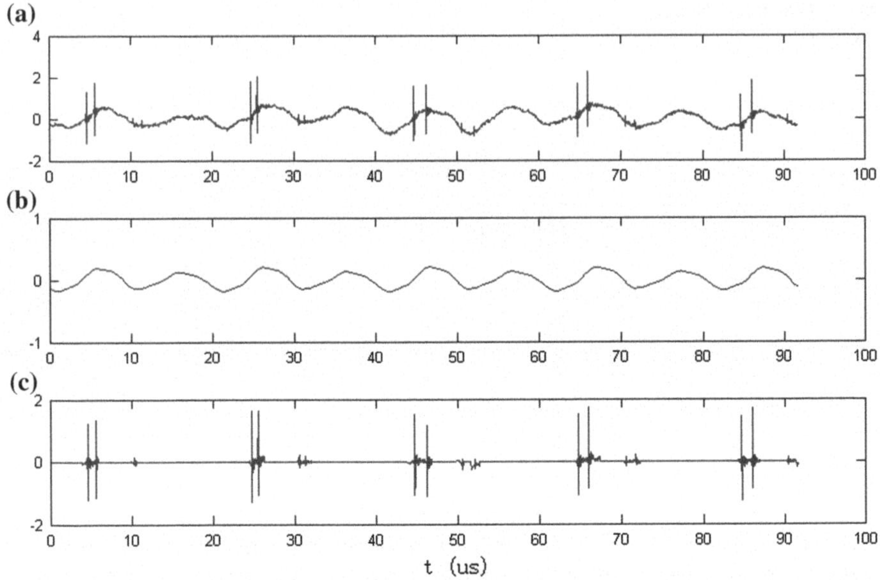

Fig. 4 Test of separation algorithm for corona discharge noise

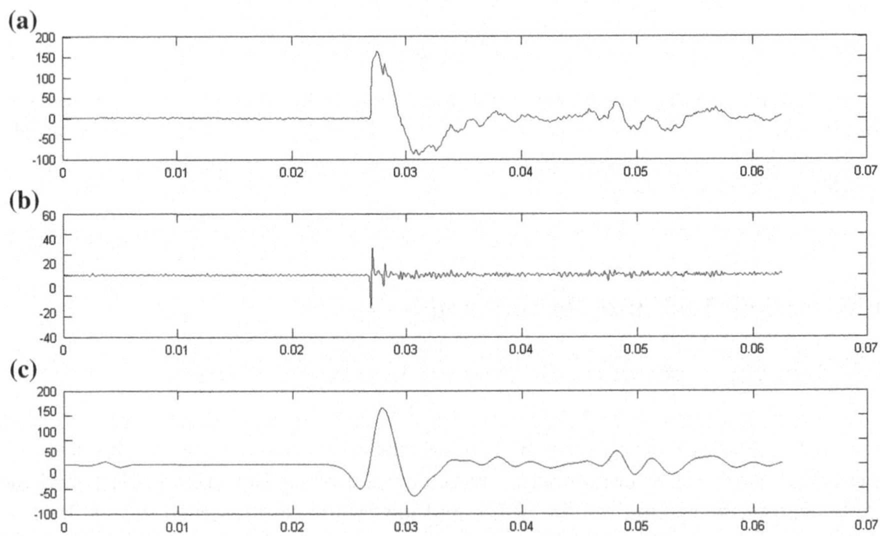

Fig. 5 Test of detection method for corona discharge noise

3.2 Detection of Corona Discharge Noise

After the procedure of noise separation, then the last step for the detection of corona discharge noise should be implemented. A period of the separated sound wave that contains the pulse signal of a distorted N-type wave together with the signals of a series of disturbance is used as shown in Fig. 5a. The DoG method is applied for the noise detection, and the characteristic noise signals resulted from the detection method are presented in Fig. 5b, c.

Obviously, as shown in Fig. 5b, c, the characteristic signal of a N-type wave is quite effective, of which the features are easier to be utilized for detecting corona discharges.

4 Conclusion

In this work, the classic audible noise due to corona discharge is fundamentally studied by giving rise to an acoustic model, which describes the major acoustic character of the N-type wave considering the gradual transformation with the sound propagation.

Further, a named difference of Gaussians method combined with frequency comb filter is adopted as the detection method of corona discharge noise. The last part of this work confirms the effectivity of the proposed detection method for corona discharge noise in more practical conditions.

On the whole, the study of our work gives rise to a new angle of sight that may be helpful to detecting and monitoring corona discharges on power transmission lines in real on-site cases.

References

Comber MG, Nigbor RJ (1976) Audible noise performance of the first three-phase ultra-high voltage transmission test line at EPRI's project UHV. IEEE Trans Appar Syst 95(4):1105–1114

Comber M, Nigbor RJ, Zaffanella LE (1987) Transmission line reference book—345 kV and above. Electric Power Research Institute EPRI, Palo Alto, pp 267–318

Corrant R et al (1976) Supersonic flow and shock waves. Beijing Science Press, Beijing, pp 376–377 (in Chinese)

Lowe DG (1999) Object recognition from local scale-invariant features. In: IEEE international conference, on computer, vol 2, pp 1150–1157

Perry DE (1972) Analysis of transmission line audible noise levels based upon field and three phase test line measurements. IEEE Trans Power Appar Syst 91(3):857–865

Tanabe K, Takebe T, Isozaki M (1996) Reduction of audible noise using asymmetrical bundles for 1,000 kV transmission lines: full-scale test results of Akagi test line. IEEE Trans Power Deliv 11(3):1482–1491

Tang J, Yang YJ, Li YS (2010) Prediction of corona effects generated from UHV AC transmission lines: I Audible noise. High Volt Eng 36(11):2679–2686

Taylor ES, Chartier VL (1969) Audible noise and visual corona from HV and EHV transmission lines and substation conductors—laboratory tests. IEEE Trans Power Appar Syst 88(5):666–679

A Case Study on Sustainable Development of Dianchi Lake Wetland

Jun Wang, Ming-quan Wang, Wei Zhao, Xi Chen and Chen Liu

Abstract This paper begins from the discussion on the definition of urban wetland and then depicts the current situation and problems associated specifically with Chinese urban wetlands. The case study of the Dianchi Lake wetland is introduced to illustrate a successful example of Chinese systematic urban wetland conservation and development. From wetland plants guidance and water quality controls, increasing the self-purifying ability to construction guidance and controls, the paper provides a comprehensive process of urban wetland management. From the case study, this paper discusses the successful experience of wetland conservation and development as well, providing a strong guideline for corresponding cases across China and the world.

Keywords Urban wetland · Conservation and development · China

J. Wang · M. Wang (✉)
Clean Energy Research Centre, Shanghai Advanced Research Institute, Chinese Academy of Sciences, No. 99 Haike Rd, 201210 Pudong Shanghai, People's Republic of China
e-mail: wangmq@sari.ac.cn

W. Zhao
Department of Urban Planning, College of Architecture and Urban Planning, Tongji University, No. 1239 Siping Rd, Yangpu, 200092 Shanghai, People's Republic of China

X. Chen
Shanghai Yi Hai Planning and Architecture LTD, Shen Zhen New Land Tool Consultants PTE.LTD, 4F, Building B, No. 605, Song Hu Rd, Yangpu, 200433 Shanghai, People's Republic of China

C. Liu
Shanghai Potevio-CAS Energy Technology Co., Ltd, Room 303, 3F, No. 700 Yishan Rd, 200233 Shanghai, People's Republic of China

S. Feng et al. (eds.), *Low-carbon City and New-type Urbanization*,
Environmental Science and Engineering, DOI 10.1007/978-3-662-45969-0_10

1 Introduction

Urban wetland is a new conception which has only undergone systematic research since the end of 1990s. Although there has been no formal definition of urban wetland to date, Yu (2001), Yang (2002), and Sun et al. (2004) all regard it as a transitional ecosystem between land and water located in urban areas, including coastal and estuarine areas, river banks, shallow lakes, water conservation areas, and natural or artificial ponds. Over the last twenty years, research on conservation and the use of urban wetlands has become increasingly important alongside the increase of worldwide concerns over urban problems such as air pollution, flood risks, and ecology unbalance. Jackson (2003) and Ehrenfeld (2000) reported a key function of wetlands among urban ecosystems as restoring water resource, relieving urban hot island effects and providing space for a wide range of animals and plants. Additionally, Boyer and Polasky (2004) and Nassauer (2004) hypothesized that urban wetlands have significant applicability in environmental control and natural disaster prevention, and treated wetlands as a part of the urban ecological infrastructure. Increasingly, nations are realizing the importance of urban wetlands in the construction of ecological cities, resulting in a new swathe of research projects and organizations focused on wetland protection, restoration, and reconstruction.

China joined the 'Ramsar Convention' in 1992, since when the state forestry administration set up the 'Ramsar Compliance Office' to perform wetland conservation and management. Chinese wetlands now account for approximately 66 million hectares, ca. 10 % of worldwide wetlands, and the highest in Asia. The wetlands in China are quite disparate in nature, from temperate to tropical, coastal to landlocked, and sea level to mountainous.

Urban wetland is much vulnerable to violation and inappropriate development due to its proximity to and inclusion in urban areas. The serious degradation of Chinese urban wetlands in recent years has been exacerbated in the following ways:

Firstly, urban wetland areas have been quickly reduced alongside rapid urbanization. Sediment deposition, lake beach reclamation, garbage disposal, and city construction all contributed to wetland transformation. Many studies have shown that curing revetment of rivers, lakes, and wetlands causes detrimental effects (Booth and Reinelt 1993; Arnold and Gibbons 1996). Chinese urban water curing is very common, with a direct consequence of wetland degradation, reducing the ability of environment adjustment and disaster prevention. Since the 1950 in Beijing, for example, the wetland area declined from 2,568.23 km^2 to only 526.38 km^2 in 2009.

Secondly, pollution of urban wetland significantly worsened. Untreated and slightly treated industrial waste and domestic sewage emissions increased each year continuously reducing the natural wetland capacity of self-purification for water. Increase of urban impervious surface and river channelization reduces the purification function of river ecosystem, making the water quality even worse. According to the 'Report on Water Resources in China' (The Ministry of Water Resources of the People's Republic of China 2004), the national industrial wastewater and urban sewage emissions increased from 62 to 69.3 billion tons from 2000 to 2004, of which 70–80 % is untreated emissions (Ran 2001; Hong 2003).

The third and the final confounding factor has been a lack of research into the supervision and potential protection of urban wetlands in comparison with wetlands as a whole. To date, research has mainly concentrated on urban sewage treatment and there has been a lack of systematic research on urban wetland protection and development. Simultaneously, there is no monitoring, protection, and development mechanisms at a government level, creating a lack of institutional guarantees for the appropriate utilization of urban wetland (Scholz and Xu 2002; Song et al. 2003).

2 Case Study—The Case of the Dianchi Lake Wetland

Dianchi Lake is the largest freshwater lake on the Yunnan–Guizhou plateau in Southwest China, located in the center of the Yunnan province approximately 5 km from the capital city Kunming. Due to the proximity of Kunming city, the lake area is a popular tourist and vacation destination. The basin area of Dianchi Lake is 2,900 km^2, with a water surface area of 300 km^2, a coastline of 163.2 km, and a water capacity of 15.6×10^9 m^3. The Dianchi wetland is one of the most important urban wetlands in South West China in terms of its rich ecological resources and variety of functions.

Over the past 50 years, the Dianchi Lake has experienced a sequence of degradation, treatment, recovery, and development, analogous to that experienced by many Chinese urban wetlands. This work focuses on the successful, comprehensive conservation and restoration of the Dianchi wetland, which can be used as a basis for the comprehensive management of urban wetlands in general.

2.1 Improvement of the Water Quality

Until the 1970s, the major problem affecting the Dianchi Lake was agricultural pollution. This was superseded by industrial pollution following the late 1970s, all due to the rapid development of industry in surrounding towns. Since then, the water quality has continued to drop rapidly, and degraded the whole ecological system in the surrounding environment.

The specific problems faced by the Dianchi wetland are as follows:

2.1.1 Excessive Land Reclamation on Wetland

Excessive farmland reclamation, changing the natural use, and occupying wetland for urban development directly caused the decline of ecological function of the Dianchi wetland. The destruction activities began from 1950s and massively broke

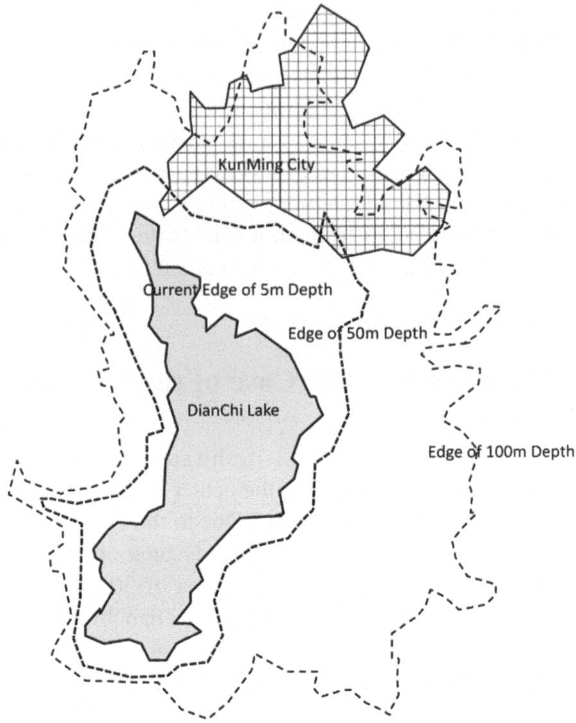

Fig. 1 Map showing the reducing lake area with depth

out after 1990s. On ca. 2000, the lake basins became shallow, water capacity was reduced with many aquatic species, and wildlife habitat sites were destroyed. Then, the water quality deteriorated and outbreak of cyanobacteria was on a massive scale. According to existing research, the purification capability of water depends on the area of the wetlands. Guo and Su (2010) calculated that a lake of 300 km^2 such as Dianchi will need about 2,912–5,814 km^2 wetland to keep its ecology balance. But following the destruction of lakeside, there are only 5,187 ha of natural marshes and only 20,017 ha of aquatic plants remain. The lack of wetland areas not only reduces the purification ability of lake itself, but also seriously influences the surrounding ecological environment.

Based on historical data, the water surface area of the Dianchi Lake is continuously shrinking, and its environmental capacity is also weakening. The Lake was about 1,000 km^2 10,000 years ago, and today, only 300 km^2 remains. Along with the significant reduction in area, the water depth also declined from 100 to 5 m (Fig. 1).

2.1.2 Excessive Destruction of Biological Resources

According to records, aquatic plants in Dianchi Lake were abundant until the 1950s. Vegetation accounted for more than 90 % of the lake, with as many as 100 species of aquatic plants, including 42 submerged species. By the 1970s, there were 46 species of aquatic plants, with a huge reduction in the distribution area from the original 90 to 12.6 %. According to a survey between 1995 and 1997, there remained only 22 species of aquatic plants, with the distribution area rapidly reduced to 1.8 %. Around 2,000, many aquatic plants and wildlife disappeared including Chara, Acuminata, Hydrilla verticillata, Small Carp, and Black Acrossocheilus yunnanensis, and some others reduced drastically (Xiao 2005). The coastal plant zone vanished creating a species imbalance among aquatic plants and causing an outbreak of Cyanobacteria in later years.

2.1.3 Increased Pollution and Water Quality Decline

The pollution to the Dianchi Lake resulting in significant damage mainly comes from the residential sewage of surrounding towns and industrial and agricultural waste water. In 1995, the volume of wastewater discharged into the Dianchi Lake was 1.85×10^8 m^3, among which the total nitrogen was 8,981 ton, total phosphorus was 1,021 ton, and chemical oxygen demand (CODcr) was 41,674 ton. Among the waste, there is 45–58 % of residential sewage and 11–32 % of industrial. In 2000, the wastewater emission had risen to 2.4×10^8 m^3, of which total nitrogen was 14,155 ton, total phosphorus 1,487 ton, and CODcr 62,449 ton. Under normal circumstances, the Dianchi Lake's environmental capacity for nitrogen, phosphorus, and CODcr is about 4,000, 400 and 11,000 ton, respectively. Such high emissions and rapid growth in pollution made the water quality inferior to the index of national quality standards with over 70 % of water areas undergoing eutrophication (Xiao 2005).

2.1.4 The Shortage of Water Resources and the Vicious Cycle of Water Usage

At present, Kunming city on the north shore of the Dianchi Lake covers ca. 180 km^2 with a population of 2 million. The city's expansion not only exacerbated pollution, but also inevitably created a shortage of water resources. The average water resource of Kunming is less than 270 m^3 per person, only equals to 4 % of the world's average rate and 50 % of the UN water standard rate.

To solve the problem of water shortage, in recent years, the local government has constructed a number of utilization facilities within the Dianchi Lake Basin. Eight large and medium-sized reservoirs have been built with a total capacity of 3.4×10^8 m^3, combined with nearly 140 small reservoirs with a total capacity of nearly 1×10^8 m^3. The reservoirs control a 1,651 km^2 runoff area, accounting for 57 % of the basin area. The annual average water supply of the Dianchi Lake Basin is more than 10×10^8 m^3, but according to calculations of surface evaporation and runoff, the actual water resource volume is less than 6×10^8 m^3. That is to say, there remains a 40 % gap of water needs to be allocated from the outside. This will also instigate negative effects to the ecological environment which in turn will form a vicious cycle.

2.2 Dianchi Wetland Conservation and Restoration

Wetland conservation and restoration includes wetland plants guidance, water quality controls, and increase in the self-purifying ability.

2.2.1 Wetland Plants Guidance

Wetland plants play an important role in Dianchi Lake's ecological environment, with a variety of benefits for recovering wetland functions. So it is emphasized that a wetland plant guide was necessary for Dianchi wetland management. Ornamental plants such as lotus, water lily, hornwort, willow flower, paper grass, Scirpus, Barracuda grass, Canna, water bamboo, and Zhongshan Chinese fir formed a unique wetland landscape. Researchers recommended plants which could thrive in the Dianchi wetland (Chen 2011), and the local authority also published the recommended list of 212 species of plants for the Dianchi Lake Basin (Administration of Kunming Dianchi Lake 2013). Furthermore, during the renovation process of the Dianchi wetland, the authorities put efforts into protecting the native natural plants on the lakeshore and riverside, taking effective measures to implement the growth guide of wetland plants.

Riverside zone vegetation protection and restoration is also an important wetland restoration method. It was recommended to retain the natural shale slope of the rivers without flood discharge usage and to recover the natural curved coastline and gentle slope by planting trees and grass. The floating leaves include Gordon Euryale and grass, while the submerged plants include eel grass, Potamogeton crispus, and Myriophyllum verticillatum, and the floating plants include duckweed and Salvinia natans.

2.2.2 Water Quality Control

One pollution control solution is to build a sewage system surrounding the lake to intercept the discharged waste (Fig. 2). The Dianchi Lake sewage system was divided into four different levels: district sewage, river sewage, rural sewage, and main tube sewage, which all link directly to different disposal stations. The sewage system stopped the polluted water and early rainfall from pouring into the Dianchi Lake. The main sewage canal is about 96 km long and capable of disposing of

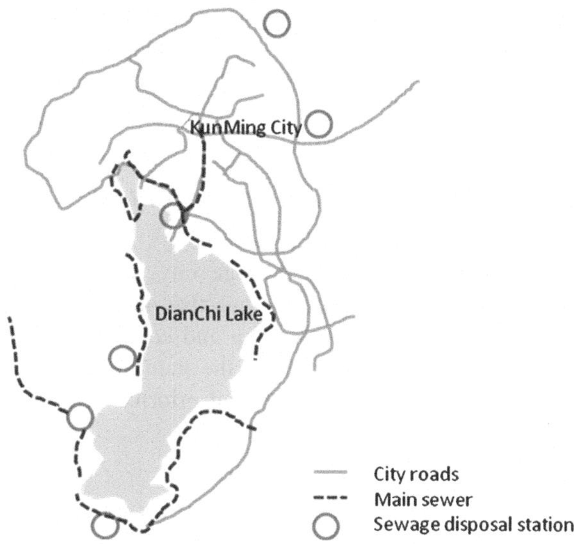

Fig. 2 Main sewage pipes surrounding the Dianchi Lake

172,500 m^3 sewage and 277,500 m^3 of early rainfall per day. The total investment in this system was 5.5 billion RMB (0.8 billion USD), and it was completed by January 2013.

Additionally, several artificial means were used to promote water circulation in order to enhance the water's self-purification ability. One was to pump pure water through plants to various bays. By this means, the water became 'living water' as opposed to 'black water.' An alternative method was to install water-cycling devices into lakes which continuously circulate surface and underlying waters, increasing dissolved oxygen concentrations and avoiding deposition of nutrients. The other method was to set rainwater biological purification canals in east and south shores with rich root plants to absorb suspended solids in intercepted rainfall runoff, before discharge into the lake.

2.2.3 Increase Self-Purifying Ability

To aid the recovery of the lake ecosystem, the authorities dismantled all existing concrete dams, using 1,887.5 m (the normal maximum water elevation) as the scope for ecological restoration. A gentle slope with a tree and shrub belt, an emergent plant belt, and a submerged plant belt was constructed at the land/water interface in order to build a complete ecological system.

According to research, a water flow rate under 0.7 m s^{-1} enhances water interaction with microorganisms and plant substrates for further purification. To reduce the flow rate to this level, the following main three measures were used: Plant species were selected to vary depth and density to form a flow-limiting barrier; secondly, islands, pools, and beaches were formed, to reduce flow through varying elevation; and thirdly, a circuitous water route was formed to increase the retention time of water in wetlands.

2.3 Construction Guide and Control

Due to the proximity of a large city, the Dianchi wetland not only bears an important ecological function but also accommodates significant leisure and tourism. Therefore, overall planning and control for the future development and construction activities must reflect the natural properties of wetland and follows the principles of feasibility, ecological priority, scarcity, priority, and aesthetics.

2.3.1 Landscape

Due to its proximity to Kunming city, the landscape building around Dianchi Lake was very important. The landscape combined natural and artificial styles for sightseeing and leisure for Kunming's citizens. Thus, at the same time as wetland ecological construction, the authorities must pay attention to combining sewage treatment with natural processes and landscape art, increasing the sustainability of the wetland itself.

2.3.2 Partition and Activity Guidance

The following five partitions and activity controls were dictated by the importance of wetland ecological function:

1. Core wetland conservation area: In this area, all activities were under the strict governance including development, sightseeing, and traffic. Pollution and aquatic breeding control alongside comprehensive management of rural sewage would guarantee the water quality in this area.
2. Ecological buffer area: This area was the buffer zone between conservation and development. Necessary waterflow shelters and wild food supply sites were set up in animal gathering areas, especially for migratory birds, to maintain the open water area and lakeshore grassland and island.
3. Wetland exhibition area: By returning farmland to lakes and reconstruction of wetland vegetation systems to recover wetland ecological function, reduce the water quality pollution by natural wetland purification. At the same time, build wetland landscape and animal habitats.
4. Tourism activity area: This area aimed to provide facilities such as exhibition hall and botanical garden for tourism use. At the same time, opportunities for

close contact between wetlands and tourists were created by water traffic routes, alongside construction of piers, bird-watching facilities, and viewing platforms.
5. Conservative agriculture areas: In this area, the ecological agriculture and fisheries were developed to become a part of experience-based tourism.
6. Tourism service area: Shopping, dining, and entertainment districts for tourism were developed with characteristics of 'Ancient Dianchi' and 'Yunnan' cultures. This area would include an appropriate proportion of residential and resort facilities.

2.3.3 Multi-ring Development

A multi-ring development model was established according to the importance of ecology environment. The inner ring contains the core natural wetland with a complete ecological system (The Primary Grade Conservation Zone according to the 'Dianchi Conservation Regulations of the Yunnan Province') (Yunnan Province Government 2012). The middle ring contains the wetland functional zone for tourism, leisure, exhibitions, new-type rural industry, etc. (The Secondary Grade Conservation Zone Dianchi). The final, outer ring contains the wetland area coordinated with the city construction area (The Tertiary Grade Conservation Zone Dianchi).

Visitors and non-conservation construction activities were forbidden in the Primary and Secondary conservation zones which had great influence on the ecological environment. For development activities in the secondary and tertiary zones to proceed, they must clearly involve 'positive protection' in the form of wetland parks. At the same time, development was guided to reflect the historical and local folk culture, for the purpose of folk culture continuance and educational function.

2.3.4 The Ecological Compensation Mechanism

The wetland ecological compensation mechanism was a coordination mechanism of wetland protection and utilization for stakeholders, comprising of the following two aspects:

The most important was to compensate for the cost of ecological restoration. As the restoration project combined a variety of local authorities like Dianchi Administration Bureau, local communities, and independent firms, the compensation was financed by the provincial government to encourage the ecological restoration. The compensation funds were concentrated on the animal and plant gathering area, important river channels, and main water districts.

The second was to compensate for the local residents for their loss of farmland and chances to do development construction. According to the strict plan, many farmlands would be reclaimed and restored into nature area. Meanwhile, all constructions which do not conform to the ecological protection were determined; thus, the opportunity cost of the residents needs to be compensated.

2.3.5 Establishment of a Dianchi Wetland Conservation and Development Committee

The Dianchi wetland conservation and development committee allows coordination between different administrations and institutions regarding the management of wetland conservation and development. The aims of this body are to do the following:

Integrate existing ecological laws and regulations, and set the lake basin management mechanisms. Formulate unified regulations in Dianchi Lake wetland conservation and restoration by implementation of 'Dianchi Conservation Regulations of the Yunnan Province' and other ecological regulations related to current policy. Expand the management power of the lake basin administrative department and give the corresponding legislative power.

To increase the budget and the ability to use economic means to control and balance the cost between protection and development. To establish collaboration between governmental macroeconomic regulation, democratic consultation, and market-oriented operation with a target of unified management of water resources.

To innovate fiscal and taxation policies, introducing the multi-channel funds into wetland protection and development to overcome the current lack of public investment. Encourage private investment in the ecology via cheap loans, extending maturities, tax returns, accelerated depreciation of fixed assets, and other preferential policies. Meanwhile, to use the ecological capital markets to gather funds, considering the long-term environmental ecological lottery or bonds, or to provide a variety of preferential policies to encourage enterprise to protect the environment.

3 Conclusion

At present, rapid urbanization is occurring in China; this causes urban wetlands to suffer greatly by construction. The natural resources of urban wetlands are decreasing and their ecological function degrading Research concerning China's urban wetland is still in its infancy, exacerbated by a lack of national policy guidance and corresponding management mechanisms for wetland conservation and development. This situation reflects the complexity of Chinese urban wetland management.

From 2010 to 2015, Kunming municipal government plans to invest 42.8 billion RMB (7 billion USD) in the management of the Dianchi Lake wetland. By the end of 2013, the water quality of the Dianchi Lake remained stable with an obvious cessation in the deterioration trend and an improvement in the eutrophic index from severe to moderate. The total phosphorus index of the water dropped by 36.3 %, and the total nitrogen dropped by 34.9 % till 2013. This indicates an obvious improvement in the ecological environment.

The comprehensive management of the Dianchi wetland is typical of multiple cases in China. It covers water area control, water quality improvement, animal and plant guides, pollution control, self-purification control, landscape building, construction control, capital operation, and many other innovations. These will provide excellent reference for similar urban wetland in China, helping to promote the urban wetland protection and development across China.

Acknowledgments This study was supported by Chinese National Natural Science Foundation of China (Grant No. 51208304 and Grant No. 51308525).

References

Administration of Kunming Dianchi Lake (2013) The Dianchi lake ecological plant species recommended list

Arnold CL, Gibbons CJ (1996) Impervious surface coverage: the emergence of a key environmental indicator. J Am Plann 62:243–258

Booth DB, Reinelt LE (1993) Consequences of Urbanization on aquatic systems—measured effects, degradation thresholds, and corrective strategies. In: Proceedings of watershed: a national conference on watershed management. Environmental Protection Agency, Alexandria

Boyer T, Polasky S (2004) Valuing urban wetlands: a review of non-market valuation studies. Wetlands 24:744–745

Chen Yi R (2011) The Dianchi wetland aquatic plant combination. Environ Sci Tribune 30(5):21–26

Ehrenfeld JG (2000) Evaluating wetlands within an Urban context. Ecol Eng 15:253–265

Guo YS (2010) The ancient Dianchi Lake's change and it's effects. Yun Nan Daily 22 Oct 2010

Hong Y (2003) Thoughts on Chinese wetland conservation. Wetland Sci 1:68–72

Jackson LE (2003) The relationship of urban design to human health and condition. Landscape Urban Plann 64(4):191–200

Nassauer J (2004) Monitoring the success of metropolitan wetland restorations: cultural sustainability and ecological function. Wetlands 24:756–765

Ran XY (2001) The discussion on management of Urban water environment. Beijing Water 4:12–14

Scholz M, Xu J (2002) Comparison of constructed reed beds with different filter media and macrophytes treating urban stream water contaminated with lead and copper. Ecol Eng 18:385–390

Song ZW, Bi XJ, Cao J et al (2003) Artificial wetland and its application in the small town wastewater treatment in our country. Ecol J 22(3):74–78

Sun GY, Yu SP, Wan ZJ et al (2004) On the nature, structure and leading edge of innovation of wetland science, geography front. Business Press, Beijing

The Ministry of Water Resources of the People's Republic of China (2004) Report on Water Resources in China. http://www.mwr.gov.cn. Accessed 11 Mar 2014

Xiao Q (2005) Dianchi lake water environment present situation, causes and countermeasures. Water Conservancy Sci Technol Econ 11(1):26–28

Yang YX (2002) The main character, progress and forecast on international wetland study. Prog Geogr Sci 21(2):111–118

Yu GY (2001) Discussion on some basic scientific problems of wetland research. Prog Geogr Sci 20(2):177–183

Yun Nan Province Government (2012) Dianchi Conservation Regulations of Yunnan Province

The Environmental Issues in Textile and Garment Trade

Lei Yao and An-yi Geng

Abstract At present, new energy and environmental protection as the theme of "low-carbon economy" are becoming the world's economic development objectives; the relationship between trade and environment has become increasingly close. China's textile and garment industry is one of the most polluting industries, and the environmental problems will become an important issue to the whole industry. This chapter aimed to discover the current environmental issues in textile and garment trade, in order to optimize the development of textile and garment industry and find solutions to these problems. This chapter mainly discussed four issues that were environmental pollution, green barriers, environmental labeling system, and low carbon.

Keywords Environment · Low-carbon economy · Textiles and garment · Trade

1 Introduction

After Copenhagen Climate Conference, climate change has aroused global attention. Specifically, environmental issues as well as low-carbon economy, low-carbon life style, and low-carbon consumption now are prevailing, and the low-carbon production is regarded as the major economic engine driving the sustainable growth for various industries. Among China's major industries, textiles and garment industry is notoriously known as a pollution-intensive industry, with alarming emissions of industrial effluence and waste gas. In the context of low-carbon economy, textile and garment industry is among the "ranking high" industries that call for immediate change in relation to environmental problems. Some foreign countries set up strict environmental protection standards to build green trade

L. Yao (✉) · A. Geng
Business School, Beijing Institute of Fashion Technology,
Beijing 100029, People's Republic of China
e-mail: yaolei276@sina.com

S. Feng et al. (eds.), *Low-carbon City and New-type Urbanization*,
Environmental Science and Engineering, DOI 10.1007/978-3-662-45969-0_11

barrier, which generates certain impact on China's export trade as China's reforming and opening accelerates. At the same time, China's environmental problems are becoming increasingly serious (Zhang 2010). As the pillar industry in China's export trade, textile and garment industry is facing more critical environmental problems. These problems would only become even more paramount as the overall environmental protection awareness increases. This study aims to review and outline the overall condition of environmental problems challenging textile and garment trade, thus providing beneficial basis for finding a way out in dealing with environmental protection issues related to China's textile and garment trade.

Currently, there are two factions according to the environmental problems which are related to international trade: free trade upholders and environmentalists. The upholders of free trade highlight the maximization of individual interests, attributing the world's economic progress to the increment of individual fortune, whose realization is best achieved through the permission of free trade. From their point of views, trade liberalization can benefit the environment and the environmental damage is the consequence of improper environmental policies instead of the trade institution itself. Porter hypothesis is the representative of the belief of this faction (David and MÓNica 2008). In contrast, the environmentalists' top concern is the ecosystem, and they emphasize on the common interests of humans and long-term interests of the individuals. In some cases, objectives of environmental protection are achieved at the cost of immediate interests of individuals. Environmentalists stress the needs to strengthen the rights of member states in trade in formulating environmental policies and propose to address the environmental problems related to trade by means of international environmental agreement and restrictive trade measures. The environmental race-to-the-bottom theory and the pollution haven hypothesis support the claims of the environmentalists.

2 Environmental Behaviors of China's Textile and Garment Industry

Textile and garment industry has long been China's traditional pillar industry as well as typical export-oriented industry, with trade surplus reaching over hundreds of billions of dollars. However, the environmental problems incurred are particularly severe. The pollution caused by textile and garment industry is specifically restraining the development of China's textile and garment trade.

Due to the influence of multiple factors during the earlier development stage of China's textile industry, including low technical level and backward production process and equipment, the environmental pollution that occurred in the production process was spectacularly prominent. At present, the emission of waste gas and water in the production process of textile industry is the major environmental problem (Wu 2011).

Table 1 Industrial waste water emission of textile and garment industry

Year	Total industrial waste water emission of textile industry (ten thousand tons)	Perception of total industrial waste water emission of nation (%)
2004	165,270	8.35
2005	181,417	8.4
2006	211,619	10.17
2007	239,663.1	10.86
2008	245,606	11.3
2009	253,843.2	12.14
2010	257,508.1	12.15

Source China Statistical Yearbook

2.1 Waste Water

Industrial waste water is the No. 1 environmental problem faced by textile and garment industry as one of the several industries characterized by enormous water utilization and industrial waste water emission. Table 1 shows the industrial waste water emission of textile and garment industry in recent years.

It can be seen from Table 1 that the total industrial waste water emission of textile industry was over 1.6 billion tons annually, accounting for 8.35 % of the total industrial effluent emission in 2004 and 12.15 % of the total in 2010. This figure constantly increased on a yearly basis, ranking among the top five in terms of the industrial waste water emission.

2.2 Waste Gas

The waste gas is mainly generated by several ten thousands of boilers used to supply energy to textile and garment industry, which discharge large quantities of combustion waste gas, SO_2, and smoke dust, causing serious pollution. Another major emission source is the production process of the textile industry. Due to its technical defect or incomplete process control, the waste gas emission occurs.

Table 2 shows the increasing emission in waste gas of textile and garment industry during the period 2004–2006. The total waste gas emission peaked in 2006—for 466,000 tons. After that, some improvement took place until 2010.

From the point of trade, with other factors unchanged, the growth rate of export trade volume of textile industry showed the same variation trend with that of the industrial effluent emission. That is, the high growth rate of export trade volume came at the expense of high emission load. In other words, low trade growth rate naturally follows if the emission load is kept at low level.

Table 2 Waste gas emission of textile and garment industry

Year	Total waste gas emission of textile industry (ten thousand tons)	SO_2 emission (ten thousand tons)	Smoke emission (ten thousand tons)	Dust emission (ten thousand tons)
2004	44.11	30.66	12.22	1.23
2005	46.25	31.15	13.52	1.58
2006	46.6	32.4	13.5	0.7
2007	42.18	28.82	13.2	0.16
2008	41.37	27.58	13.49	0.31
2009	40.38	26.85	13.33	0.2
2010	38.75	25.84	12.77	0.14

Source China Statistical Yearbook

Generally speaking, China's textile export trade is closely associated with the environmental performance. We believe that the environmental regime is the crucial factor in textile and garment trade.

3 Analysis on Environmental Regime in Textile and Garment Trade

As the public awareness of environmental protection keeps increasing, practicability, esthetics, and durability which used to be the major criteria of the products can no longer completely satisfy the consumer requirements. Instead, the customers now have a growing demand on the security and sanitation of the products. As a response, some regimes and measures related to international textile and garment trade are gradually improving and developing. The next section is devoted to the introduction of common regimes and measures adopted in textile and garment trade, including green barrier, environmental indication system, carbon footprint, carbon labeling, and carbon tariff (Jiang et al. 2013).

3.1 Green Barrier

There are mainly two categories of green barrier in textile and garment trade.

The first category of green barrier is set against the environmental impact produced during the entire process, including design, production of garments, the scrapping, and recycling of the garments. For instance, garments only manufactured by enterprises which are certified to ISO 9000 standards are permitted to enter the trade exhibition held by America and European Union. It is also required by European Union that any garments sold in EU countries should meet ISO 14000 standards regarding to the pre-production, manufacture, sales, use, and post-use treatment. ISO 9000 refers to the quality management system standard and is a collective term of a family of standards. International standard organization (ISO) then issued ISO 14000 series environmental management standard.

The second category of green barrier is set against the negative impact of the products on the security and health of the customers (Lin 2010). That is, the textile and garments manufactured are not allowed to produce negative impact on the health of the customers. For instance, European Union prohibits the use of pesticide containing toxic metallic compounds in cotton planting since 1997; the ban on decabrominated diphenyl ether as a brominated flame retardant (BFR) in textiles, which came into force on January 1, 2010, is stipulated in the amendment to eco-label.

Hong Kong (China), Japan, America, and European Union have long been the four major markets for China's textile and garments export. A growing environmental and security concern in these countries directly results in increasingly sound and stringent green trade barrier system and standards. The implementation of green trade barrier system involves all of the aspects of textile and garment industry, from the production and sales to the scrapping and after-use process. And high expenses of equipment and certification lead to increased cost, posing great challenge to China's textile and garment export trade.

3.2 Environmental Indication System

Environmental indication system for textiles consists of two forms, i.e., indication system for the life cycle of textiles and single-aspect environmental indication system (Li 2013). The former, also known as eco-indication system for textiles, requires that every aspect in the textile trade, including research and development, production, transportation, use, and recycling, brings no damage to both the environment and human health. The environmental indicator is only issued when all of the environmental protection requirements are met. Single-aspect environmental indication system reflects the environmental impact of textiles in a specific aspect, e.g., "no damage to human skin," rather than provide information regarding the relationship of textiles to other environmental aspects. The second system does not require complex and strict analysis as the first one does, and the cost incurred by the obtainment of this certification for the manufacturers is lower.

Environmental indication system produces both positive and negative impacts on China's textile and garment trade. The positive impact is that the environmental indication system can increase the brand awareness of the enterprises which have been certified and urge the enterprise to save resources, recycle the waste products, and reduce the management cost. As a consequence, the product structure of China's textile and garment export can be adjusted and the international competitiveness of the textiles will be enhanced, with better capacity to circumvent the green trade barrier set by the developed countries. The negative impact includes the increased cost for the export-oriented enterprises and the weakened price advantage of China's textile and garment export products, which might impede the entry of China's textile and garment products into the international market.

3.3 Carbon Footprint, Carbon Labeling, and Carbon Tariff

With the global expansion of the influence of low-carbon economy, "low carbon" has also become one of the mainstream business values in international supply chain, which gives rise to the conceptualization and implementation of carbon footprint, carbon labeling, and carbon tariff (Wen 2013).

1. Carbon footprint

   Carbon footprint refers to the influence of one's resources awareness and behaviors on the natural world. To put it simply, it is the carbon consumption of individuals and enterprises (Sun et al. 2011).

   However, the carbon footprint during the entire life cycle of the textiles is difficult to calculate at present. Since many uncertain factors are involved in the carbon footprint of plant fibers during its planting and growth and the life cycle and washing methods for the textiles are diversified, an effective method for calculating the carbon footprint during the use and recycling of the products is still unavailable. Additionally, as the chain from agricultural production to industrial production and to the use and discarding of the products is particularly long, the calculation precision of the carbon footprint of the final products is hard to guarantee. In spite of this, the concept of carbon footprint along with its calculation method which is being constantly improved provides a solid technical support for the sustainable development of textile and garment industry.

2. Carbon labeling

   Carbon labeling refers to the label of quantitative index of the greenhouse gas emissions during the entire production process of the products on the product label, with the aim to mitigate climate change, reduce greenhouse gas emissions, and to promote the low-carbon emissions technology. In this way, the customers are informed of the carbon information of the particular product that is purchased. Carbon labeling is mainly used for export products.

   So far, China has not yet adopted carbon labeling, and the researches and reports relating to carbon labeling for textile and garment industry are rare. Most of the textile spinning enterprises have weak awareness in this respect and are rather passive. However, the external pressure of low-carbon economy is now growing for the textile and garment enterprises, and the irreversible trend is that the textile spinning, printing and dyeing, and garments enterprises in developing countries would be required by developed countries to attach carbon labeling on the products, to specify the carbon emissions of the products during the entire production process. Carbon labeling would become an effective approach to judge whether the products meet the ecological and environmental requirements in the era of low-carbon economy (Wu et al. 2011).

3. Carbon tariff
   Carbon tariff refers to the special tariff imposed if the products fail to meet the standards set by the import countries relating to energy saving and emission reduction. As low-carbon economy is an irreversible trend, carbon tariff is also an avoidable reality. According to the report released by the World Bank, if the carbon tariff is to be implemented on the international market, China's manufacturing industry would be confronted with a tariff of 26 % on average, and as a consequence, Chinas' export volume would drop by 21 %. Or if the carbon tariff is implemented by America and Europe, the new trade barrier established would directly impair the global competitiveness of China's textile and garment export. A more harsh result will be that more than 20 million workers in textile and garment industry would be faced with unemployment.
   The implementation of carbon tariff is undoubtedly a lethal blow to China's textile and garment industry. Increased cost combined with other adverse factors would result in the industry transfer of some foreign textile processing enterprises into countries with lower costs. If that is exactly what happens, the yearly decreasing trend for China's textile industry is unavoidable. Moreover, as an additional tariff, carbon tariff would be another disaster for the recession of China's textile and garment industry.

4 Conclusion and Prospect

Textile and garment industry is China's traditional pillar industry in national economy which is closely associated with people's livelihood, and it is also the industry with significant competitive advantages on the international market. On one hand, textile industry is the typical pollution-intensive industry, whose expansion in production scale as a result of trade growth has brought enormous damage to China's environment. On the other hand, the succession of environmental trade barrier set by foreign countries raises new challenges to China's textile and garment export. The contradiction between economic growth and environmental deterioration is becoming more prominent than ever during the production of textile products.

The major reason for numerous trade restrictions encountered by China's textile and garment export trade is that China's textile and garment products cannot satisfy the environmental and health requirements raised by foreign countries. Green trade barrier, environmental indication system, and carbon emission can be used as subterfuges by developed countries to turn down China's products. However, the environmental problem is indeed the un-negligible problem troubling China's textile and garment trade. To better resolve this problem, we should attempt to gear our awareness to international conventions while internationalizing the economic development. It is the government's responsibility to promote the understanding of environment-related problems and to introduce advanced measures from the foreign countries, so that these measures can be adapted to China's conditions. The environmental problems should be analyzed from a long-term perspective, and the attempt to increase the environmental standards of China's products to the internationally accepted level should be made. The strict legislation is also needed to avoid some international disputes, so as to eliminate the barrier to China's textile and garment export trade.

Acknowledgments Supported by National Social Science Fund (Grant: 14BJL056), Beijing Planning Project of Philosophy and Social Science (Grant: 13JDJGB045), Beijing Funding Project of Talents (Grant: 2013D005001000004), Project of selecting and cultivating of innovative talents team in Beijing Institute of Fashion Technology (Grant: PTTBIFT_TD_007).

References

David O, MÓNica K (2008) Industrial development for the 21st Century. Palgrave Macmillan, USA, pp 183–234

Jiang G, Chen J, Wang D (2013) Basic theory and system design of low-carbon economy. In: 2013 international conference on public administration 9:872–876

Lin Y (2010) Conflicts and coordination path of China's textile export trade and environmental. China Ocean University, Beijing

Li Y (2013) Transformation of green trade policy in the context of low-carbon economy. Shandong University of Finance and Economics, Shandong

Satoshi W, Makoto N (2013) Japan's experience of energy saving and business model for low-carbon economy. Chin J Popul Resour Environ 11:288–298

Sun Q et al (2011) Carbon footprint and textile industry. Text Lead 3:15–18

Wen S (2013) The factor analysis of international trade investment based on green carbon emissions. In: Proceedings of 2013 2nd international conference on social science and education, vol 47, pp 304–308

Wu D et al (2011) Development and prospects of textile carbon labelling. Text Lead 3:19–22

Wu S (2011) A study on the foreign environmental regulation effect on China's export trade—based on the cases of EU and Asean. Services and Management Engineering, 2011 International Conference on Information, pp 557–561

Zhang Y (2010) Study on relationship between international trade and environmental pollution. Henna University, Henna

A Calculation Model for Greenhouse Gas Emission Impacts of Electric Vehicle Carsharing

Liang Hu, Ming-quan Wang and De-yang Kong

Abstract To deal with global warming and maintain the harmonious relationship between people and the nature, low-carbon city and transportation have become the important strategies of sustainable development. Carsharing is a new means of transportation in China, and some research has indicated that it holds considerable benefits to both society and environment. In recent years, electric vehicle (EV) has become the trend of automotive industry and some EV carsharing programs have been launched; therefore, it is necessary to study their greenhouse gas (GHG) emission impacts on cities. By comparing traditional carsharing with EV carsharing, the authors analyze the households' travel behaviors before and after participating in EV carsharing, which are described by 12 travel behavior parameters. Adding three energy consumption parameters and three emission parameters, the authors propose a calculation model for the GHG emission impacts of EV carsharing. The impacts on low-carbon cities and transportation can be assessed based on this model after acquiring specific data of those parameters by conducting questionnaire surveys, recording with instruments, or referring to research papers.

Keywords EV carsharing · Low-carbon · Greenhouse gas (GHG) emissions · Calculation model

L. Hu · D. Kong (✉)
School of Automotive Studies, Tongji University,
201804 Shanghai, People's Republic of China
e-mail: kongdeyang@126.com

M. Wang
Chinese Academy of Sciences, Shanghai Advanced Research Institute,
201210 Shanghai, People's Republic of China

S. Feng et al. (eds.), *Low-carbon City and New-type Urbanization*,
Environmental Science and Engineering, DOI 10.1007/978-3-662-45969-0_12

1 Introduction

With the development of Chinese economy and the increase in disposable income per capita, a rising number of Chinese households can afford a car. Chinese car ownership was only about 9.7 million in 2002, but this figure had grown to about 88.4 million by 2012 (National Bureau of Statistics of China 2013), a 24.7 % annual increasing rate on average. However, the popularity of private cars and the increasing number of vehicle kilometers traveled (VKT) have intensified the greenhouse gas (GHG) emissions. According to some research, the amount of CO_2 emitted by the transportation area in China represented around 8 % of all CO_2 emissions nationwide (Wu 2007), and the petroleum consumption of transportation accounted for 45.6 % of total consumption in China in 2011 (National Bureau of Statistics of China 2013).

Carsharing originates in Switzerland and Germany. It is an innovative means of transportation between private cars and public transportation. The members should make reservations by telephone or the Internet before using the cars, and they pay a membership fee in some combination with per-hour and per-kilometer charges rather than the costs of private vehicle ownership. By the July of 2012, the United States (USA) had 26 carsharing programs, 806,332 members, and 12,643 vehicles (Carsharing 2014). However, carsharing is still young in China and the first carsharing company was registered in 2010 (Introduction to Edoauto carsharing 2014).

In recent years, electric vehicles (EVs) are more widely used in carsharing service due to their low operation costs and emissions. 'Autolib,' which is the largest EV carsharing program in the world at present, was launched in December 2011 in Paris. This program plans to offer 3,000 shared EVs and to establish charging pillars by the road (2014). China has also initiated several EV carsharing programs such as 'EV Beijing' in Beijing, 'EV CARD' in Shanghai, and 'Weigongjiao' in Hangzhou.

Research shows that carsharing may offer significant social and environmental benefits. Cervero and Tsai studied a carsharing program in Los Angeles called 'City CarShare' (Cervero et al. 2007; Cervero and Tsai 2004), and the results indicated that the members had reduced their VKT after joining carsharing. Martin and Shaheen (2011) studied the GHG emission impacts of traditional carsharing in North America. They conducted an online questionnaire survey among the members in 11 carsharing programs and acquired 6,281 effective samples. The results showed that every carsharing household can reduce 0.58 t (observed impact) and 0.84 t (full impact) CO_2 emissions per year on average. Xia et al. (2007) researched into an informal carsharing project in Beijing from the perspective of economic and ecological efficiency, concluding that the efficiency of shared cars is 2.34 times as many as that of private cars. Zhang et al. (2012) conducted a questionnaire survey among 271 residents in Shanghai, and the results suggested that the VKT dropped by 228,520 km per year and the GHG emissions fell by 61,035 kg annually.

However, almost all the above research focuses on traditional carsharing. EV is the trend of global automotive industry, so it is necessary to study the environmental impacts of EV carsharing. By studying the households' travel behaviors before and after joining EV carsharing and the differences between traditional and EV carsharing, the authors propose a calculation model for the GHG emission impacts of EV carsharing.

2 Households' Travel Behaviors

This research first studies the change in households' travel behaviors before and after EV carsharing and then proposes a calculation model for the GHG emissions of every household. The unit of analysis is household because an individual's carsharing use can affect the travel decisions of all household members (Martin and Shaheen 2011).

2.1 Households' Travel Behaviors Before EV Carsharing

Families can be categorized into three types, which are one-car family, several-car family, and no-car family, based on the number of cars owned by families. Zhang et al. (2012) conducted a questionnaire survey about carsharing among Shanghai residents, showing that one-car family accounted for 44.6 %, several-car family only 8.5 %, and no-car family 46.9 %.

Car-owning families travel mainly by private car. Statistically, 78 % of all private cars travel within 64 km daily on average in the USA (Mineta et al. 2003) and 80 % within 49.2 km in Beijing (Wu 2013). Getting the data of total VKT of several-car families requires the total travel distance of all their cars. Meanwhile, families also plan to purchase one more car if they have higher income or existing cars cannot meet their demands, bringing about potential emissions subsequently.

No-car families travel mainly by bicycle, bus, and subway before joining EV carsharing. At present, more Chinese families plan to purchase a car and this phenomenon is illustrated by a study showing that 72.4 % of no-car families in Shanghai plan to buy a private car within 5 years (Zhang et al. 2012). Apart from traditional internal combustion engine vehicles (ICEVs), hybrid electric vehicles (HEVs) and EVs are gradually accepted by consumers. Qin, a plug-in HEV designed by BYD, sold at 6,457 in total between January and July of 2014, has become the best selling new-energy vehicle in China (2014). Our survey shows that 46, 35, and 11 % of 1015 respondents plan to buy a gasoline ICEV, HEV, and EV, respectively (see Fig. 1). Only a few people plan to buy a natural gas ICEV or a diesel ICEV.

As a result, the households' travel behaviors before EV carsharing can be described by six parameters—'number of cars owned,' 'annual VKT of every car,' 'plan to buy a car or not,' 'plan to buy an ICEV or not,' 'plan to buy a HEV or not,' and 'plan to buy an EV or not.' The last three parameters are mutually exclusive.

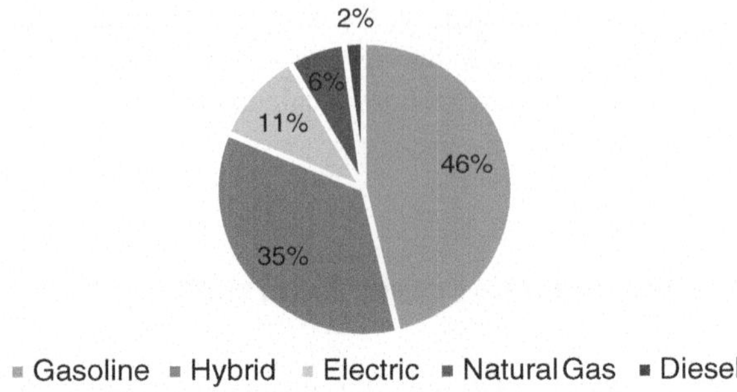

Fig. 1 Fuel/Energy types of vehicles planned to purchase

2.2 Households' Travel Behaviors After EV Carsharing

The 1,015 respondents expect to share different types of vehicles (see Fig. 2). HEV gains most popularity and next comes gasoline ICEV. About 20 % of respondents think EV is most appropriate for carsharing. However, carsharing programs in China only provide gasoline ICEVs and EVs at present.

Some research indicates that households' travel behaviors have changed after traditional carsharing (Cervero et al. 2007; Glotz-Richter 2012): No-car family travels more distance by car; car-owning families not only sell some or even all their private cars, but also drive less. So the changing amount of GHG emissions can be calculated based on the total change in VKT.

However, the differences of EV carsharing are that the members travel by both ICEV and EV, and the two have different emission impacts. The reasons are as follows. First, no-car families' VKT increases because a part of public transport travel is replaced by driving shared vehicles. Second, car-owning families reduce their car ownership, and they use EV carsharing as well as drive private cars. Third, car-owning families even sell all their cars and only use EV carsharing with less VKT. As for the purchasing plan, this paper assumes that all potential car consumers abandon their plans after joining EV carsharing.

Consequently, households' travel behaviors after joining EV carsharing can be described by three parameters—'number of cars owned,' 'annual VKT of every car,' and 'annual VKT of using EV carsharing.'

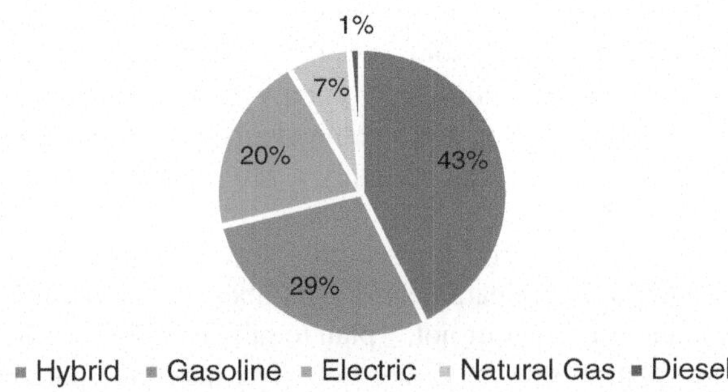

Fig. 2 Fuel/Energy types of vehicles expected for carsharing

3 Calculation Model and Its Parameters

3.1 Calculation Model

Before participating in EV carsharing, every household's travel behaviors can be described by six parameters—'number of cars owned,' 'annual VKT of every car,' 'plan to buy a car or not,' 'plan to buy an ICEV or not,' 'plan to buy a HEV or not,' and 'plan to buy an EV or not.' Besides, the fuel economy parameters of ICEVs, HEVs, and EVs as well as the emission parameters of gasoline and electricity are needed to calculate the corresponding amount of GHG emissions.

Under such scenario, the GHG emissions come from four aspects: first, emissions from driving private cars; second, potential emissions from purchasing ICEVs in the future; third, potential emissions from purchasing HEVs in the future; and fourth, potential emissions from purchasing EVs in the future. The second to the fourth aspects have not occurred yet, so they should be calculated based on the projected values of households' VKT.

As a result, the amount of annual GHG emissions on average of every household before joining EV carsharing is given in (1).

$$
\begin{aligned}
f = {} & N_b \cdot L_b \cdot C_{\mathrm{ICE}} \cdot \mathrm{GHG}_{\mathrm{ICE}} \\
& + D \cdot D_{\mathrm{ICE}} \cdot L_{\mathrm{E-ICE}} \cdot C_{\mathrm{ICE}} \cdot \mathrm{GHG}_{\mathrm{ICE}} \\
& + D \cdot D_{\mathrm{HEV}} \cdot L_{\mathrm{E-HEV}} \cdot C_{\mathrm{HEV}} \cdot \mathrm{GHG}_{\mathrm{HEV}} \\
& + D \cdot D_{\mathrm{EV}} \cdot L_{\mathrm{E-EV}} \cdot C_{\mathrm{EV}} \cdot \mathrm{GHG}_{\mathrm{EV}}
\end{aligned}
\tag{1}
$$

In this equation, the unit of f is kg; N_{b} is the number of cars owned by families before joining EV carsharing, and the value is 0 for no-car family; L_{b} is the annual VKT on average of every private car (in km); $L_{\mathrm{E-ICE}}$, $L_{\mathrm{E-HEV}}$, and $L_{\mathrm{E-EV}}$ are the projected households' annual VKTs on average of driving ICEVs, HEVs, and EVs, respectively (in km); C_{ICE} is the fuel economy parameter of ICEVs (in L/100 km); C_{HEV} is the fuel economy parameter of HEVs; C_{EV} is the fuel economy parameter of EVs (in kWh/100 km); $\mathrm{GHG}_{\mathrm{ICE}}$ is the GHG emission parameter of ICEVs (in kg/L); $\mathrm{GHG}_{\mathrm{HEV}}$ is the GHG emission parameter of HEVs; $\mathrm{GHG}_{\mathrm{EV}}$ is the GHG emission parameter of EVs (in kg/kWh); D, D_{ICE}, D_{HEV}, and D_{EV} stand for 'plan to buy a car or not,' 'plan to buy an ICEV or not,' 'plan to buy a HEV or not,' and 'plan to buy an EV or not,' respectively. Value 1 means 'Yes,' while value 0 means 'No.'

After joining EV carsharing, households' travel behaviors will change and can be described by three parameters—'number of cars owned,' 'annual VKT of every car,' and 'annual VKT of using EV carsharing.'

Since this paper assumes that all potential car consumers will abandon their purchasing plans after joining EV carsharing, the emissions under this scenario come from two aspects: first, emissions from driving remained ICEVs, and second, emissions from using EV carsharing. So the amount of annual GHG emissions on average of every household after joining EV carsharing is given in (2).

$$q = N_a \cdot L_{a-ICE} \cdot C_{ICE} \cdot GHG_{ICE} + L_{a-EV} \cdot C_{EV} \cdot GHG_{EV} \quad (2)$$

In this equation, the unit of q is kg; N_a is the number of cars owned by families after joining EV carsharing; L_{a-ICE} is the annual VKT on average of remained ICEVs (in km); L_{a-EV} is the annual VKT on average of using EV carsharing (in km).

Combining Eqs. (1) and (2), the annual changing amount of GHG emissions on average brought by all households who join EV carsharing is given in (3).

$$\Delta_{GHG} = \sum (q_i - f_i) \quad (3)$$

In this final equation, f_i and q_i are the ith household's emissions before and after EV carsharing, respectively. Minus Δ_{GHG} means that EV carsharing can cut down GHG emissions.

3.2 Travel Behavior Parameters

In the calculation model, there are 12 travel behavior parameters—N_b, N_a, L_b, L_{E-ICE}, L_{E-HEV}, L_{E-EV}, L_{a-ICE}, L_{a-EV}, D, D_{ICE}, D_{HEV}, and D_{EV}. For every family who joins EV carsharing, the data of L_b, L_{a-ICE}, and L_{a-EV} can be recorded by instruments, and the data of the remaining 9 parameters can be acquired by conducting questionnaire surveys.

Take L_{E-ICE}, L_{E-HEV}, and L_{E-EV} for instance. In our survey, the 1,015 respondents predict their annual VKT of driving new cars. Some respondents plan to buy diesel or natural gas cars but they are hardly sold in Chinese market. After deleting them, there are 899 effective samples remained. Figure 3 shows that people who plan to purchase ICEVs expect to drive the longest distance, and next come HEVs and EVs. It indicates that people who travel longer distance tend to buy ICEVs, while people who travel shorter distance tend to purchase EVs.

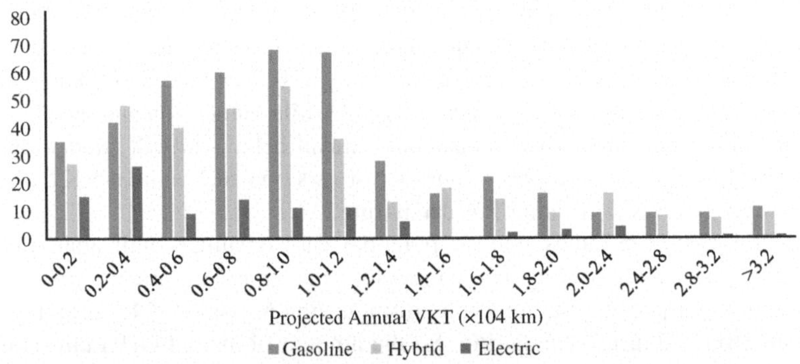

Fig. 3 Projected annual travel distance of driving new cars

3.3 Energy Consumption Parameters

In the calculation model, C_{ICE}, C_{HEV}, and C_{EV} are the energy consumption parameters.

C_{ICE} is the gasoline consumption of ICEVs every 100 km. Referring to the Chinese official limit (8.6–9.5 L/100 km) for passenger vehicles (2004), the value of C_{ICE} is set to be 9 L/100 km in this paper.

C_{EV} is the electricity consumption of EVs per 100 km. Now, there are a few EV carsharing programs in China. For example, 'EV Beijing' is a platform for Beijing Municipality to promote the use of EVs. This program provides BAIC E150 EV as the shared vehicle and has constructed several stations such as one in TusPark (2014). Another example is 'EV CARD' located in Jiading District, Shanghai, initiated by Shanghai International Automobile City Company and Tongji University. The target customers are university teachers and students, and the shared cars are ROEWE E50 and SHANGHAI GM SPRINGO (2014). Related parameters of the above three EVs are listed in Table 1.

The electricity consumption per 100 km of the three models is all 15 kWh, so the value of C_{EV} is set to be 15 kWh/100 km in the calculation model.

There are many different types of HEVs, and each has different driving range, so the value of C_{HEV} is set to be the average number of C_{ICE} and C_{EV} values which means HEVs consume 4.5 L gasoline and 7.5 kWh electricity per 100 km.

3.4 Emission Parameters

In the calculation model, $\mathrm{GHG}_{\mathrm{ICE}}$, $\mathrm{GHG}_{\mathrm{HEV}}$, and $\mathrm{GHG}_{\mathrm{EV}}$ are the emission parameters, representing the GHG emissions of ICEVs, HEVs, and EVs, respectively.

Gasoline produces GHG such as CO_2 during the process of not only combustion but also exploitation and refinement. Similarly, although EVs only consume electricity and produce no emission during running, the emissions produced during the process of electricity generation using fossil fuels such as coal and crude oil cannot be ignored. Argonne National Laboratory in the USA proposes a well-to-wheel (WTW) life cycle assessment system (see Fig. 4) and an assessment model called GREET (Wang 1996). The WTW can be divided into two stages—well-to-tank

Table 1 Related parameters of three shared EV models

EV model	Battery capacity (kWh)	C_{EV} (kWh/100 km)
BAIC E150 EV	25.6	15
ROEWE E50	18	15
SHANGHAI GM SPRINGO	21.4	15

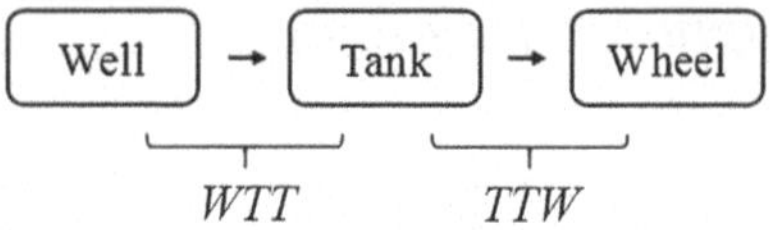

Fig. 4 Well-to-wheel life cycle assessment system

(WTT) and tank-to-wheel (TTW), evaluating the emissions of gasoline and electricity during the whole process of production and usage (Wang 2002).

Therefore, the model in this paper calculates the life cycle GHG emissions of ICEVs, HEVs, and EVs. Based on the GREET model, Liu et al. (2007) concluded that consuming 1 L gasoline releases 3.03 kg CO_2 and consuming 1 kWh electricity releases 1.10 kg CO_2. The amount of CO_2 emissions is the equivalence of all GHG such as CO_2, CH_4, and NO_2 based on their greenhouse effects. The emission intensity of HEVs is set to be the average level of ICEVs and EVs.

4 Conclusion

Before participating in EV carsharing, the households' travel behaviors can be described by six parameters—'number of cars owned,' 'annual VKT of every car,' 'plan to buy a car or not,' 'plan to buy an ICEV or not,' 'plan to buy a HEV or not,' and 'plan to buy an EV or not,' while after joining EV carsharing, the households' travel behaviors change to be described by three parameters—'number of cars owned,' 'annual VKT of every car,' and 'annual VKT of using EV carsharing.' Therefore, there are 12 travel behavior parameters such as the number of cars owned by families before EV carsharing (N_b) in the calculation model.

Apart from those, there are also three energy consumption parameters and three emission parameters—gasoline consumption per 100 km of ICEVs (C_{ICE}), energy consumption per 100 km of HEVs (C_{HEV}), electricity consumption per 100 km of EVs (C_{EV}), life cycle emissions of ICEVs ($\mathrm{GHG}_{\mathrm{ICE}}$), life cycle emissions of HEVs ($\mathrm{GHG}_{\mathrm{HEV}}$), and life cycle emissions of EVs ($\mathrm{GHG}_{\mathrm{EV}}$).

Before joining EV carsharing, driving private cars produces GHG emissions. Meanwhile, there are also potential emissions from purchasing ICEVs, HEVs, or EVs in the future. After joining EV carsharing, households abandon their purchasing plans, so the potential emissions are avoided in fact; however, driving remained private cars and using EV carsharing still bring about emissions. In particular, no-car families drive more due to EV carsharing, which is the increasing impact of EV carsharing on GHG emissions, whereas in contrast, the decreasing impact is in two aspects: First, the VKT of driving ICEVs declines; second, the potential emissions brought by purchasing plans are avoided. The total GHG emission impacts are the combination of increasing impact and decreasing impact.

After acquiring the data of every household's travel behavior parameters by conducting questionnaire surveys or recording with instruments, the GHG emission impacts can be calculated with Eqs. (1)–(3).

However, there are three major limitations in this research. First, there is no empirical calculation. Since there have not been large-scale carsharing programs in China, we are not able to conduct surveys among carsharing members. Besides, using instruments such as GPS to record members' travel behaviors may invade their privacy, so the related data are difficult to acquire. Second, the annual VKT of plan-to-buy cars is just estimated by the respondents; therefore, the data are not accurate enough and we need better prediction methods. Third, the values of energy consumption parameters and emission parameters are different in different research, so a reasonable range is needed.

Acknowledgments Financial support: National Science and Technology Support Plan (2013BAG06B00, 2015BAG11B01); Chinese Academy of Sciences Innovation Engineering Project (KGCX2-EW-321); National Natural Science Foundation of China (51308525); International Foundation: Drexel-SARI Co-Research; and Education on Low Carbon and Healthy City Technology and Evaluation (Drexel-SARI 2014F003).

References

Autolib' [EN/OL]. WIKIPEDIA (2014-8-10) [2014-8-20]. http://en.wikipedia.org/wiki/Autolib%27

BYD new-energy vehicles had been sold from January to July (EN/OL). (2014-8-11) [2014-8-20]. d1ev.com. http://www.d1ev.com/news/company/20140811133538.html (in Chinese)

Carsharing (EN/OL). Innovative Mobility Research (2014-8-6) [2014-8-20]. http://www.innovativemobility.org/carsharing/index.shtml

Cervero R, Tsai YS (2004) City Carshare in San Francisco, California: second-year travel demand and car ownership impacts. Transp Res Rec: J Transp Res Board 1887:117–127

Cervero R, Golub A, Nee B (2007) City CarShare: longer-term travel demand and car ownership impacts. Transp Res Rec 1992:70–80

EV Beijing [EN/OL] [2014-8-20]. http://www.evbeijing.cn (in Chinese)

Glotz-Richter M (2012) Car-Sharing—Car-on-call for reclaiming street space. In: Conference on transport research arena. Athens, pp 1454–1463

Introduction to Edoauto carsharing (EN/OL) (2014-8-20). http://www.eduoauto.com/guanyugongsi.php (in Chinese)

Liu H, Wang H-w, Luo Q, Wang Y, Ou Yang M-g (2007) Energy, environment and economic assessment of life cycle for electric vehicle. Technol Econ Areas Commun 6:45–48 (in Chinese)

Martin EW, Shaheen SA (2011) Greenhouse gas emission impacts of carsharing in North America. IEEE Trans Intell Transp Syst 12(4):1074–1086

Mineta NY, Jackson MP et al (2003) Highlights of the 2001 National Household Travel Survey. US Department of Transportation, Washington, DC

National Bureau of Statistics of China (2013) China Statistical Yearbook 2013. China Statistics Press, Beijing (in Chinese)

Standardization Administration of the People's Republic of China (2004) GB 19578-2004 Limits of fuel consumption for passenger cars. Standards Press of China, Beijing (in Chinese)

Vehicle information of EV CARD [EN/OL] [2014-8-20]. http://www.evcardchina.com/car_rental/vehicle_Information.aspx (in Chinese)

Wang M (1996) Development and use of the GREET model to estimate fuel-cycle energy use and emission of various transportation technologies and fuels, Center for Transportation Research. Argonne National Laboratory, ANL/ESD-31

Wang M (2002) Fuel choices for fuel-cell vehicles: well-to-wheels energy and emission impacts. J Power Sources 112(1):307–321

Wu W-h (2007) Comparisons of energy consumption and emission in transport field between China and typical countries. Energy China 29(10):19–22 (in Chinese)

Wu L-w (2013) A GPS-based research on the driving patterns of private passenger vehicle in Beijing. MS program, Tsinghua University Beijing, China (in Chinese)

Xia K-x, He M-s, Zhang H (2007) Study on car sharing service based on theory of economic ecological efficiency. Manag World 11:152–153 (in Chinese)

Zhang M, Hui Y, Wang M-q (2012) Urban greenhouse gas emission of car sharing. China Popul Resour Environ 22(9):48–53 (in Chinese)

The Analysis of Guangxi Environmental Problems of Urbanization and Low-carbon Development Strategy

Li-hong Zhang, Lv-xuan Zhao, Xiao-wei Song, Nan-jun Qin and Jun Zhang

Abstract The urbanization of Guangxi starts late, and it has developed rapidly, but in the process of urbanization, industrial and energy structure is irrational, resulting in the growing problem of cities and towns pollution. So changing the traditional development model of urbanization, implementing low-carbon development path is necessary. Based on the status research of Guangxi urbanization, this article analyzes the main features of urbanization in Guangxi, finding the problems of increasing environmental pollution, the construction of environmental protection facilities lag, lack of green assessment system, and other issues caused by the irrational industrial structure. This article puts forward four low-carbon development proposals: first, establishing a green GDP examination and evaluation system, strengthening supervision and restraint effect; second, adjusting industrial layout, enhancing the level of industrial development; third, improving the environmental management system, and strengthening environmental infrastructure; and fourth, using low-carbon development and sustainable development as guidance, carrying out scientific urban planning. The purpose is to promote the healthy development of urbanization in Guangxi and explore a new and sustainable, low-carbon development-oriented urbanization road.

Keywords Environmental issues · Industrial structure · Low-carbon urbanization · Pollutant emissions

1 Introduction

Now, Guangxi is in the stage of rapid development of urbanization, with an average growth rate of urbanization rate reaching 1.73 %, higher than the national average. However, the overall urbanization rate of Guangxi is lower than the national

L. Zhang · L. Zhao (✉) · X. Song · N. Qin · J. Zhang
Scientific Research Academy of GuangXi Environmental Protection, 530022 Nanning, China
e-mail: 358056078@qq.com

S. Feng et al. (eds.), *Low-carbon City and New-type Urbanization*, Environmental Science and Engineering, DOI 10.1007/978-3-662-45969-0_13

average (Xu 2010). Statistics shows that from 1953 to 2012, the urbanization rate increased from 8.52 to 43.53 % in Guangxi, and the urbanization rate is always lower than the national average urbanization rate of about 10 %. Guangxi urbanization rate in 2012 ranked 26th in 31 provinces and municipalities, just ahead of Henan, Gansu, Yunnan, Guizhou, and Tibet provinces.

Though it can stimulate domestic demand and promote economic growth by urbanization in Guangxi, there are also high energy consumption, high waste, high pollution, and many other hazards and environmental issues (Li 2014). Therefore, using the guidance of ecological civilization (Pan 2011), taking a low-carbon urbanization, and handling the relationship between economic development and environmental constraints has become an urgent task for us (Pan et al. 2014).

2 The Main Feature of Urbanization in Guangxi

2.1 Urbanization Has Changed Guangxi Industrial Structure, and Industry Has Become the Leading Industry

Accompanying urbanization, the industrial structure of Guangxi has changed every year; Fig. 1 summarizes the industry contribution to economy from 1978 to 2011 in Guangxi.

As can be seen from the figure, the primary industry contribution ratio to GDP is showing lower trends in Guangxi. The contribution rate was 40 % in 1978, and it dropped to 17 % in 2011; while the second and tertiary industries' contribution rate is increasing, especially the contribution rate of the secondary industry increased by about 15 % since 2002–2011. With the start and carrying out the country's Western development strategy, Guangxi has been undertaking industrial transfer from the eastern region, constructing a lot of industrial parks, and the industrial output increasing significantly; thus, the industry has been becoming the dominant

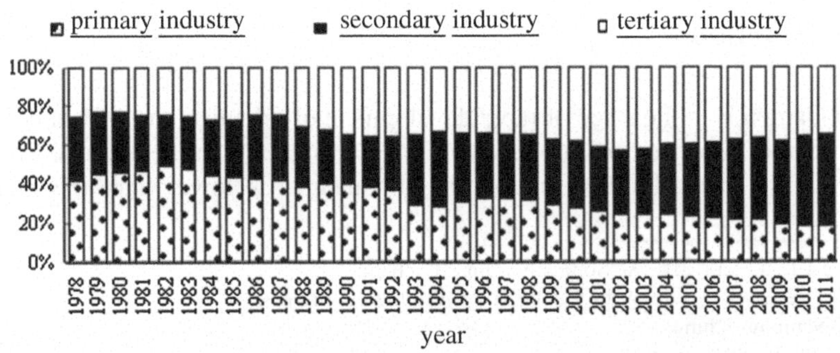

Fig. 1 Schematic of industry contribution to economy 1978–2011 in Guangxi

industry in Guangxi. Though the tertiary industry growth rate has declined slightly in the recent 10 years, the varying trends of proportion of tertiary industry are relatively stable, and the tertiary and second industry still has promoted jointly regional economic development in Guangxi.

2.2 *Urbanization Changes Energy Use Patterns, Industrial Coal Increases Year by Year*

1. Coal Consumption
   Guangxi is still in the early stage of development of urbanization and industrialization; as technology advances, energy consumption indicators show an upward trend over a period of time, and due to a faster pace of development than the more developed regions, the total growth trend is very clear. Statistics of the rate of urbanization and energy use structure from 2003 to 2012 in Guangxi (Table 1) shows that Guangxi urbanization rate was 43.53 % in 2012, compared with 2003 increased by 54 %; the total coal consumption was 70,749,600 t, compared with 2003 increased by 2 times. Among them, due to the accelerate of growth rate of industrial output, the industrial coal consumption indicators continued to rise, and the industrial coal consumption increased by 2.2 times from 21,606,300 t in 2003 to 68,773,400 t in 2012; the life coal consumption grew by only 46 %, and basically kept stable since 2005, which is because the life coal consumption of urban residents of Guangxi used mostly liquefied petroleum gas, coal gas, natural gas, and used the lower proportion of direct fire coal.
2. Electricity Consumption
   With the improvement of living standards of urbanization, per capita electricity consumption continues to increase, and with urban population growth, the total amount of electricity grows substantially. As can be seen from Table 1, with the improvement of the urbanization rate, the per capita electricity consumption of urban, rural residents in Guangxi showed a significant increasing trend. Per capita electricity consumption of urban and rural in 2012 was 434.11 and 202.13 kw h/a, respectively, just increased doubly compared with 2003.

2.3 *Urbanization Changes Quantity of Pollutant Discharged, and the Environmental Pressures Continues to Increase*

Table 2 summarizes the behavior of industrial pollutants emissions from 2003 to 2012 in Guangxi. By analyzing, we can see that the contradiction between urbanization development and environmental pollution was still very prominent, especially the total amount of industrial waste gas emissions continued to rise, and the regional atmospheric environment was depredating; since 2013, the winter haze

Table 1 The changes of rate of urbanization and energy use structure in Guangxi

Statistical year	Guangxi urbanization rate (%)	Coal consumption (10,000 t)			Electricity consumption (kw h/a)		
		Total coal consumption	Life coal consumption	Industrial coal consumption	Total electricity consumption	Town	Country
2003	28.30	2,295.84	135.21	2,160.63	274.84	183.67	91.17
2004	29.06	2,828.30	150.97	2,677.33	256.29	159.38	96.91
2005	31.70	3,040.49	158.55	2,881.94	295.24	190.30	104.94
2006	33.62	3,602.76	194.83	3,407.93	330.65	217.81	112.84
2007	34.64	4,165.68	190.78	3,974.9	358.80	231.68	127.12
2008	36.24	4,511.61	168.06	4,343.55	410.42	266.24	144.18
2009	38.16	4,687.10	189.54	4,497.56	446.84	290.23	156.61
2010	39.20	5,345.55	194.99	5,150.56	462.10	300.9	161.20
2011	40.60	6,104.52	196.90	5,907.62	563.20	377.53	185.67
2012	43.53	7,074.96	197.62	6,877.34	636.24	434.11	202.13

Table 2 Guangxi industrial pollutants emissions

Year	2003	2004	2005	2006	2007	2008	2009	2010	2011	2012
Industrial waste gas emissions (hundred million standard m^3)	6,635.3	10,656.5	8,338.3	8,969	12,723.5	11,643.0	1,318.1	14,519.6	29,853.5	27,610.7
Discharge of industrial waste water (hundred million t)	11.9	12.2	14.6	12.9	18.4	20.6	16.2	16.5	10.1	11.1
Industrial solid waste emissions (10,000 t)	108.71	131.83	110.48	22.74	10.34	8.66	7.97	8.12	2.57	0.41
Sulfur dioxide emissions (10,000 t)	72.0	89.7	97.5	94.4	92.6	87.0	83.5	84.8	48.9	43.1
Ammonia nitrogen emissions (10,000 t)	3.1	5.0	5.7	3.6	2.5	2.1	1.4	1.4	0.85	0.86
Chemical oxygen demand emissions (10,000 t)	62.5	69.3	66.4	67.9	60.8	56.0	51.9	49.3	20.5	20.0

phenomenon occurred frequently, and the environmental pressures continued to increase. The total amount of industrial wastewater discharge showed a downward trend after the first rise, industrial solid waste decreased year after year, and emissions of sulfur dioxide, ammonia nitrogen, and chemical oxygen demand experienced a trend of decrease after the first increase. The reason of the above changes was that along with urbanization and rapid industrial development, industrial "three wastes" emissions increased at the same time. With the development of science and technology, "three wastes" handling capacity enhanced, and the emissions of pollutants decreased gradually which was brought from industrial development during the process of urbanization.

3 The Major Environmental Problems of Urbanization Exist in Guangxi

3.1 The Urbanization Quality is Low, Without the Establishment of Ecological Assessment System

Guangxi urbanization rate was rapid, but the urbanization quality was low, showing in 4 parts that economic targets over the desired, environmental targets was difficult to complete; development targets over the desired targets of people's livelihood were difficult to complete; targets of number over the desired targets of quality were difficult to complete; area of land used for construction over the desired; The urban population was difficult to complete. Moreover, with only economic objectives as the focus of assessment, and the lack of a scientific and rational evaluation system, the development of environmental protection was slower than economic development (Li and Xia 2012). The phenomenon of pollution accidents occurred frequently in the process of urbanization, which has become the key factor that restricted the development of urbanization in Guangxi.

3.2 Industrial Structure is Unreasonable, and Structural Pollution is Serious

Whether the industrial structure is reasonable will directly impact the rationalization level of urbanization and the entire economic structure. In the process of economic growth, the first industry and third industry produced less pollution, and the second industry especially was the main pollution source (Li and Jiang 2009).

Urbanization and economic development of Guangxi mainly relies on industrial production. The changes of Guangxi industrial coal consumption from 2003 to

2012 display that industrial development still mainly relies on coal. Guangxi industrial base was low and no reasonable planning and control of the industrial production scale in the process of development, because that coal consumption is increasing year by year. Depending on coal resources excessively will bring the ecological problems of resource depletion, the environmental problems of atmospheric pollution, and other problems. It will restrict the development of Guangxi urbanization.

3.3 Construction of Environmental Protection Facilities was Backward, and Capital Investment was Insufficient

The urbanization of Guangxi was the traditional urbanization, the extensive development way; economic growth is too dependent on the consumption of resources and energy. The rapid development of population and industrial production, limited environmental pollution control technology and capital, and lack of environmental protection facilities have resulted in a significant increase in emissions of pollutants and urbanization rates. With the increase of the national and Guangxi emission reduction efforts, enhancing environmental protection policy, and financial and human inputs, the environmental pollution control capacity of new construction projects has improved, but the industrial pollution problems left over by history are still not resolved thoroughly.

Environmental pollution control investment accounted 1.71 % for the proportion of Guangxi GDP, which ranked 8th of nation. But according to the experience of developed countries, environmental protection investment accounted 3 % or above for GDP, and environmental quality will improve. Apparently, the Guangxi environmental protection investment is far for this target, and Guangxi environmental governance and supervision level is far below national leading level.

3.4 Lack of Technological Innovation Driving Force, the Implementation of the Strategy is not in Place

Economic transformation is the internal motivation to make the traditional urbanization to shift to a low-carbon urbanization, by promoting the application of the technology innovation in urbanization construction, which will lead to the transformation of the economy, and the new urbanization pattern will inevitably arise. Otherwise, the traditional economic growth mode which is of high cost, high energy consumption, high pollution, and resource-based will be with the traditional urbanization patterns. As a western province, relying too much on the eastern industry transfer to promote urbanization is not conducive to cultivate inner innovation drive and hinder the economic development and transformation.

Guangxi has been prepared with urban agglomerations and major cities development planning and master planning, but there are inconsistencies in goal setting and urban development orientation, which cause confusion and even obstacles to the development of urbanization. And it is lack of sufficient communication between departments, and in the urbanization strategy formation process, it has not yet considered resources and environmental factors, and the guide and binding functions from environmental objectives.

4 Countermeasures and Suggestions

4.1 Establish Green GDP Examination Evaluation System, Strengthen the Effect of the Supervision and Constraint

Based on the requirements of optimizing the land space, promote resource conservation, strengthening environmental protection and system construction; it should build low-carbon urban ecological civilization construction assessment system which includes economic development, social development, and environmental development, also it should bring into the annual performance evaluation content of government and relevant departments (Zhang 2012). In the process of assessment of urbanization, it should include considering resource and environment elements, combined with the relevant requirements of the environmental protection strategy setting, attaching great importance to the accessibility of environmental protection goal, and establishing the target system, assessment method, rewards, and punishment mechanism which meet the requirements of ecological civilization.

4.2 Adjust the Industrial Layout, Enhance the Level of Industrial Development

To adjust the industrial layout, it should be on the basis of the national energy-saving goal, establishing more stringent regional energy conservation policy (Zhou 2012). By carrying out the total target control in advance, it will implement that the total energy consumption declines ahead of time, it will reduce the dependence on the resource especially energy from development, speed up the transform and upgrade traditional industries, and limit the development of industries which make large resource consumption and serious pollution. Encourage the development of high and new technology industry and the development of low consumption and high-value-added products, actively developing service industry, and promote the development of the financial industry, insurance industry, modern circulation industry, and new services which has a significant impact on social economic life.

To synchronize to promote agricultural modernization and new rural construction, to improve the efficiency of agricultural production, to liberate rural labor will lay a solid foundation of urbanization. Attaching great importance to the integration of urban and rural development, strengthening the construction of equal social service facilities, improving the rural public service and ecological service level, and enhancing rural and agricultural ecological infrastructure construction, it will realize the organic combination for the new rural construction and urbanization. Adjusting measures to local conditions to develop small towns, it will raise the level of small town life service (Cao et al. 2011). At the same time, the feedback effect of the township enterprise, which used small towns as the carrier on the agriculture, will promote the development of agricultural modernization and industrialization, and increase farmers' income and living standard.

4.3 Improve the Environmental Management System, to Strengthen Environmental Infrastructure Construction

To strengthen the construction of township-level environment management institutions, take the environmental planning as an important part of urban planning, and use the ecological civilization to guide the urbanization development will fully embody the development of the holistic and strategic; to use the policy such as fiscal, financial, and land, strengthen the supervision and management of the environment; to carry out and comply with the environmental protection laws and regulations will enhance the awareness of the environment of the rule of law.

To increase the construction of urban environmental infrastructure, especially investment of environmental protection facilities and pipe network in medium and small towns and the old city, and improve the environmental governance ability construction of Guangxi, it will alleviate the environmental pressure of the urbanization process. To improve the disposal ability of sudden environmental accidents, and establish a perfect environment monitoring and early warning system, it will fully reflect the status of environmental monitoring quality and trends, accurately warning of various environmental emergencies.

4.4 Guided by the Low-carbon Sustainable Development, Carry out Scientific Urban Planning

Scientific and rational pattern of urban space can enhance the comprehensive carrying capacity of cities and prevent the excessive expansion of the main city, and also effectively promote the coordinated development of medium cities and towns, and to realize the division of various functional areas and its role in function (He 2013).

1. Principles of conservation, rational use of land, and space resources to implement in urban planning, construction, and management, and demonstrate the overall coordination of resources, environmental infrastructure, and urban development, it will realize for optimal allocation of resources in a larger range;
2. Reduce environmental pollution by optimization of urban space and the industry layout, optimizing space layout and industrial structure on the basis of different environmental function districts and the resource environmental bearing capacity requirements;
3. Implement the higher environmental standards in the new urban district and new town construction, so as to realize the harmonious development of economic and social development and ecological environment;
4. Increase the investment of urban environmental infrastructure and the proportion in GDP (Xiu et al. 2007); it will guarantee the healthy and fast development of new urbanization.

5 Conclusions

The central economic work conference has put forward "fully integrated the ecological civilization concept and principle into the whole process of urbanization, taking a new urbanization road of intensive, smart, green and low-carbon" in December 2012. Therefore, the development of urbanization in Guangxi should be the perspective of ecological civilization and environmental protection; optimize the industrial layout, enhance the level of industrial development, and improve the rational use of resources and energy; perfect environment management system, enhance the level of pollution control, reduce the negative effects of urbanization development on the environment, and make the harmonious development of the urbanization of population agglomeration, space utilization, economic development, and social urbanization, to avoid the one-sided pursuit of space expansion, city scale, and only on the GDP; with the economic and social development, strictly implement the national environmental policy and the related policies and requirements of constructed ecological civilization, increase the intensity of environmental protection, to avoid the load that "treatment after pollution" and at the expense of the environment for economic growth, and it will realize the healthy and low-carbon development of the urbanization construction in Guangxi.

References

Cao LP, Zhu DJ, Yi H (2011) Research on low carbon service industry concept, classification and social and economic effect. Shanghai J Econ 08:3–10

He W (2013) Study on the structure of urban and rural and urban China. Nankai University, Tianjin

Li XJ (2014) New urbanization: can not engage in zaochengyundong. Chin People Mag 05:38–39
Li MG, Jiang ZM (2009) Review of the development of the theory of Industrial economics. Chin Bus Mark 04:30–32
Li SH, Xia YY (2012) The development of low carbon economy under the background of new urbanization road. J Soc Sci Hunan Norm Univ 03:84–87
Pan JH (2011) City development in the transition to low carbon. Economic Daily, 30 Mar 2011, 005
Pan JH, Liang BF, Xiong N, Qi GZ (2014) The path of low carbon urbanization. Environ Prot 01:33–36
Xiu RX, Wu G, Ceng XA, Shun JG, Yu DY (2007) Research progress of green GDP accounting target. Chin J Ecol 07:1107–1113
Xu GY (2010) Research on the development of low carbon agriculture Chinese. Economist 10: 72–78
Zhang J (2012) The UK low carbon economic policy and practice and to China enlightenment. East China Normal University, Shanghai
Zhou N (2012) Connotation and evaluation of low carbon industry. J Poyang Lake 01:24–30

Traffic Safety Evaluation in the Port Roadway Based on BP Neural Network

Li-shan Sun, Wei Luo, Wei Zeng, Cong-ying Qiu and Li Cui

Abstract On the basis of comprehensive analysis of traffic safety evaluation studies abroad, this paper presented the evaluation index system composed of four areas, such as people, traffic flow characteristics, traffic environment, and traffic management, consisting of a total of 15 indexes for special situation and problems in port road traffic safety. Each index has definite index classification criteria, evaluation requirements, and investigation methods. This paper also built evaluation model based on BP neural network and analyzed its calculation method in detail. This method has a guiding role for the management and decision levels of the port road traffic management department.

Keywords BP neural network · Port road · Safety evaluation · Traffic safety

1 Introduction

In recent years, with the rapid economic growth and with the increasing number of motor vehicle ownership, the number of accidents and deaths has been growing. Related to traffic safety, researchers conducted extensive research, and it has made some progress. But in the current feasibility study of highway and the preliminary design stage, the evaluation and optimization of overall road safety are still relatively lacking. In particular, research work in the port area and other aspects of road accidents is not an in-depth investigation. A typical set of port traffic brings about

L. Sun (✉) · W. Luo · C. Qiu · L. Cui
Key Lab of Traffic Engineering, Beijing University of Technology,
No. 100 Pingleyuan, Chaoyang District, Beijing 100124, People's Republic of China
e-mail: lssan@bjut.edu.cn

W. Zeng
Tianjin Municipal Engineering Design and Research Institute, Tianjing 300051
People's Republic of China

S. Feng et al. (eds.), *Low-carbon City and New-type Urbanization*,
Environmental Science and Engineering, DOI 10.1007/978-3-662-45969-0_14

adverse phenomenon, such as overload, overloaded weight, and large traffic volume, because of its special functions. The traffic environment is harsh.

Based on the above background, for the special status and existing problems of port road traffic, this paper established a comprehensive index system for special port road traffic conditions, constructed the evaluation model based on BP neural network, and introduced the realization of the model. This method can play an important role in improving the traffic administrative department of the management and decision-making level.

2 Literature Review

Evaluation of road traffic safety is the evaluation of a particular area, route, road, or place (section), which can provide an important basis for the objective analysis of road safety conditions (Chen et al. 2008). It analyzes the road traffic safety situation, development process, and trends and then estimates the level of road safety in traffic safety evaluation. Traffic safety evaluation provides an effective basis for studying traffic safety management systems and traffic safety technology. It can improve traffic safety measures and planning and management level. It also can effectively control the growth trend of traffic accidents and reduce the loss of traffic accidents.

The main method of traffic safety evaluation is the probability and statistics and incremental analysis abroad. Analysis of these methods is too simple and not taking into account other factors from the root. It is difficult to make an accurate assessment, and therefore, it does not have the versatility. Hakkert et al. (1996) used the CART method to analyze the security level for intercity main road. But the road conditions are variable and limited. So their research only has a certain representative. USA scholar Perkins (Safety Campaign Annual Seminar 1999) put forward the traffic conflict technique (TCT), but it is fairly difficult to realize. Because it requires auxiliary technique, and the observation of traffic conflict is complex.

Scholars at home have studied much more about traffic safety, but most of the studies concentrated in these two areas too. Zhang (1998) proposed a Chinese TCT, but this method is more complicated to operate. Qing (2002) used the grade coefficient method system to study the state of road safety systematically. This method is simple, practical, and easy to accept by road base management personnel. But it also has shortcomings, which are lack of accuracy and a practical improvement on application. Zufeng (2008) set up a reasonable evaluation index system which consists of four aspects: people, vehicles, road, and environment. But the human factors are fairly heavy when scaling the ratings.

The artificial neural network has some similar characteristics with human brain neural networks, such as self-learning, self-organization, nonlinear dynamic processing, distributed knowledge storage, and memory. It provides a powerful tool for the study and to deal with uncertainty phenomenon (Yang et al. 2005). Because of its characteristics, neural network has been applied to the traffic field in recent years. Pan et al. (2005) put forward a comprehensive evaluation method based on the gray cluster theory and neutral network technology. But the precision of the model still have much room for improvement. Chen (2006) has used BP neural network method on Beijing city road to traffic safety level evaluation and got a higher forecasting accuracy. According to the actuality of freeway traffic safety, Li and Wang (2010) have proposed the model of BP neural network. Application of BP neural network in traffic field gradually perfected.

Many factors influence the road traffic safety in port. The relationship is complex, the noisy data is huge, and factors are difficult to determine, in line with the characteristics of neural networks. Therefore, it is entirely scientific to use artificial neural network study traffic safety evaluation in port area.

3 Establish Traffic Safety Evaluation Model in the Port Roadway Based on BP Neural Network

Based on determining the port complex traffic state the evaluation system, it confirms object vector and sample data. Then, it conducts the BP neural network algorithm and the specific evaluation process graph model as shown in Fig. 1.

3.1 The Establishment of Index System

There are many impacts on traffic safety indicators. It is necessary to consider the characteristics of indicators, logical relationships between indicators, targets, and quantization, and other issues. Only in this way, it can put amount of integrated indicators together. Otherwise, it will lose its reasonableness and the true value of traffic safety comprehensive evaluation (Weimin and Yi 1999).

From the harbor complex state highway starting, traffic safety evaluation system is divided into four main factors: characteristic aspects of the driver, traffic flow characteristics, and traffic management.

Drivers' characteristics factors. The human factor is the most important cause to induce road traffic accidents. As subject of traffic behavior, human is one of the most important factors in the road traffic accident cause. The first element of human factors is motor vehicle drivers. Their professional skill, driving experience, safety awareness, and drunk driving are closely related to the occurrence of traffic accidents. According to the survey on the cause of the accident in some of the major countries, the average proportion of the driver factors is 70.8 %. In addition, non-motorized vehicle drivers and pedestrians who lack awareness of traffic safety, poor in self-awareness, and ignore the traffic rules cause many traffic accidents too. Therefore, the decisive factor to improve traffic safety and reduce traffic accidents is improving the people's safety awareness and ameliorating safety behavior.

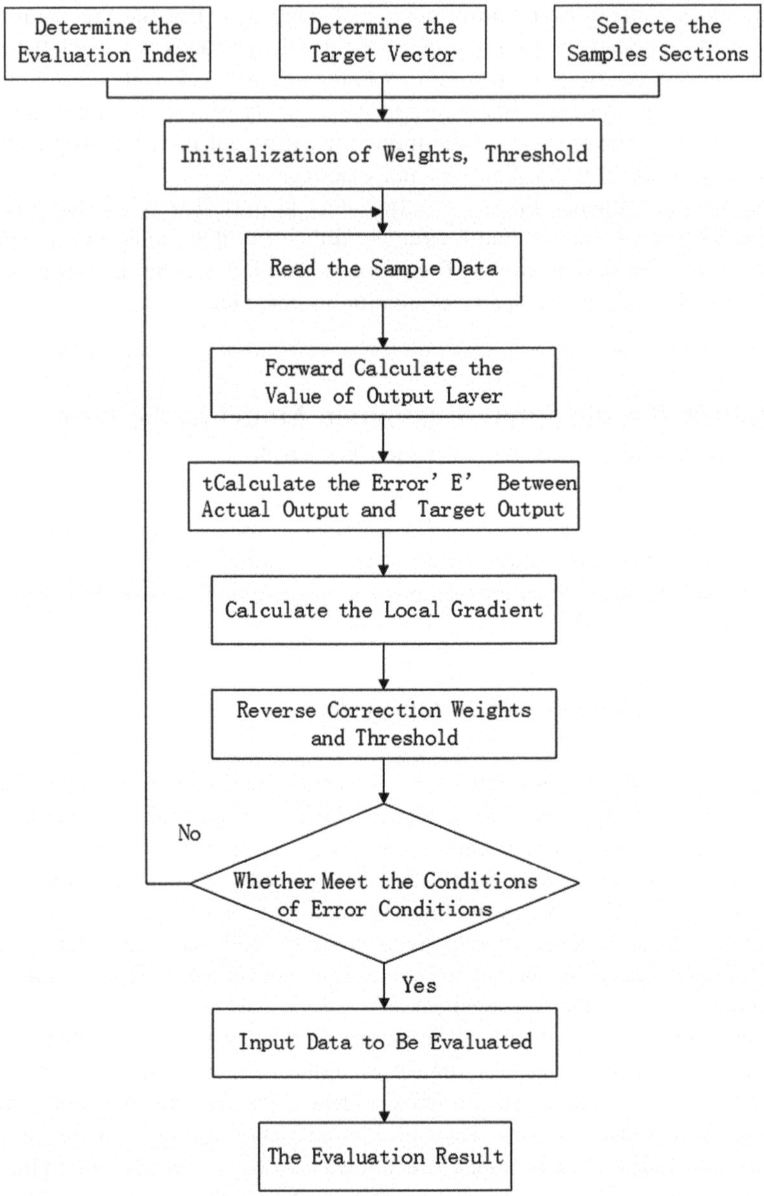

Fig. 1 The process of BP neural network model

Traffic flow characteristics factors. Traffic flow factors include traffic volume, speed and density, and extracting metrics, such as saturation, design speed, and average speed. Different types of vehicles are running on the road. Their running state is having different changes not only with the traffic environment and driver's characteristics, but also with the characteristics of road.

Traffic environmental factors. Environmental factors include road alignment of flat, vertical, and horizontal, traffic signs, safety facilities, and weather. There are no complete traffic signs and safety facilities in some areas. Thus, it causes the driver lax security awareness. So it causes more serious accidents. Noise and weather also have an important impact on traffic safety, for example, in the rain, snow, fog, and other inclement weather conditions, and driving safety factor will come down.

Traffic safety management factors. Traffic management systems include monitoring and management. Traffic management and monitoring system use the intelligent transportation systems (ITS) in traffic management system to establish integrated transportation management system, which can round play a role, real-time, accurate, and efficient. Research shows that the implementation of ITS can make the road capacity increased by 2–3 times. When the vehicles travel on intelligent roads, the traffic accidents can be reduced exponentially.

To sum up, in line with the scientific objectivity, availability, completeness, and feasibility, the evaluation index system is established, as shown in Table 1. The evaluation index system has four categories, fifteen indicators. Ten indicators are quantitative index. And five indicators are qualitative indexes (P12, P32, P33, P41, and P42). Each indicator is determined based on the index grading, evaluation requirements, and survey methods, and divided into excellent, good, medium, and poor, in the 4 grades.

Table 1 Traffic safety evaluation index system of port road

First-grade indexes		Second-grade indexes	
Names of indexes	Coding	Names of indexes	Coding
Drivers	P1	The proportion of drivers who have less than 35 years of driving experience	P11
		Safety awareness of drivers	P12
Traffic flow	P2	Saturation	P21
		The size of the proportion of vehicle traffic	P22
		Design speed	P23
		Average speed	P24
		The average speed difference between carts and trolley	P25
		Headway	P26
		The ratio of lane change	P27
Traffic environment	P3	The bad weather days in a year	P31
		Safety induction facility	P32
		Safety protection facilities	P33
		The alignment of flat, horizontal, and vertical	P34
Traffic safety management	P4	Monitoring system	P41
		Management	P42

3.2 The Model Construction of the Evaluation Index System of BP Network

Based on the evaluation index system, the construction of BP network input layer is 15 nodes corresponding to the 15 indexes. The corresponding index which the qualitative index value converted into quantitative index as input data into the BP network to calculate.

3.2.1 The Number of Hidden Nodes in Middle Layer

In the engineering case, it usually uses the following formula to determine the intermediate layers (Zhang 1993):

$$n_1 = \sqrt{n + m} + a \tag{1}$$

where
n_1 is the node number of hidden layers;
n is the node number of input layer;
m is the node number of output layer; and
a is constant between 1 and 10.

In this paper, it uses the above formula to calculate the initial value of hidden layer nodes. On this basis, it uses the test method. The procedure is as follows: It continuously changes the hidden layer node in the training process, through the evaluation and comparison of the network training and test errors, training step, network structure, and other standards in different number of hidden nodes to select the optimal number of hidden nodes. After the spreadsheet, finally, it selected twenty-two nodes.

3.2.2 The Division of Evaluation Grade

This study will divide the evaluate level into 4 levels. The expected output value is shown in Table 2.

Corresponding to the four grades, the output layer has four nodes. It constructs a "15-22-4"—three layered BP network as the network model evaluation system. After calculating the output layer data, the results are compared with the expected output value. Then, there got the specific evaluation grade, as shown in Fig. 2.

Table 2 The division of evaluation grade

Security classification	Desired output
Very safety	0.9, 0.1, 0.1, 0.1
Relatively safety	0.1, 0.9, 0.1, 0.1
General safety	0.1, 0.1, 0.9, 0.1
Unsafety	0.1, 0.1, 0.1, 0.9

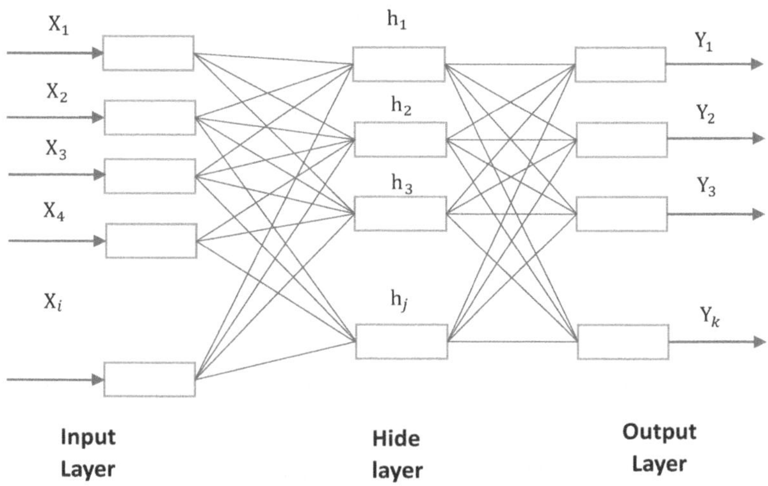

Fig. 2 The schematic diagram of *three layers* of BP neural network model

4 The Specific Realization of the BP Neural Network Model

4.1 Setting Network Parameters

The training parameters are an important index to determine the network performance, including the initialization of weights and thresholds, net.trainParam parameters (Liu et al. 2010). The system will automatically initialize the given weights and thresholds. And the specific net.trainParam parameter settings are shown in Table 3.

4.2 The Training Model

When setting the network parameter, it can enter the network sample data files and the weight matrix, beginning to training network.

In the process of realizing the traffic safety evaluation method of neural network, the key is determining the value of input and output in sample model. This study

Table 3 The settings of evaluation network parameter

Code of parameters	Meanings of parameters	Setting value
net.trainParam.show = 1	Display interval in training	1
net.trainParam.time = 500	Time allowed in training	500
net.trainParam.lr = 0.3	Learning step	0.3
net.trainParam.mc = 0.95	Momentum parameter	0.95
net.trainParam.epochs = 1,000	The maximum number of training	1,000
net.trainParam.goal = 0.001	The minimum mean square error	0.001

selected five typical port road data as the sample data. The results in network training are as shown in Figs. 3 and 4.

As shown in Fig. 3, the training results and expected output value reached 0.89631. It is good fitting. As shown in Fig. 4, the neural network comes to complete the training when iterates 901 times. The error reached its target goal. Therefore, the training effect is good, and it accepts the neural network model.

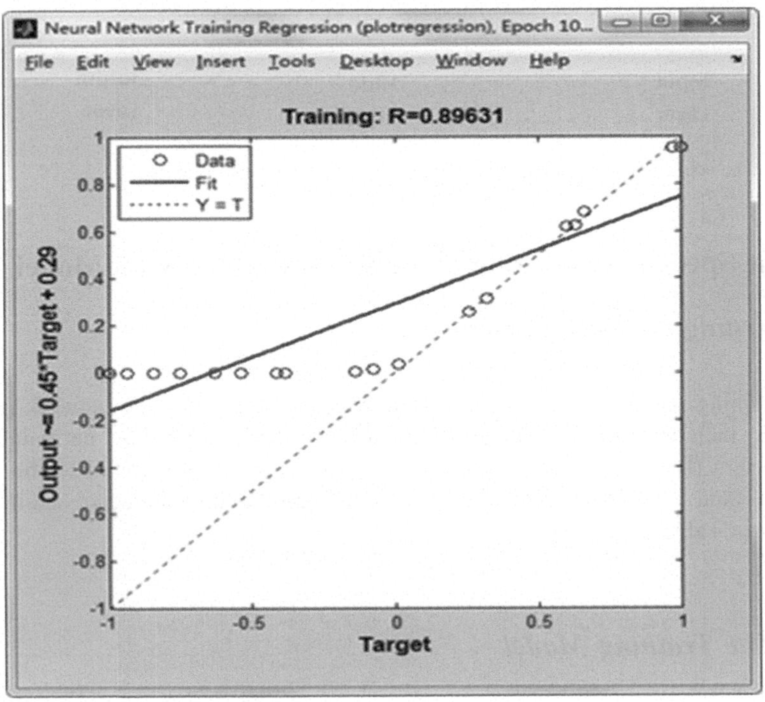

Fig. 3 The results of training network

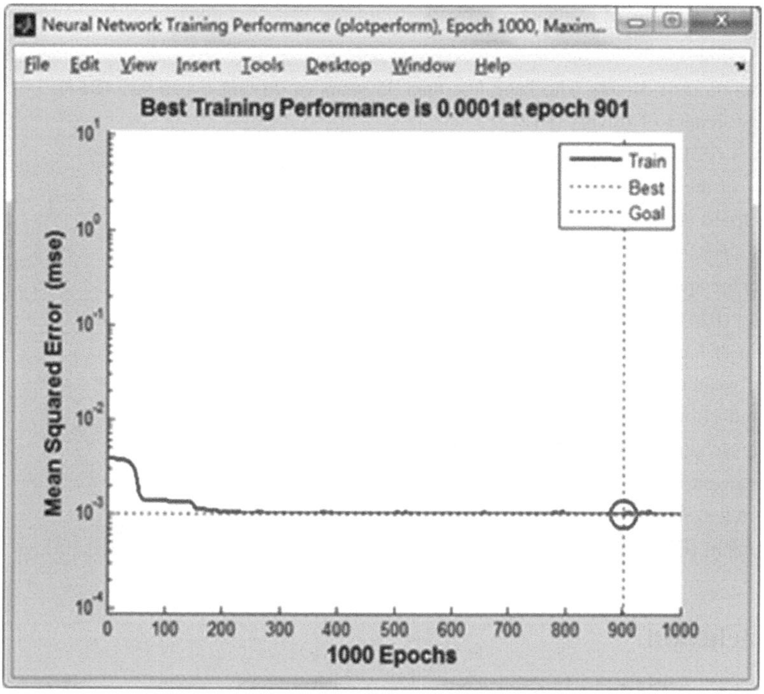

Fig. 4 The error log curve in the process of network training

4.3 Example Analysis

As development of a key city, The Binhai New District of Tianjin highway transportation development is extremely rapid.

With the rapid development of Binhai New District and a large number of motor vehicle soared, it aggravates the structural contradictions between growth of motor vehicle and private facilities lag, causing the seaside Avenue Road in Binhai New District' traffic conditions are more complex. Combined traffic environment research, questionnaires of traffic participants, and other forms, it conducted an traffic investigation analysis for the seaside Avenue Road. The evaluation index values are shown in Table 4.

Put the evaluation index values obtained from investigation into the BP neural network evaluation model that has been built. The result is "a = (0.003, 0.0397, 0.0790, 0.9931)". From the evaluation results, the fourth data are the maximum—0.9931, closing to 0.9. The other data are in the vicinity of 0.1. The evaluation result is convergent, so it can determine that the freeway traffic safety situation is not safe. According to the result of the evaluation, the traffic environment of seaside Avenue Road is not good, needing for traffic optimization and strengthening traffic safety management.

Table 4 The settings of evaluation network parameter

Names of index	Score
The proportion of drivers who have less than 35 years of driving experience (P11)	81
Safety awareness of drivers (P12)	79
Saturation (P21)	52
The size of the proportion of vehicle traffic (P22)	75
Design speed (P23)	120
Average speed (P24)	70
The average speed difference between carts and trolley (P25)	29
Headway (P26)	9
The ratio of lane change (P27)	53
The bad weather days in a year (P31)	49
Safety induction facility (P32)	82
Safety protection facilities (P33)	81
The alignment of flat, horizontal, and vertical (P34)	72
Monitoring system (P41)	85
Management (P42)	83

5 Conclusion

This paper presented the evaluation index system suitable for port special traffic conditions. On this basis, the port road safety evaluation model based on BP neural network was designed and developed. After the model was applied to the seaside Avenue Road, the conclusion is that the seaside Avenue Road is in unsafe traffic condition. BP neural network safety evaluation model proposes a new way for solving port safety evaluation of the uncertainty and the dynamic complexity and also avoids the difficulties encountered as the establishment of a complex mathematical model to describe the nonlinear relationship.

Acknowledgments Supported by the Science and Technology Projects of Tianjin Municipal Highway Administration (NO. 2012-33), Science and Technology Projects of Beijing Municipal Commission of Transport (NO. kj2013-2-32), Beijing Nova program (Grant No. Z141106001814110), Specialized Research Fund for the Doctoral Program of Higher Education (20121103120025), Funding Project for Academic Human Resources Development in Institutions of Higher Learning Under the Jurisdiction of Beijing Municipality, Tianjin Binhai Avenue Construction & Development Co. Ltd and Tianjin Municipal Engineering Design & Research Institute.

References

Chen D (2006) Application of BP neural network in Beijing road safety assessment. China Water Transp 12

Chen J, Li C, Ding G (2008) Traffic safety assessment of freeway based on BP neural network. J Tongji Univ: Nat Sci 36(7):927–931

Hakkert AS, Hocherman I, Mensah A (1996) Levels of safety on interurban roads. Transp Res Rec 1553:95–102

Li C, Wang Z (2010) Design and realization of the assessment system of freeway traffic safety based on bp neural network. J Wuhan Univ Technol: Transp Sci Eng 34(3):476–479

Liu Q, Chen G, Liu X, Zhan J (2010) Research on initialization algorithms of weights and biases of BP neural network. J Southwest China Normal Univ: Nat Sci Ed 35(006):137–141

Pan Y, Zhai C, Zhu S (2005) An evaluation of the road traffic safety based on the grey cluster and neural network. J Chongqing jiaotong Univ 24(2):101–104

Qing L (2002) Road traffic safe grade appraisement and example analysis. Transp Sci Technol 3:44–46

Safety Campaign Annual Seminar (1999) The Roarders

Weimin C, Yi W (1999) Some problems and their methods in comprehensive safety assessment. China Saf Sci J 9(4):75–78

Yang TJ, Zhang XC, Yang XG, Pan HZ (2005) Research on the evaluation of city road traffic safety based on BP artificial network. J China Univ Min Technol 34(1):37–40

Zhang L (1993) Artificial neural network model and its application. J Fudan Univ P13–47

Zhang S (1998) Chinese traffic conflict technique. Southwest Jiaotong University, Chengdu

Zufeng S (2008) Synthesis evaluation on urban traffic safety level. Urban Transp China 2:22–24

Carbon Emission of Guangxi's Major Industries and Measures for Low-carbon Economic Development

Lv-xuan Zhao, Li-hong Zhang, Xiao-wei Song, Nan-jun Qin and Jun Zhang

Abstract Guangxi's economic development mainly relies on major industries such as electric power, steel, cement, nonferrous metallurgy, papermaking, sugar, starch, and glass industry. With continuous development of the economy, the situation of carbon emission reduction of these industries is not optimistic. In this paper, carbon emission of Guangxi's major industries in 2012 was calculated using Oak Ridge National Laboratory (ORNL) and Intergovernmental Panel on Climate Change (IPCC) method, and then, the situation of carbon emissions was analyzed. The results showed that carbon emission of Guangxi's major industries mainly came from energy consumption, but carbon emissions in the technological process also should not be ignored; carbon emissions were mainly from industries such as steel, cement, thermal power, and nonferrous metallurgical and chemical industry where energy consumption is high and energy efficiency is low. The key factors that influence carbon emissions of these industries included irrational energy consumption structure and energy processing, the overall low utilization efficiency, and the increasing demand with the rapid economic growth. In order to realize a low-carbon economy and sustainable development in these major industries, the following measures were proposed such as formulation of low-carbon development planning and regulatory standards in these industries, management strengthening of the energy consumption in existing enterprises, optimization of the industrial and energy structure, and actively carrying out industrial upgrade, vigorous promotion of the use of new energy, establishment of the financing and policy guaranteeing system, and improvement of the processing and utilization of energy resources.

Keywords Carbon emission · Low-carbon economy · Guangxi · Measures

L. Zhao · L. Zhang (✉) · X. Song · N. Qin · J. Zhang
Scientific Research Academy of GuangXi Environmental Protection,
530022 Nanning, China
e-mail: 85385220@qq.com

S. Feng et al. (eds.), *Low-carbon City and New-type Urbanization*,
Environmental Science and Engineering, DOI 10.1007/978-3-662-45969-0_15

1 Introduction

The global warming caused by greenhouse gas emissions is a major environmental problem facing the present humanity. In general, the greenhouse gases associated with climate change are mainly referred to six gases which are identified by the Kyoto protocol, including carbon dioxide, methane, nitrous oxide, hydrofluorocarbons, fluorinated carbon, and sulfur hexafluoride. Among these, carbon dioxide has attracted the most attention due to the longest survival time and the largest stock in the atmosphere. The Intergovernmental Panel on Climate Change (IPCC) points out that in all kinds of greenhouse gases blamed for climate change, carbon dioxide accounted for more than 50 and 70 % carbon dioxide of human activities is from fossil fuel combustion (IPCC 2006). Therefore, reduced carbon emissions have become the key measure in response to global climate warming (Schimel 1995). At present, China's economic development is faster, but it is the extensive economic growth mode which is based on a large number of resource consumption. According to the statistics, China is one of the world's largest carbon emitters; China's annual greenhouse emission is by nearly half of all the emissions in developing countries and is close to 15 % of the world's total emissions (Energy research institute national development and reform commission 2003); and high CO_2 emission to make carbon reduction in China is facing greater pressure (Zhu et al. 2010; Zhu and Qiu 2010; Zhao et al. 2009; Li and Xian 2014; Tang et al. 2010).

Guangxi is located in the fringe of the southern and the southwestern, it is the important gateway for the association of Southeast Asian nations and frontal zone of China, and it is also the most convenient sea channel in southwest of China, especially the Beibu Gulf Economic Zone will become new growth pole of coastal economic development in China. In recent years, the key industries of Guangxi such as electric power, steel, cement, chemical industry, nonferrous metallurgical, papermaking, sugar, starch, lime, and glass developed rapidly; in the meanwhile, the energy consumption is also increasing, and the reduce situation of carbon emissions is not optimistic. Therefore, to calculate the carbon emissions in Guangxi key industries, analysis of carbon emission characteristics of the present situation and key factors that influence the carbon emissions have been made, and based on this, measures to develop low-carbon economy and sustainable development in the key industry of Guangxi, have been put forward, which has an important practical significance on the carbon emission reduction work and healthy development of the social economy in Guangxi.

2 Carbon Emission of Guangxi's Major Industries

2.1 Calculation Method of Carbon Emissions

Carbon emissions of the key industries in Guangxi were mainly from energy consumption and technological process.

Due to the energy consumption of clean energy such as water and electricity, nuclear power, wind power, solar, and biomass energy does not produce carbon dioxide, carbon emissions of energy consumption mainly consider the burning of fossil fuels. The carbon emissions of the energy consumption were calculated using the Oak Ridge National Laboratory (ORNL) (1990), and its general formula is as follows:

$$C = Ekn \tag{1}$$

where C is carbon emissions, E is energy consumption, k is effective oxidation scores, and n is carbon content of every tons standard coal.

The carbon emissions of the technological process were calculated using the carbon emissions calculation method proposed by 2006 IPCC guidelines for national greenhouse gas inventories (IPCC 2006).

Therefore, the total amount of carbon emissions of the key industries in Guangxi were added carbon emissions from technological process and energy consumption, whose main calculation formula is shown in Table 1.

All kinds of fossil fuels are converted to standard coal coefficient by using numerical reference in Chinese Energy Statistics Yearbook 2013 (Table 2). The energy consumption, industrial production, product type, product yield, and electric power consumption data of the key industries were from the Guangxi statistical yearbook 2013.

2.2 Status Quo of Carbon Emissions

1. Carbon emissions of the energy consumption in the key industries
   The usage of fossil fuels in the key industries of Guangxi in 2012 was shown in Table 3. The main fossil fuel in Guangxi is coal, of which the usage of the raw coal is 47.7453 million tons, washed coal is 7.8523 million tons, and coke is 6.8943 million tons. As for the usage of fuel oil, the usage of diesel is 0.0869 million tons and gasoline is 0.0085 million tons. The usage of fuel gas in the major industries in Guangxi was almost zero. The energy consumption and carbon emissions of the key industries are shown in Fig. 1. As can be seen from the figure, energy consumption of the thermal power industry was the largest (17.166 million tons), followed by steel industry (6.8527 million tons), cement industry (6.3784 million tons), nonferrous metallurgical industry (3.5728 million tons),

Table 1 Calculation formula of the key industries

Industry	Carbon source		Total carbon emissions
	Energy consumption	Technological process	
Power	1. Coal: $M_1 = C \times 0.982 \times 0.73257 \times 44/12$ 2. fuel: $M_2 = C \times 0.982 \times 0.73257 \times 0.813 \times 44/12$ 3. fuel gas: $M_3 = C \times 0.982 \times 0.73257 \times 0.561 \times 44/12$ $M_{energy} = M_1 + M_2 + M_3$ Where C is standard coal equivalent; 0.982 is effective oxidation scores; 0.73257 is carbon content of every tons standard coal; 44/12 is conversion ratio of C and CO_2.	/	M_{energy}
Steel		[a]E_{CO_2} = Production × Emission factor	$M_{energy} + E_{CO_2}$
Cement		E_{CO_2} = Cement clinker × 0.52	$M_{energy} + E_{CO_2}$
Chemical		E_{CO_2} = Synthetic ammonia production × 2.1-urea production × 12/60	$M_{energy} + E_{CO_2}$
Glass		E_{CO_2} = Glass production × 0.17	$M_{energy} + E_{CO_2}$
Lime		E_{CO_2} = Lime production × 0.75	$M_{energy} + E_{CO_2}$
Nonferrous metallurgical		/	M_{energy}
Sugar		/	M_{energy}
Papermaking		/	M_{energy}
Starch		/	M_{energy}

Note [a]E_{CO_2} = (BOF × 1.46 + EAF × 0.08 + OHF × 1.72) + (Pig iron production × 1.35) + (Direct reduced iron production × 0.70) + (Slag production × 0.20) + (Pellet production × 0.03), where BOF is the steel production of the alkaline oxidation of converter, EAF is the steel production of the electric arc furnace, and OHF is the steel production of the open hearth

Table 2 The fold of standard coal reference coefficient between different fossil fuels and standard coal

Name	The fold of standard coal reference coefficient (tons/standard coal)	Name	The fold of standard coal reference coefficient (tons/standard coal)
Raw coal (tons)	0.7143	Oil field gas (million cu. m)	13.300
Cleaned coal (tons)	0.9000	Crude oil (tons)	1.4286
Other cleaned coal (tons)	0.3571	Gasoline (tons)	1.4714
Briquette coal (tons)	0.6000	Kerosene (tons)	1.4714
Coke (tons)	0.9714	Diesel (tons)	1.4571
Other coke (tons)	1.3000	Fuel oil (tons)	1.4286
Coke oven gas (million cu.m)	6.143	Liquefied petroleum gas (tons)	1.7143
Blast furnace gas (million cu.m)	1.286	Refinery dry gas (tons)	1.5714
Other gas (million cu. m)	3.570	Other petroleum products (tons)	1.2000

chemical industry (2.5882 million tons), and papermaking industry (2.2341 million tons). The energy consumption of the six industries accounted for 95.96 % of the major industries' overall energy consumption. These results show that steel industry, cement industry, nonferrous metallurgical industry, chemical industry, and papermaking industry were the main energy consumption industries in Guangxi. As for carbon emission, the more the energy consumption, the more the carbon emission, and the main energy consumption industry is also the major carbon industry. Carbon emission of the thermal power industry is 45.2794 million tons, followed by steel industry (18.0757 million tons), cement industry (16.8245 million tons), nonferrous metallurgical industry (9.4238 million tons), chemical industry (6.8375 million tons), papermaking industry (5.893 million tons), glass industry (1.6774 million tons), sugar industry (1.4149 million tons), lime industry (0.9111 million tons), and starch industry (0.3039 million tons).

2. Carbon emissions of the technological process in the key industries
   As for carbon sources, in addition to energy consumption, the technological process also produces CO_2. Carbon emissions of the technological process in the

Table 3 The usage of fossil fuel in the major industries of Guangxi in 2012 (million tons)

	Steel	Thermal power	Nonferrous metallurgical	Cement	Chemical	Sugar	Papermaking	Lime	Glass	Total
Washed coal	7.5765	–	0.1209	3.83×10^{-2}	8.14×10^{-2}	–	–	0.038	–	7.8523
Coke	6.7555	–	0.1118	3.7×10^{-3}	1.12×10^{-2}	1.21×10^{-2}	–	–	–	6.8943
Raw coal	2.5987	25.3409	3.0988	7.9387	3.4702	0.7294	3.8352	0.4054	0.328	47.7453
Gasoline	9.2×10^{-5}	9.4×10^{-5}	0.7×10^{-3}	6.7×10^{-5}	0.1×10^{-3}	0.5×10^{-3}	0.1×10^{-3}	6.7×10^{-3}	0.1×10^{-3}	8.5×10^{-3}
Diesel	2.9×10^{-3}	2.17×10^{-2}	9.8×10^{-3}	2.96×10^{-2}	5.3×10^{-3}	0.2×10^{-2}	2.8×10^{-3}	1.21×10^{-2}	0.7×10^{-3}	8.69×10^{-2}

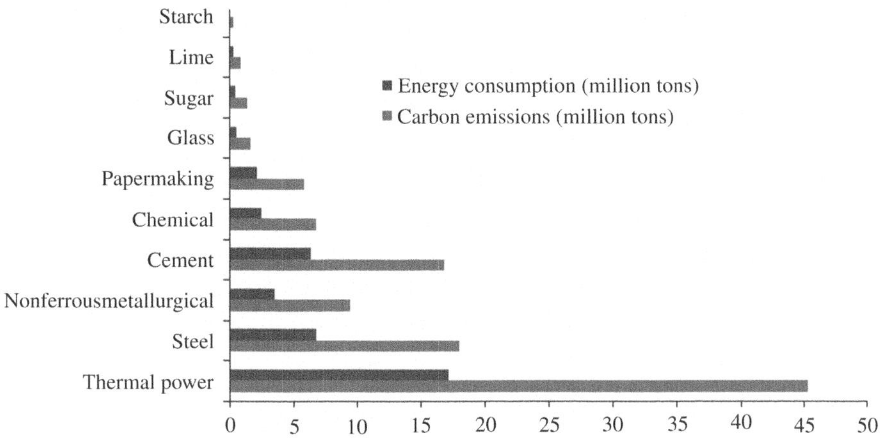

Fig. 1 Energy consumption and carbon emission of energy consumption in key industries in Guangxi in 2012

key industries in Guangxi were calculated according to Table 1, and the results are shown in Fig. 2. As shown in Fig. 2, the total carbon emission of the technological process in Guangxi was 94.6372 million tons in 2012. Among them, carbon emission of the technological process in the steel industry was the largest (60.5344 million tons), which accounts for 63.96 % of the overall technological process carbon emissions, followed by the cement industry (29.2159 million tons), which accounts for 30.87 % of the overall technological process carbon emissions, synthetic ammonia industry (2.2127 million tons), lime industry (1.6413 million tons), and glass industry (1.0329 million tons), and carbon emissions of the technological process in the synthetic ammonia industry, lime industry, and glass industry only account for 5.17 % of the overall technological process carbon emissions. These results show that carbon emission of the technological process was mainly from steel industry and cement industry.

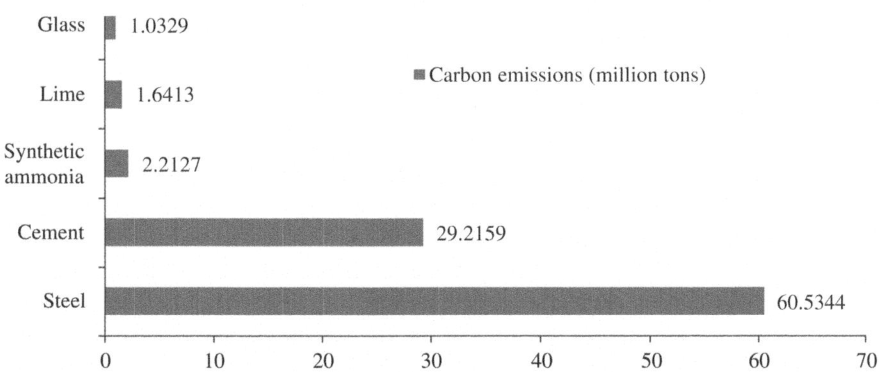

Fig. 2 Carbon emissions of the technological process in the key industries of Guangxi in 2012

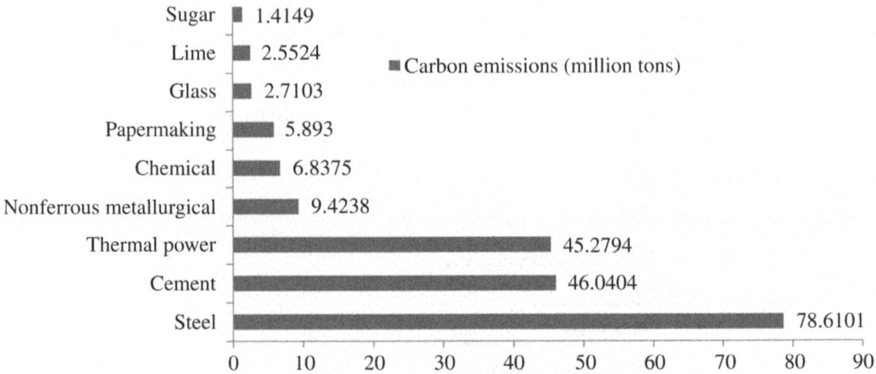

Fig. 3 Overall carbon emissions of the key industries in Guangxi in 2012

3. Overall carbon emissions of the key industries in Guangxi
   The overall carbon emissions of the key industries were calculated based on the carbon emissions from the energy consumption and technological process in the key industries in Guangxi in 2012. Carbon emissions of the steel industry were the largest (78.6101 million tons), which accounts for 37.38 % of the overall carbon emissions of the key industries in Guangxi. The second is the cement industry and thermal power industry, carbon emissions by 46.0404 million tons and 45.2794 million tons account for 21.90 and 21.53 %, respectively, followed by other key industries such as nonferrous metallurgical industry (9.4238 million tons), chemical industry (6.8375 million tons), and papermaking industry (5.893 million tons). The six industries are the most important carbon emissions industry, and carbon emissions of the six industries accounted for 96.64 % of the overall carbon emissions of the key industries in Guangxi (Fig. 3).

3 Analysis of Influence Factors for Carbon Emissions of the Key Industries in Guangxi

3.1 Energy Consumption Structure

According to the energy consumption of the key industries in Guangxi, the energy consumption structure of the key industries of Guangxi was based on coal consumption; raw coal consumption accounted for 76.3 % of fossil fuel consumption in 2012. Coal is the main source of carbon emissions, and carbon emissions of the energy consumption in Guangxi were mainly from the key industries such as thermal power, steel, cement, nonferrous metallurgical, chemical, and papermaking industry. Therefore, Guangxi's economic growth is based on huge coal consumption and is highly dependent on coal resources, but Guangxi is not a major coal production area of coal resources; coal resources are mainly transferred from other provinces. Therefore, improving the clean energy utilization and optimizing the energy structure will become the key to reducing carbon emissions in Guangxi.

3.2 Energy Efficiency

The key industries such as steel, nonferrous metallurgical, papermaking, cement, chemical, and thermal power industry of Guangxi all belong to high energy consumption industries. According to statistics, the GDP of the key industries in Guangxi was 439.645 billion yuan and total energy consumption of the key industries was 420.381 million tons, and per 10,000 yuan GDP energy consumption is 0.96 tons which is significantly higher than the national average (0.7 tons). This result has shown that the key industries in Guangxi still belong to the extensive economic growth mode. Economic growth mainly depends on the large consumption of energy, but the efficiency of energy processing and utilization is low; most of the energy resources are lost and wasted in the resource exploitation, processing, conversion and utilization process, which leads to large carbon emissions.

3.3 Economic Development

In the "twelfth five-year" period, with the "several opinions on further promote economic and social development in Guangxi," "development planning of Guangxi beibu gulf economic zone," "development planning of Pearl River-Xijiang River economic zone," and other policy documents to implement, the key industries of Guangxi faced with an excellent opportunity for development, the GDP of Guangxi has grown significantly, however, the energy consumption demand of Guangxi is also increasing; energy consumption growth will remain at about 11 %, while the absolute increment will continue to increase (Su and Wu 2008). It can be foreseen that in the quite a long period of time, carbon emissions of the key industries in Guangxi will increase with the development of economy, and the reduction in carbon emissions is not optimistic.

4 Measures to Develop a Low-carbon Economy and Sustainable Development in Guangxi

4.1 Optimize the Industrial Structure, Actively Carry Out Industrial Upgrading

The key industries in Guangxi are mainly high energy consumption industries, and the new low-power industrial proportion is low. To achieve a low-carbon economy, on the one hand, it need to optimize the internal structure of the industries, reduce the proportion of high energy consumption industries, and foster emerging low-carbon industry, especially to speed up the development of the emerging strategic industries such as new energy, energy conservation and environmental protection, new materials, and information industry; on the other hand, to the existing traditional high energy consumption, key industries such as steel, electric power, and nonferrous metallurgical industry can, through the enterprise implementation of technological transformation and clean production, improve the efficiency of the processing and utilization of energy resources and take a new road to industrialization.

4.2 Optimize the Energy Structure; Vigorously Promote the Use of New Energy

The energy consumption structure of the key industries in Guangxi is based on fuel coal; the proportion of using natural gas and other clean energy is less. This energy consumption structure objectively causes lower energy efficiency, serious environmental pollution, and higher carbon emissions in Guangxi. Therefore, it should step up the development and use of clean energy, and strive to reduce the proportion of coal in the energy consumption structure, and accelerate energy consumption structure mainly from fuel coal change to oil and natural gas. In the meanwhile, new energy is the direction of the future energy development and also the only way for developing low-carbon economy, and Guangxi is one of the largest sugar cane and cassava production bases in China. It should actively implement the key technology of new energy in the biological chemical and biomass gasification fuel industry and speed up the industrialization of biomass energy development. At the same time, it should, according to the geographical position advantage of Guangxi, promote new energy such as solar energy, oceanic energy, and wind energy development and utilization.

4.3 *Strengthen the Energy Consumption Management of Existing Key Industrial Enterprise, Full Implementation the Energy Assessment of the Key Industries*

Formulating of energy consumption management measures of key enterprises in Guangxi as soon as possible, establish energy consumption management database of the key enterprises, key industries enterprise should execute mandatory energy evaluation, and be assessed by the government regularly. Strict implementation of national industry access system, control unreasonable growth of energy consumption timely, improve energy assessment based on results of shutting down the exit mechanism for elimination, close or move high energy consumption and high pollution enterprises step-by-step, eliminate inefficient equipment and backward production capacity.

4.4 *Formulate the Low-carbon Development Planning and Regulations Standard of the Key Industries and Improve the Fund and Policy Security System*

According to the economic development of the key industries in Guangxi, and the status quo of the energy consumption and demand, combined with the concept of low-carbon economy development, formulate the low-carbon economy development plan of the key industries in Guangxi, establish reasonable low-carbon development goals and tasks of the key industries in Guangxi, determine the key areas of conserving energy and emission reduction, propose the key projects and implementation plan of carbon reduction, and issue the standard system, technical specification and other policy documents, promoting the low-carbon development of the key industries as soon as possible.

Increase the funding support for the low-carbon development of key industries, improve the financial input and subsidy policy of low-carbon development. Through the economic means, give financial support and preferential tax policy incentives to the enterprise which get obvious effect of energy saving and emission reduction by implementing clean production and circular economy, give moderate fiscal subsidies to the enterprise which implements low-carbon key technology research and development and promote achievement, provide strong policy and financial security for the key industries to achieve low-carbon development.

5 Conclusions

Guangxi is located in the fringe of the southern and the southwestern of China. It is a gateway to Southeast Asian nations and frontal zone of China and also the most convenient sea channel in southwest of China, especially the Beibu Gulf Economic Zone will become new growth pole of coastal economic development in China. In this paper, based on the analysis on the status quo of carbons emissions in the key industries such as power, steel, cement, chemical, nonferrous metallurgical, papermaking, sugar, starch, lime, and glass industry, it points out the key factors that influence the carbon emissions of the key industrial, and put forward measures to realize the low-carbon economy and sustainable development.

The results showed that the carbon emission of major industry in Guangxi was mainly from energy use, but carbon emissions of the technological process also cannot be ignored; the economic development of Guangxi was mainly rely on the steel industry, cement industry, and so on, and these industries were high energy consumption industries, and mainly, energy was coal-fired, energy efficiency was low, and carbon emissions were large. Quite a long period of time in the future, economy of the key industries in Guangxi will grow rapidly and seriously in the form of carbon emissions reduction, and to realize the low-carbon economy and sustainable development of the key industries, the following measures must be taken: formulation of low-carbon development planning and regulatory standards in these industries, management strengthening of the energy consumption in existing enterprises, optimization of the industrial and energy structure, actively carrying out industrial upgrade, vigorous promotion of the use of new energy, establishment of the financing and policy guaranteeing system, and improvement of the processing and utilization of energy resources.

References

Energy Research Institute National Development and Reform Commission (2003) Study on the sustainable development of China's energy and carbon emissions scenario. Energy research institute national development and reform commission, Beijing, pp 20–39 (in Chinese)

IPCC (2006) IPCC guidelines for national greenhouse gas inventories: vol II (EB/OL). The Institute for Global Environmental Strategies, Japan 2008. http://www.ipcc.ch/ipccreports/Methodology-reports.htm

Li YY, Xian YN (2014) The current situation and prediction of energy consumption and carbon emissions, and way choice of low-carbon development in Beijing. J Central Univ Finance Econ 6:105–112 (in Chinese)

ORNL (1990) Estimate of CO_2 emission from fossil fuel burning and cement manufacturing. ORNL/CD IAC-25. Carbon Dioxide Information. Analysis Center, Oak Ridge National Laboratory, Oak Ridge, Tennessee, USA

Schimel DS (1995) CO_2 and carbon cycle. In: Climate change 1994: radioactive forcing of climate change (IPCC). Cambridge University Press, Cambridge, pp 35–71

Su QY, Wu SX (2008) Some thinking of the energy situation and energy industry in Guangxi. Energy Environ 2008(6):20–22 (in Chinese)

Tang YQ, Chen J, Ran T (2010) Study on the status quo of carbon emission and the strategy to develop low-carbon economy in Chongqing. Sichuan Environ 29(1):87–90 (in Chinese)
Zhao M, Zhang WG, Yu LZ (2009) Carbon emissions from energy consumption in Shanghai city. Res Environ Sci 22(8):984–989 (in Chinese)
Zhu Y, Qiu SF (2010) Calculation and analysis of carbon emissions from energy consumption in Fujian province. Fujian BBS, Humanit Soc Sci 10:145–148 (in Chinese)
Zhu Y, Zhou H, He XW (2010) Study on CO_2 emission investigation and reduction measure in Liaoning Province. J Anhui Agric Sci 38(23):12357–12360 (in Chinese)

The Influence of Seaport Operations on the Coastal City Environment

Mamatok Yuliya and Chun Jin

Abstract The main objective of this research is to explore the influence of activities of a seaport on the carbon pollution of the surrounding city and suggest the adequate measures for its reduction. This paper focuses on the land-side area of the seaport by considering three kinds of emissions sources: cargo handling equipment, heavy-duty vehicles, and locomotives, as they pollute the seaport land-side area and are included into the transport infrastructure of the city. The multiple methodologies were used in this research to confirm the environmental effectiveness, including the total frame of three approaches, the multiple case study to choose suitable evaluation methods, case study for verification of the proposed method, and the quantitative method to calculate the emission and inventories. A case study was carried out on the influence of the operations of Qingdao Port on Qingdao city. As a result, some measures of emission reduction are proposed for Qingdao Port to decrease CO_2 produced by CHE, HDV, and RL through improving operation, such as using clean fuels, controlling emissions, encouraging the choice of the rail transportation, and developing dry ports.

Keywords Coastal city environment · Emissions estimation · Qingdao port · Seaport carbon emissions

1 Introduction

The development of port facilities and their associated operations contributes significantly to the growth of maritime transport and the economic development of coastal regions (Puig et al. 2014). However, the pollution from ports' operations has

M. Yuliya · C. Jin (✉)
Faculty of Management and Economics, Dalian University of Technology, No. 2 Linggong Road, Ganjingzi District, Dalian 116024, Liaoning Province, People's Republic of China
e-mail: jinchun@dlut.edu.cn

S. Feng et al. (eds.), *Low-carbon City and New-type Urbanization*, Environmental Science and Engineering, DOI 10.1007/978-3-662-45969-0_16

been remained the key problem concerned by proximity of the ports (Acciaro et al. 2014). The most common pollutants associated with port-related operations are the following: carbon dioxide (CO_2), methane (CH_4), nitrous oxide (N_2O), particulate matter (PM) (10, 2.5 μ), diesel particulate matter (DPM), oxides of nitrogen (NO_x), oxides of sulfur (SO_x), and carbon monoxide (CO) (IAPH Tool Box for Port Clean Air Programs 2008). For example, heavy-duty vehicles account for 10 % of diesel PM, 36 % of NO_x, and 1 % of SO_x (Giuliano and O'Brien 2007). Port operations impact on air, water, and soil, affecting both the terrestrial and marine environments. The situation with seaside mobile emissions was investigated by a lot of researches (Qi and Song 2012; Schrooten et al. 2008, 2009; Miola and Ciuffo 2011; Adamo et al. 2014) and that with the land-side has not been studied thoroughly. Considering the seaside emissions, the precise calculation methodologies in this area are of high demand. Automatic identification system (AIS) is regarded as a very effective tool to create a state-of-the-art inventory of marine vessel emissions. The data of vessel identification, position, course, and speed are transmitted continuously, providing a comprehensive and detailed data set for individual vessels, which can be used to estimate and allocate emissions based on improved traffic pattern data (Perez et al. 2009).

This paper focuses on emission estimates of greenhouse gases (GHGs) from the land-side mobile emissions of the seaport. Land-based emission sources include cargo handling equipment (CHE), heavy-duty vehicles (HDVs), and rail locomotives (RLs) which are operating within a port area (World Ports Climate Initiative 2010). The contribution of this research involves the following points:

- Establishment on the emissions estimation methodologies by analyzing the most progressive experiences of seaports;
- Empirical study on the emissions estimation in Qingdao Port to evaluate the carbon emissions pressure on the city environment;
- Recommendations for the seaport executives in land-side emissions mitigating strategies.

The rest of this paper is arranged as follows: Sect. 2 represents the literature review. Section 3 describes the methodology applied in this research. Section 4 discusses the results from the emissions estimation of the case study. The Sect. 5 provides some recommendations for the seaport operations. Finally, the conclusions are summarized.

2 Literature Review

The problem of seaport land-side emissions was developed according to the following aspects.

The first aspect is the calculation methodology applied for the estimation of land-side emissions. Two general approaches could be represented as the solution of this problem. The first one is a detailed approach, which uses source-specific data. It makes use of the greatest levels of detail and provides the highest levels of accuracy. The second one is a streamlined approach, where land-side emissions are estimated from other detailed inventories, using the surrogates to estimate their activities. The streamlined approach requires a lower level of detail, but it is based on assumptions limiting accuracy (Browning 2009). This approach was tested to calculate air emissions in Saigon Port in Vietnam. The results showed that the total emissions of all pollutants were dominated by cargo handling equipment (Ho 2013).

The second aspect of the land-side emissions is terminal operations and associated energy consumption. By providing the insight into the energy of the terminal processes, the model of bottom-up calculation could be suggested. This model included a bottom-up calculation of the amount of work supplied by machines, not using the amount of fuel as input, but as the result of the model. This model was applied in Rotterdam (Geerlings and van Duin 2011). In addition, an operational activity-based method was developed in contrasts to the traditional aggregated activity-based method. The results showed that the aggregated method could better overestimate CO_2 emissions and the operational activity-based method is more appropriate (Dong-Ping and Jingjing 2012).

The seaport environment influences the city environment to a big extent. The monitoring of the city and seaport pollutions is the vital problem, which could be solved starting from the development of a system of sustainable environmental management indicators. The attempt to analyze a total of 17 pressure/state indicators with the potential environmental impacts was made in the Port of Valencia in Spain (Peris-Mora et al. 2005). In addition, estimating the emissions due to seaports in such a way that they can be included as part of the cities inventory or be used by the port itself has a great optimization potential. This methodology was adapted in the Port of Barcelona in Spain and helped to find out that the highest polluters were auto carriers (Villalba and Gemechu 2011).

Many studies were devoted to the problem of the sustainability of the coastal cities. They described the methodologies and data used to determine GHG emissions, attributable to the following cities or city-regions: Greater Toronto, New York City, Greater London, Barcelona, Cape Town, and Bangkok. The problem of CO_2 pollution is urgent for China as well. Some researches tried to investigate the scale of pollution in several Chinese cities and regions such as Hong Kong (Harris et al. 2012), Nanjing (Bi et al. 2011), Macao (Li et al. 2013), and Suzhou (Wang et al. 2014).

The majority of the observed researches developed the aspects of the seaport emissions inventories designing. In addition, the problem of the cities emissions inventories from the global perspective gains the growing interest. The observation of the provided papers also represents the scientific research situation in different world regions.

3 Methodology

The multiple methodologies were used in this research to confirm the environmental effectiveness, including the total frame of three approaches, the multiple case study to choose suitable evaluation methods, case study for verification of the proposed method, and the quantitative method to calculate the emission and inventories.

Case studies have been used extensively in port economics and management. Considering the exploratory nature of the paper, which investigates the influence of seaport land-side operations on the city environment, the application of case studies is particularly appropriate.

In this research, the multiple case study method was used to investigate the experience of some seaports, which are progressive in the low-carbon port construction and emissions inventories development (Port of Los Angeles (USA), Port of Gothenburg (Sweden), Port of Jurong (Singapore), and Port of Rotterdam (The Netherlands)). The investigated ports were chosen because of the following criteria: They are leaders in their regions; they are on the cutting edge of the environmental policy solutions implementation; their carbon emission impacts on the city environment are significant; and their achievements are representative and should be studied thoroughly to be applied by others.

An emission inventory is a quantification of all emissions of criteria and other pollutants (including toxics and greenhouse gases) that occur within a designated area by their source. An emission inventory is necessary for port authorities, state and local entities as well as other interested parties to understand and quantify the air quality impacts of current port operations and to assess the impacts of port expansion projects or growth in port activities. Emission-producing activities for ports are grouped into three scopes, represented in Table 1. These scopes include port direct, port indirect, and port tenants' emissions according to the ownership and responsibility aspects (World Ports Climate Initiative 2010).

When developing carbon footprint inventory, the detailed approach is highly preferable (Browning 2009). Three data elements are critical in this approach. These elements include the following:

Table 1 Port-related emission sources

Type	Style	Source examples
Scope 1	Port indirect	Purchased electricity for port-owned buildings and operations
Scope 2	Port direct	Port-owned fleet vehicles, buildings, stationary sources
Scope 3	Port tenants	Ships, trucks, cargo handling equipment, rail, harbor craft, port employee vehicles, buildings, purchased electricity

1. Source data (detail the emissions source characteristics, which include size or rating of the engine, type of fuel consumed, engine technology information, age of the engine, manufacturer, model);
2. Activity data (detail how the source operates over time and how engine loads and/or fuel consumption change by mode of operation, distance traveled by speed, power production rates);
3. Emission factors (provide the means to convert the estimates of energy output or fuel consumption into the pollutant emission rates).

Detailed approach closely models actual port operations and can provide emission reduction strategy progress but requires significant time to conduct first inventory. The attempt to adapt the detailed approach was made is this research based on the Qingdao Port case.

4 Seaport Land-Side Carbon Emissions

4.1 Land-Side Mobile Emission Sources Structure and Estimating Methodologies

Land-based emission sources include CHE, HDVs, and RLs, operating within a port area (World Ports Climate Initiative 2010). CHE might consist of terminal tractors, top and side loaders, forklifts, wharf cranes, rubber tire gantry cranes, and skid loaders. HDVs include on-road trucks, buses, and other port vehicles. RLs are divided into line haul locomotives and switch yard locomotives.

The methodologies of seaport land-side emissions calculation are based on WPCI recommendations. The input parameters are collected in Table 2.

1. Cargo Handling Equipment

   The calculation of emissions from CHE could be made for each piece of equipment or for the fleet of equipment as a whole. Estimates for each piece of equipment are preferable because that method helps point out potential targets for emission reduction efforts. It is important for both fuel-based and energy-based calculations to calculate the emissions generated from equipment, using different fuels separately, because the emission factors of each fuel are different.

$$\mathrm{TE_{CHE}} = \sum_{i=1}^{n} \mathrm{TE_{CHE(\mathit{i})}} \tag{1}$$

   where, is the particular type of cargo handling equipment (e.g., forklift).

   Energy-based method

Table 2 Input parameters and symbols interpretation

Parameter	Interpretation	Dimension
Cargo handling equipment		
TE_{CHE}	The sum of emissions from all the cargo handling equipment per year	tonnes CO_2E
$TE_{CHE(i)}$	Total emissions from the cargo handling equipment of particular type	tonnes CO_2E
$R_{p(i)}$	Rated power (for a particular type of equipment)	kW
$L_{f(i)}$	Load factor	dimensionless
$O_{t(i)}$	Operating time per year	h
$E_{fe(i)}$	Emission factor of the equipment engine	gCO_2/kW-h
$F_{c(i)}$	Fuel consumption per year	l/year
$E_{ff(i)}$	Emission factor of the particular fuel consumed	$kgCO_2$/l
Heavy-duty vehicles		
TE_{HDV}	The sum of emissions from all types of HDVs per year	tonnes CO_2E
$TE_{HDV(j)}$	Total emissions from a particular type of HDV per year	tonnes CO_2E
$E_{iter(j)}$	Emissions caused by the vehicle idling mode in the terminal	tonnes CO_2E
$E_{rter\ (j)}$	Emissions caused by the vehicle running mode in the terminal	tonnes CO_2E
$E_{rreg(j)}$	Emissions caused by the vehicle running mode in the region	tonnes CO_2E
$P_{(j)}$	Population of HDV type j in the port truck fleet	dimensionless
$N_{t(j)}$	Number of truck trips per year	dimensionless
$I_{t(j)}$	Average idle time	min per truck trip
$E_{fi(j)}$	Emission factor in idling	gCO_2/h
D_{term}	Average trip distance on-terminal	km
$E_{fr(j)}$	Emission factor in running	gCO_2/h
D_{reg}	Average region trip distance	km
$F_{c(j)}$	Average fuel consumed per trip	gallons
$E_{ff(j)}$	Emission factor of the particular fuel consumed	$kgCO_2$/gallon
Locomotives		
TE_{LOG}	The sum of emissions from all the locomotives per year	tonnes CO_2E
$TE_{LOG(i)}$	Total emissions from the locomotives of particular type	tonnes CO_2E
$R_{p(i)}$	Rated power (for a particular type of locomotive)	hp
$L_{f(i)}$	Load factor	dimensionless
$O_{t(i)}$	Operating time—time in notch per year	h/year
$E_{fe(i)}$	Emission factor of the locomotive engine	gCO_2/hp-h
$F_{c(i)}$	Fuel consumption per year	gallons/year
$E_{ff(i)}$	Emission factor of the particular fuel consumed	$kgCO_2$/gallon

$$\mathrm{TE}_{\mathrm{CHE}(i)} = \sum_{i=1}^{n} \left(R_{\mathrm{p}(i)} \times L_{\mathrm{f}(i)} \times O_{t(i)} \times E_{\mathrm{fe}(i)} \right) \tag{2}$$

Fuel-based method

$$\mathrm{TE}_{\mathrm{CHE}(i)} = \sum_{i=1}^{n} \left(F_{c(i)} \times E_{\mathrm{ff}(i)} \right) \tag{3}$$

2. Heavy-duty Vehicles
   When estimating emissions, generated from heavy-duty trucks, two modes of operation are considered: (1) idle emissions occur, when the engine is on yet the vehicle is not moving; (2) running emissions occur, when the engine is on and the vehicle is also in motion. Greenhouse gas emissions from trucks can also be classified by truck operation area: "on-terminal," where trucks traverse the terminals with their loads, idling to pick up or drop off cargo; "on-port," where trucks enter or exit port property or travel between terminals; and "regional," where trucks operate outside of port property to deliver goods.

$$\mathrm{TE}_{\mathrm{HDV}} = \sum_{j=1}^{n} \mathrm{TE}_{\mathrm{HDV}(i)} \tag{4}$$

where, j is the particular type of heavy-duty vehicle.
Energy-based method

$$\mathrm{TE}_{\mathrm{HDV}(j)} = \sum_{j=1}^{n} \left(E_{\mathrm{iter}(j)} + E_{\mathrm{rter}(j)} + E_{\mathrm{rreg}(j)} \right) \tag{5}$$

$$\begin{aligned} \mathrm{TE}_{\mathrm{HDV(j)}} = \sum \Big(\big(P_{(j)} \times N_{t(j)} \times I_{t(j)} \times \frac{1\,\mathrm{h}}{60\,\mathrm{min}} \times E_{\mathrm{fi}(j)} \big) \\ + \big(P_{(j)} \times N_{t(j)} \times D_{\mathrm{term}} \times E_{\mathrm{fr}(j)} \big) + \big(P_{(j)} \times N_{t(j)} \times D_{\mathrm{reg}} \times E_{\mathrm{fr}(j)} \big) \Big) \end{aligned} \tag{6}$$

Fuel-based method

$$\mathrm{TE}_{\mathrm{HDV(j)}} = \sum_{j=1}^{n} \left(P_{(j)} \times N_{t(j)} \times F_{c(j)} \times E_{\mathrm{ff(j)}} \right) \tag{7}$$

3. Locomotives
   Unlike heavy-duty diesel trucks, engine load for locomotives is not a direct function of vehicle speed. The activity of locomotives tends to be expressed in terms of "time in notch" or "throttle position" which ranges from idle to one of eight different operating modes, each of which represents successively higher

average engine load. Only the emissions associated with the combustion of diesel fuel would be considered in estimating greenhouse gases from these engines.

$$\mathrm{TE_{LOC}} = \sum_{i=1}^{n} \mathrm{TE}_{\mathrm{LOC}(i)} \tag{8}$$

where, *i* is the particular type of locomotive (line haul or switchers).

Energy-based method

$$\mathrm{TE}_{\mathrm{LOC(i)}} = \sum_{i=1}^{n} \left(R_{p(i)} \times L_{\mathrm{f}(i)} \times O_{t(i)} \times E_{\mathrm{fe}(i)}\right) \tag{9}$$

Fuel-based method

$$\mathrm{TE}_{\mathrm{CHE}(i)} = \sum_{i=1}^{n} \left(F_{c(i)} \times E_{\mathrm{ff}(i)}\right) \tag{10}$$

4.2 Qingdao Container Terminal Case Study

In this research, the attempt to apply and adopt the practice and best experience of the worldwide seaports in CO_2 emissions calculation was taken for Qingdao Port in China. The case study consisted of several steps. First, all available relevant documents (annual reports, strategic plans, and statistics) were analyzed. Second, interviews with the responsible managers of Qingdao Qianwan Container Terminal Ltd. (QQCT) were conducted. These interviews were based on a list of questions, derived from the literature review and other case study documents. The results of adaptation and calculation of emissions for Qingdao Port are provided in the next subsection.

4.2.1 Qingdao City and Port Brief Introduction

As a sub-provincial city, Qingdao had permanent resident population of 8.71 million totally in 2012. Qingdao covers an area of 11.6 km^2. Qingdao city has pioneered a new way by applying the low-carbon economic mode to the features of its city development. Qingdao is the industrial center of the West Coast Economic New Zone. The planning area of this Cooperation Region is 190 km^2 with the leading and demonstrative role of Sino-German Eco-Park and Sino-Japan and Sino-South Korea innovative industrial parks, as shown in Fig. 1.

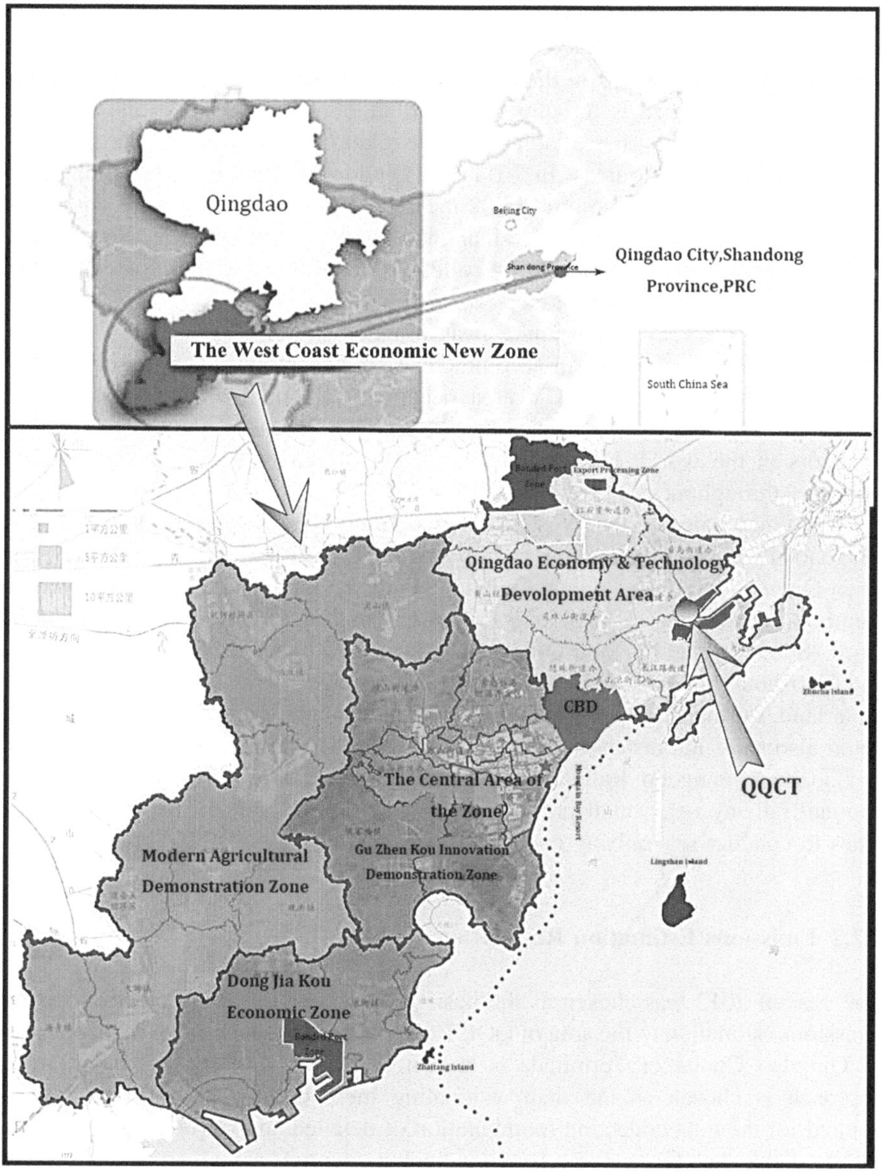

Fig. 1 The west coast economic new zone

The Sino-German Eco-Park is committed to the development of a smart city and is enthusiastically building a top ranking information base in China. The project realizes the following results as thoroughly sense: Internet with ubiquitous broadband and the application of intelligence blending. As a result, a new type of city development will be formed with new characteristics, such as the high integration of smart technologies, high-end development of smart industries, and convenient and high-efficient smart services.

Qingdao Port is located in the Yellow River Basin and on the western Pacific Rim. Port of Qingdao is an important hub of international trade and seagoing transportation. The port has established trade relations with over 450 ports in more than 130 countries and areas. In 2011, the Qingdao Port total cargo volume was 372,000 t and containers volume 13.02 mTEUs. The Port of Qingdao is composed of three areas: Qingdao Old Port, Huangdao Oil Port, and Qianwan New Port. It provides services of loading and unloading, storage and logistics for containers, coal, iron ore, crude oil, grain, etc.

Qingdao QQCT, a joint venture, was founded in July 2000. QQCT operates 11 container berths, which can accommodate six-generation mega-vessels of over 10,000 TEUs. The terminal CY area occupies 2.25 million km^2, enjoying the minimum depth draft water of up to −17.5 m. QQCT, one of the largest terminal operators in the world, has the total berth length of 3,400 m and the designed container throughput of 6.5 million TEUs.

QQCT is located in the West Coast Economic New Zone and only 68 km to Qingdao city through Jiaozhou Bay Expressway. QQCT has efficient road access to Jinan-Qingdao Highway, Yantai-Qingdao Highway, and 308 National Highway. It connects outside area in Shandong Province. Tongsan Highway goes through the North–South coastal line and Qinglan Highway. Starting from the east to the west and Jiaozhou-Huangdao Railway inside the terminal, it connects QQCT to the hinterland. Furthermore, the bridge and the sea channel, opened to traffic in 2011, could also short the distance from QQCT to Qingdao downtown.

Jiaozhou-Huangdao Railway can reach the dock area directly and link the national railway net, which provides great convenience for inland provinces and cities to conduct sea–railway combined transport through QQCT.

4.2.2 Emissions Estimation Results

The year of 2012 was chosen as the basis year of activity. We limit the scope of emissions estimation to the area of QQCT responsibility, which is the main operator of Qingdao Container Terminal, as shown in Fig. 2 Activity-based bottom-up approach is chosen as the main estimating methodology. Hybrid approach is applied for the data collecting (combination of detailed and surrogate approaches). The available data for some ports in developing countries are not sufficient to make a detailed estimation. Some surrogate data are taken from the other sources: US Environmental Protection Agency (EPA) guidence, PoLA and PoLB inventories 2012, which are considered to be the best current practice in inventory development. For CHE emissions evaluation, energy-based approach is used (Eq. 2), which is considered to be more accurate in comparison with the fuel-based approach. For HDV and locomotives, we also use energy-based approach (Eqs. 6 and 9).The boundary of the emissions estimation is limited to the in-terminal, in-port, and in-region container transportation.

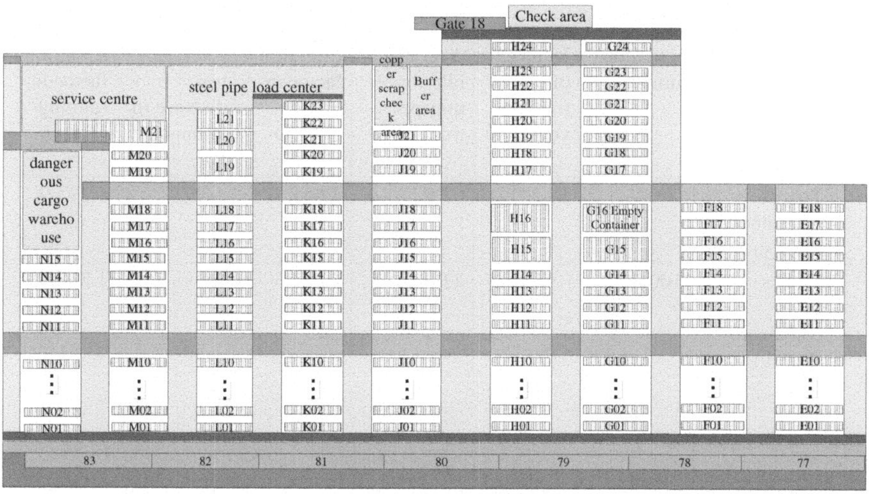

Fig. 2 Qingdao Qianwan container terminal layout

Input data collecting process generally consists of the collecting activity data, represented in Tables 3, 4 and 5, and choosing the emission factors. CHE load factors and operating time are calculated for each month and then summarized per year. Emission factors including those factors for HDVs are taken from the US EPA guidance document (Browning 2009). Locomotives emission factors, expressed in terms of grams of pollutant per kilogram of fuel consumed, are taken from both the California EPA (Air Resource Board 2013) and the US EPA (Office of Transportation and Air Quality 2009).

Figure 3a–d represents the emissions structure by the source category. Results' analysis will be provided in the next section.

Table 3 CHE input data

No.	Types	Count	Engine type	Power, kW	Annual operating hours, h	Load factor
1	RTG	106	Electric	134	3,565	0.42
2	ITV	84	Diesel	75	1,657	0.2
3	Heavy forklift	6	Diesel	450	1,880	0.22
4	Empty forklift	14	Diesel	130	3,231	0.39
5	Small forklift	27	Diesel	37	1,163	0.14
6	Crane	2	Electric	500	360	0.04

Table 4 HDV input data

No.	Heavy-duty vehicle type	Population (number)	Number of truck trips per year	Average idle time/min per truck trip	Emission factor in idling gCO_2/h	Average trip distance in terminal/km	Emission factor in running/ gCO_2/h
1	Trucks (container facility)	1,000.0	14,501.0	62.0	4,655.3	1.3	1,293.0
2	Trucks (non-container facility)	100.0	3,741.0	47.0	4,655.3	0.5	1,293.0

Table 5 RLs input data

No.	Locomotive Type	Population (number)	Rated power/hp	Load factor/-	Time in notch 4/h per year	Emission factor/ gCO_2/hp-h
1	Locomotive (line haul)	2	2,500.00	0.3	950	510.1
2	Locomotive (switchers)	3	2,500.00	0.3	580	505.5

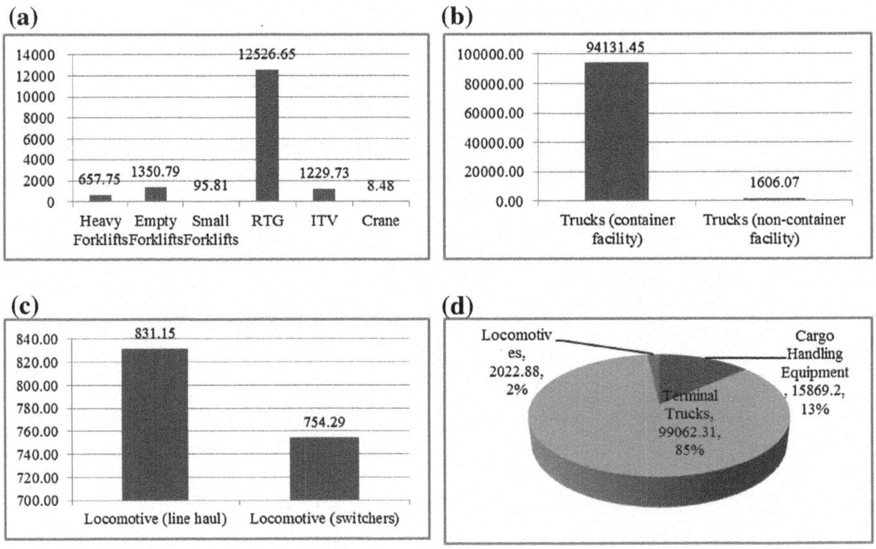

Fig. 3 Emissions structure by the source category, tCO_2E. **a** Cargo handling equipment, **b** heavy-duty vehicles, **c** locomotives, **d** total land-side emissions

5 Discussion

The main aim of the conducted research is to evaluate the CO_2 emissions from the seaport activities and their influence on the city environment as well as to propose some suggestions for the carbon emissions mitigation.

The seaport activity includes both the seaside and land-side operations. Considering the seaside area, many investigations and seaport inventories have proved that the emissions from the oceangoing vessels create the largest amount of emissions on the territory of the seaport. It goes without saying that the seaport authorities should take urgent measures for the decreasing emissions from oceangoing vessels activities.

At the same time, the analysis of the seaport land-side emissions claims for the close attention as well. Seaport land-side infrastructure is a part of urban territory, thus influencing the city ecology. The analysis of the received results shows that HDVs create the largest amount of land-side CO_2 emissions (99,062.31 tCO_2E) in comparison with CHE (15,869.2 tCO_2E) and RLs (2,022.88 tCO_2E) correspondingly. Total amount of CO_2 emissions from QQCT land-side mobile sources in 2012 was evaluated as 116,954.39 tCO_2E.

It should be noticed that the received results could become more representative if the activities of the other terminals in Qingdao Port (oil, dry bulk, and passengers' terminals) were also included in the evaluation. At the same time, the research process detected the problem of the data collection. The input parameters for the emissions evaluation need to be collected from the different sources (different terminal operators). This process supposes their close cooperation in information exchange. As the problem of carbon emissions calculation for the port activities has risen relatively recently, the seaport authorities have not created the uniform informational base used for the emissions calculation.

Apart from, the collected data accuracy also is quite a problematic question. When calculating the carbon emissions, the hybrid approach was used, and some parameters like the emissions factors were taken from the outside sources and the reports of the other seaports, which have created quite a long history of emissions inventories. Thus, one of the outcomes of the conducted research is the necessity of creating the uniform database for the CO_2 emissions calculation, assuming all the stakeholders cooperation.

Despite of the mentioned limitations, the received results could lead us to the following findings and suggestions.

1. Seaport and City CO_2 Emissions
   The diversity of greenhouse gas accounting methodologies currently utilized by cities around the world makes meaningful comparisons of their emissions and the seaport emissions almost impossible. There are no reliable sources providing with the information about the Qingdao city carbon emissions. The Governmental Report 2012 of the Qingdao City Statistical Department gives just the statistics about SO_2 (100,000 t), NOx (120,000 t), and PM (45,000 t) polluters. Among the observed literature, just the results of land-side emissions calculation

Table 6 Chinese cities emissions

City/Province	Source	Year	City emissions, millions tCO_2E	City territory, km^2	Population, millions
Nanjing/ Jiangsu	Bi et al. (2011)	2009	75,43	6,582	8
Suzhou/ Jiangsu	Wang et al. (2014)	2010	169,44	8,488.42	10.46
Shanghai	Liu et al. (2011)	2008	188,32	6,340.5	25
Hong Kong	Harris et al. (2012)	2008	42	1,104	7.18
Xiaolan/ Guangdong	Feng et al. (2014)	2010	2,072	71.96	0.317
Qingdao/ Shandong	–	–	–	11.6	8.71

of Port of Barcelona—156,206 tCO_2 (in 2008)—could be taken into consideration for the comparative purposes (QQCT—116,954 tCO_2, in 2012). In addition, the results of the carbon emissions from some Chinese cities, towns, and regions were given in Table 6 for the purpose of some general comparison. This outcome indicates the theoretical and practical gap in the methodologies of the cities emissions evaluation and the seaport emissions, caused by the increasing land-side activities. This aspect needs further development in the academic research.

2. Qingdao Seaport Case.
The results of the empirical research of QQCT indicated that the largest amount of the CO_2 emissions is caused by the HDVs activities. The terminal authorities should focus on the measures for their reduction in the first place. But the measures for the CO_2 emissions reduction from CHE and RL are necessary to be developed as well. The possible CO_2 emissions mitigating strategies could be combined into several main categories such as equipment replacement, clean fuels, emission control technologies, operational improvements, and idle reduction technologies. Equipment replacement implies replacement with engines meeting cleaner standards. Clean fuels mean low- to ultra-low-sulfur diesel fuel, emulsified diesel fuels, oxygenated fuel (O_2 diesel fuel), biodiesel, LNG, and CNG. Emission control technologies usually might involve diesel oxidation catalysts (DOCs), diesel particulate filters (DPFs) with PM emission reduction benefits, and selective catalytic reduction (SCR). Operational improvements could be implemented through the using of radio frequency identification (RFID) and optical character readers (OCR) to enhance the efficiency of gates and terminals. Idle reduction technologies include truck stop electrification (provides cab power for the truck while a truck is stationed in an area for a period of time) as well as automatic shutdown and start-up systems, battery power, auxiliary power units, diesel-driven heating systems for HDVs

Table 7 Energy use and emissions for typical transport units of different modes

Energy use/Emissions g/t/km	Rail electric	Rail diesel	Heavy truck
kWh/t/km	0.043	0.067	0.18
CO_2	18	17	50
SO*x*	0.44	0.35	0.31
NO*x*	0.10	0.00005	0.00006
PM	n/a	0.008	0.005

and automatic engine stop–start controls (AESS), auxiliary power unit (APU), diesel-driven heating systems (DDHS), shore power plug-in unit, and a hybrid switching locomotive for RLs.

These strategies could be implemented through the following instruments: lease requirements; tariff charges; voluntary programs; capital funding; and memorandum of understanding with the port, regulatory agencies, and other stakeholders.

Along with the mentioned measures for CO_2 emissions reductions from HDVs, the strategy of switching on the rail transportation could be carried out by the QQCT authorities. The rail transport is considered to be more efficient than trucks, as represented in Table 7.

Using rail transportation is closely connected with the dry port concept. For environmentally conscious shippers, it offers the possibility of using rail mode instead of road mode and thus decreases the environmental impact of their products. Rail operators obviously benefit from dry ports, especially from distant dry ports, because it increases the scale of their business in a comparatively lucrative segment. This is particularly important for rail transport depending on economies of scale and can make continental services viable although ports are reluctant to bring in containers not relating to shipping. At least, the fixed costs of the intermodal terminal itself can be distributed between transshipments when adding the dry port flows. Road transport operators do not benefit from this configuration directly since the aim is to move transport of containers from road to rail, but they are still involved in the intermodal transport chains. As they are not particularly paid for waiting in congestion or at crowded gates at the port, they can serve the dry port surroundings with shorter hauls and better total revenues.

The strategy of dry port implementation is extremely vital for QQCT as this container terminal is located on the territory of Qingdao West Coast Economic New Zone that follows the strategy of low-carbon development and smart city concept. In addition, Qingdao city plans to implement the new construction strategy and to establish an eco-coastline along Jiaozhou Bay with different functional emphases. For example, it might include conferences and exhibitions in Hongdao and logistics operation in Huangdao District. The new construction strategy is expected to fuel economic growth of Qingdao's satellite cities.

QQCT needs to take into account the mentioned development policies and work in collaboration with its tenants, terminal operators, and governmental authorities when developing its infrastructure and CO_2 emissions mitigating strategies.

6 Conclusion

The conducted research resulted in the following conclusions:

1. The seaports need to develop the comprehensive database for the precise CO_2 emissions estimation in cooperation with all the stakeholders;
2. The seaport mobile sources emissions should be included in the city inventories, and further improvement in city GHG inventory procedures is warranted;
3. The research gives the suggestions for QQCT in decreasing CO_2 from CHE, HDV, and RL via the operational improvements, using clean fuels, emissions control technologies, and idle reduction technologies;
4. The implementation of the dry port concept, supposing the modal shift from road to rail transportation, seems to be the most effective strategy, considering the governmental region development in future.

The further research might be developed in the mobile sources, used in dry bulk and oil terminal, the analysis of the dry port concept implementation, and the cities GHG inventories methodologies, including the seaport activities.

References

Acciaro M, Ghiara H, Cusano MI (2014) Energy management in seaports: a new role for port authorities. Energy Policy 71:4–12

Adamo F, Andria G, Cavone G, Capua C, Lanzolla A, Morello R et al (2014) Estimation of ship emissions in the port of Taranto. Measurement 47:982–988

Air Resource Board (2013) Estimated Annual Average Emissions, California Environmental Protection Agency. http://www.arb.ca.gov/app/emsinv/2013/emssumcat_query.php?F_YR=2012&F_DIV=-4&F_SEASON=A&SP=2013&&F_AREA=CA#8 2012

Bi J, Zhang R, Wang H, Liu M, Wu Y (2011) The benchmarks of carbon emissions and policy implications for China's cities: case of Nanjing. Energy Policy 39(9):4785–4794

Browning L (2009) Current methodologies in preparing mobile source port-related emission inventories. Final Report for US Environmental Protection Agency, http://www.epa.gov/sectors/sectorinfo/sectorprofiles/ports/ports-emission-inv-april09.pdf

Dong-Ping S, Jingjing X (2012) An operational activity-based method to estimate CO_2 emissions from container shipping considering empty container repositioning. Transp Res Part D Transp Environ 17(1):91–96

Feng Ch, Gao X, Wu J, Tang Y, He J, Qi Y, Zhang Y (2014) Greenhouse gas emissions investigation for towns in China: a case study of Xiaolan. J Cleaner Prod. In Press

Geerlings H, van Duin R (2011) A new method for assessing CO_2-emissions from container terminals: a promising approach applied in Rotterdam. J Clean Prod 19(6–7):657–666

Giuliano G, O'Brien T (2007) Reducing port-related truck emissions: the terminal gate appointment system at the ports of Los Angeles and Long Beach. Trans Res Part D Trans Environ 12(7):460–473

Harris PG, Alice SY, Chow J (2012) Symons. Greenhouse gas emissions from cities and regions: international implications revealed by Hong Kong. Energy Policy 44:416–424

Ho BQ (2013) Air emission inventories methodology for port and air quality simulation. Mod Transp 2(1):1–9

IAPH Tool Box for Port Clean Air Programs (2008) A reference guide, provided by international association of ports and harbors. http://wpci.iaphworldports.org/iaphtoolbox/
Li JS, Chen GQ, Lai TM, Ahmad B, Chen ZM, Shao L, Xi J (2013) Embodied greenhouse gas emission by Macao. Energy Policy 59:819–833
Liu Z, Geng Y, Xue B (2011) Inventorying energy-related CO_2 for city: Shanghai study. Energy Procedia 5:2303–2307
Miola A, Ciuffo B (2011) Estimating air emissions from ships: meta-analysis of modelling approaches and available data sources. Atmos Environ 45(13):2242–2251
Office of Transportation and Air Quality (2009) Emission Factors for Locomotives, US EPA. http://www.epa.gov/otaq/locomotives.htm
Perez H, Chang R, Billings R, Kosub T (2009) Automatic identification systems (AIS) data use in marine vessel emission estimation. In: Proceedings of 18th annual international emission inventory conference comprehensive inventories—leveraging technology and resources, Baltimore, Maryland, US Environmental Protection Agency
Peris-Mora E, DiezOrejas JM, Subirats A, Ibañez S, Alvarez P (2005) Development of a system of indicators for sustainable port management. Mar Pollut Bull 50(12):1649–1660
Puig M, Wooldridge C, Darbra RM (2014) Identification and selection of environmental performance indicators for sustainable port development. Mar Pollut Bull 81(1):124–130
Qi X, Song D (2012) Minimizing fuel emissions by optimizing vessel schedules in liner shipping with uncertain port times. Transp Res Part E Logistics Transp Rev 48(4):863–880
Schrooten L, Vlieger I, Panis L, Styns K, Torfs R (2008) Inventory and forecasting of maritime emissions in the Belgian sea territory, an activity-based emission model. Atmos Environ 42 (4):667–676
Schrooten L, Vlieger I, Panis L, Chiffi C, Pastori E (2009) Emissions of maritime transport: a European reference system. Sci Total Environ 408(2–20):318–323
Villalba G, Gemechu ED (2011) Estimating GHG emissions of marine ports—the case of Barcelona. Energy Policy 39(3):1363–1368
Wang H, Wang Y, Wang H, Liu M, Zhang Y, Zhang R et al (2014) Mitigating greenhouse gas emissions from China's cities: case study of Suzhou. Energy Policy 68:482–489
World Ports Climate Initiative (2010) Carbon footprinting for ports. Guidance document. Carbon Footprint Working Group, Port of Los Angeles, Lead Port

Complex-Based Gestalt Principle of Urban Visual Analysis: A Review

Ming Li and Jian-gang Xu

Abstract In the discipline of urban design application, the Gestalt principle implied in the traditional visual analysis pathway is a microscopic description of the interaction between the visual subject and object. The theory of complex adaptive system is generally a macroscopic description of the interaction between the visual subject and object. Gestalt principle in the microscopic sense, when integrated into the complex scientific viewpoint, can effectively characterize the evolution of urban visual elements and therefore enhance the performance of its functions. The focus points of this article are as follows: (1) The consistency between Gestalt principle and the integration principle in city as a complex system was reviewed. (2) By investigating the visual features of the city in various periods, the depth and width of the application of Gestalt principle of vision over time was discussed. (3) A systematic review of the existing tools for visual analysis was given. The integrated organization role and the self-similarity rule based on Gestalt principle were extracted. (4) A possibility of constructing a comprehensive tool for vision adaptability analysis was predicted from the perspective of complex system.

Keywords Complex adaptive system · Urban design · Holism · Urban image · Typology · Phenomenology

M. Li (✉)
School of Geographic and Oceanographic Sciences,
Nanjing University, 210093 Nanjing, People's Republic of China
e-mail: 573567882@qq.com

J. Xu
School of Architecture and Urban Planning, Nanjing University,
210093 Nanjing, People's Republic of China

S. Feng et al. (eds.), *Low-carbon City and New-type Urbanization*,
Environmental Science and Engineering, DOI 10.1007/978-3-662-45969-0_17

1 Introduction

Vision is the source of 83 % cognitive information for man and dominates 75–90 % of human activities. As a subsystem of city as a complex system, the city's visual system is related to nearly every domain of human being. Under the concept of subject integration, the object of visual analysis in urban planning and design has been extended from the simple urban physical environment to the interaction and mutual influence between man, nature, and society. According to the qualitative–quantitative integrated method for complex system, the comprehensive method for the analysis of the city's visual system should not only account for the intrinsic complex features of visual object, but also clarify the complex features arising from the concerns of human as the subject. However, the researches in this field are very few.

Holland John, a representative figure from Santa Fe Institute, put forward the theory of complex adaptive system (CAS) in 1996. This theory generally deals with the macroscopic description of the complex features in mutual concerns between human subject and city object. Here, the urban visual system is considered to be composed of a large amount of human and thing unit that interact among themselves. It represents the interaction between substance and psychology as an outcome of the holistic psychological progression and interaction between the object and subject of vision. Historically, the microscopic psychological progression on the part of the visual subject has attracted great attention from the time of Kant. The integrity of the perceptive act is emphasized. The later generations of philosophers directly or indirectly devote similar explanation to this topic. Although no explicit definition was given to this object of research at that time, the philosophers actually follow the Gestalt principle when reflecting on this topic. Several urban visual analysis pathways emerging under the reflection upon modernism after 1960s are all related to the Gestalt principle in one way or another. It is believed that the urban visual system is in constant evolution. The generation of visual hierarchy, the differentiation of visual elements, and the appearance of diversity are all the results of the holistic interaction between the object and subject of vision.

2 Concepts Explanation

The urban visual phenomena can be viewed from the perspective of city as a complex system. It is believed that the urban visual system is the city image acquired by visual perception and the subsequent ideological and emotional processing of the physical environment and spiritual environment of the city. This is in essence a Gestalt interaction feature based on the object and subject of vision. The Gestalt visual analysis tools in urban planning and design integrate social, psychological, geographical, and local folk characteristics as well as the available technical and engineering methods to finally give rise to perceptible urban morphology.

From the perspective of complex adaptive system, the following conclusions about Gestalt visual analysis tools can be obtained using the object–subject unification principle and object–environment integration principle. On the one hand, the urban visual system has complex features and a binary structure. Therefore, urban morphology, lifestyle, and cultural activities of human being that depend on the temporal and spatial dimension all fall within the domain of visual analysis. On the other hand, the visual adaptive feature of the city reflects the movement and interaction between the physical elements as the visual object and the human as the visual subject. If the factor of the object–subject interaction is excluded, there will be no independent rule of visual adaption. The visual analysis established without the real perception of man has no practical sense.

3 Gestalt Connotations Presented by Urban Visual Features in Different Periods

The development course and tradition of visual features of western cities is powered by two sources: one is Utopia in ideological realm, which propels the pursuit for the ideal of human habitat, and another is the strong drive of instrumental rationality in the technical realm. Therefore, the idea of urban construction in the earlier period is rich in perfection and idealism. This gives rise to the divergence of city's visual features in different time periods.

3.1 Holistic Visual Features Manifested in the Idea of "Same Construction of Family and Home" in Feudal Times

The urban visual features have the ethical element of "same construction of family and home" in the feudal times. The city's spatial structure was actually a layer-by-layer extension from the habitat place of the family to urban settlement. Accordingly, the social structure of the city has a correspondence from family to nation (city) and from nation (city) to the whole world. The historical period from ancient Greece to Renaissance witnessed the principle of limited Gestalt holism in the realm of urban planning and design. Nature, religion, and civic culture were taken as the elements of the city, conferred with human touch. Taking human body as the standard, the architectural space was built on the scale of man. The simple Gestalt principle applied to urban visual design reflects the pursuit for holism, order, and function (Table 1).

Table 1 Urban visual features in feudal times

Period	Gestalt holism	Visual features
Ancient Greece	Integration with natural environment	Taking nature as the element of city, the architectural space is conferred with the scale of man
Ancient Rome	Integration with religion	Rich in realistic and utilitarian values, the urban visual morphology presents definability
Middle Ages	Integration with civic culture	Good visual experience and the emphasize on spatial continuity on different scales
Renaissance	Integration with rationality	Emphasis on the practice and rationality, giving full play to subjective initiative of human

3.2 The Worship of the City's Functionality in the Times of Modernist Movement Caused the Breakdown of the Holistic Connection Between Object and Subject of Vision

From the end of nineteenth century to 1960s, the philosophy of urban planning in the west was dominated by physical determinism and elite determinism under functionalism. The idea of urban construction was the continuation of rationalism and physical planning. The representative theory was the "architecture determinism that supports tradition by a fragmented social science." The rapid development of science and technology powered the expansion of western cities. Since the beginning of last century, the idea of functional zoning proposed in the Athens Charter has become the cornerstone of the urban planning application and practice. At this time, the urban visual feature was rich in perfection and idealism. The visual feature of "internationalism" emerged everywhere and that of "regionalism" was rejected. The rule of organization based on the holistic interaction between the object and subject of vision, as required by Gestalt principle, broke down.

3.3 With the Revival of the "Human-Oriented" Concept in Contemporary Cities, the Visual Analysis Methods Based on Gestalt Principle Are Constantly Emerging

From 1960s to the present day, the urban planning theory based on functional zoning can no longer meet the requirements associated with the interactions between people in urban society. One of the outcomes of the prosperity of modernism was the "city disease" with declining space quality. This led to the reflection on formalism and absolutism (Yin 1999). The aggravation of urban problems and

Table 2 Visual features of contemporary cities

Important theories	Gestalt holism	Visual features
TEAM10	Advocating the harmony of city and man	The beauty of vision lies in the fusion between man, time, and space
Architectural typology	Temporary entity but enduring image	The historical prototype of building determines its visual form
City typology	Highlighting the secularity of city	The perfection and secularity of vision can coexist in the city
Townscape	Sense of place connects man and environment	Emphasizing the holistic continuity of landscape vision
City image	Identifiability of city	The image of city consists of five factors: region, road, boundary, node, and landmark
Design with nature	Urban evolution explained by nature	Ecological beauty of vision presented
Pattern language	Complex structure of urban system	Visual feature reflects the complex demands of man, with small pattern implied in large pattern
Postmodernism	City is the attention given to history	Vision has diversity, and secular visual aesthetics is more important to the cities
Neo-urbanism	Compact and mixed human community	Visual experience jointly enhanced by traditional lifestyle and modern civilization

predicament finally gave birth to a major transition in the philosophy of urban construction. Since the late 1960s, many new analysis methods of urban vision have emerged (Table 2).

4 Gestalt Principle for Contemporary Visual Analysis Pathway

After 1960s, the integrated pathway for visual analysis of urban planning and design falls into three categories generally. The first category is based on the critical heritage of the view on values of modernist cities. There is a shift from the emphasis on the functionality of vision to the holism, continuity, and popularization. This pathway is related to the "aesthetic activity" layer of the human visual perception. The "type-morphology" pathway represented by Aldo Rossi and Creil is the most notable. The second category starts from the criticism on the oblivion to human psychology by modernism, with the emphasis on the humanistic and emotional characteristics of urban environment. This pathway is related to the "environmental cognition" layer from the perspective of visual cognition. The most representative is the "cognition-image" pathway by Kevin Lynch. The third category starts from the philosophical concept, which highlights the role of direct experience in brushing aside the theoretical model in front of the visual phenomena. Then, one can more easily turn to the basis for the very existence of urban visual phenomena. This pathway is related to the “sense of existence” layer from the perspective of visual cognition. The “architecture phenomenology” pathway represented by Christian Norberg Schult is the most widely known (Fig. 1).

If viewed from the perspective of complex system, the three integrated pathways of visual analysis above showed close connections with Gestalt holism. On the one hand, they have close connections with the three elements of structuralism: holism, convertibility, and adaptability. Hence, the theoretical construction of architectural typology and city typology are affected. On the other hand, since the birth of psychology, the elementarism psychology represented by Wilhelm Wundt is weak in its explanation of “urban phenomena.” It provides the possibility and inevitability for the generation of Gestalt theory. Therefore, the achievements made in relation to this theory are also the epistemological source for the study of city image and architectural phenomenology. Although these visual analysis pathways differ in inclination, they share one thing in common: Visual phenomena come from the human cognition of the environment, which results in the psychological progression in terms of experience and understanding. From the perspective of Gestalt holism, the object of psychological progression with respect to urban visual phenomena consists of three aspects: human (visual subject), scene (visual object), and their interaction (between human and scene). The main integrated pathways for visual analysis of contemporary western cities are connected with these three aspects.

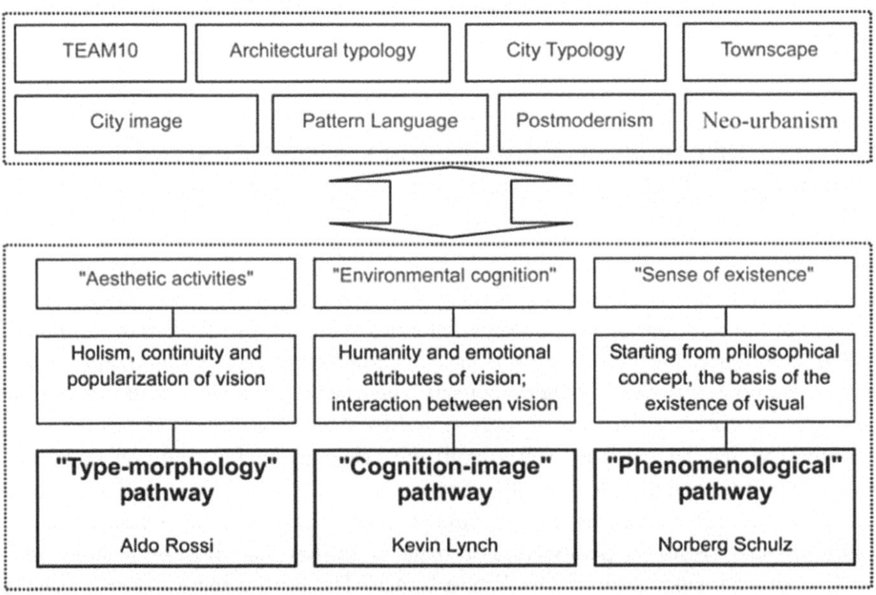

Fig. 1 Three major integrated pathways for visual analysis

Table 3 Connections between visual analysis pathways for contemporary cities and gestalt principle

Integrated pathway	Holistic organization rule	Self-similarity rule of Gestalt	Evaluation model
"Phenomenological" pathway	Holistic "meaning space"	"Place" and "meaning"	Identifiability evaluation
	The meaning of place is converted to various spatial morphologies through the composition of substance	Through the integration with place, the various meanings of specific scenarios are brought together	Positioning is carried out in the space of place, while identity is performed on the element interface of place
"Morphology-type" pathway	Holistic "form-type"	"Simplification" and "conversion"	Durability evaluation
	The conversion of "historical prototype" is the deep structure implied in the single unit and group of architecture	Through abstraction, the prototype of proportion and spatial model is obtained, and the form conversion is conducted	The continuity and durability of city and architecture are emphasized; the continuity of city morphology is acquired
"City image" pathway	Holistic "point-line-plane"	"Identification" and "image"	Identifiability evaluation
	Five factors, road, region, boundary, node, and marker, constitute the basic backbone for the cognition of the city	City is the space of visual perception and experience. Attention is paid to spatial cognition and positioning	The readability of individuality and structural features is emphasized

Gestalt principle implied in the integrated pathways for visual analysis of contemporary cities comprises of the holistic organization rule and self-similarity rule (Table 3). According to Gestalt principle, the human perception on the whole is superior to its composition, such as element, member, individuals, or parts, in terms of metaphysics, epistemology, or explanatory power. They define the basic model of visual analysis. However, Gestalt stresses the exploration of the holistic organization factors in the field, which represents the self-similar dynamic mechanism in the process of visual analysis.

5 Outlook

Among the several visual analysis tools for urban design mentioned in this article, either "phenomenological" pathway, "morphology-type" pathway, "city image" pathway, or "pattern language" pathway, they all show close connections with the holistic psychological progression in visual Gestalt. However, the visual construction tools for contemporary urban design based on traditional Gestalt principle have their own defects. Regardless of the development stage of relevant application theories in urban planning discipline, one has to make a choice between two orientations: The first is to firmly believe that the urban system can be simple, uniform, and balanced; and the second is to admit that the urban system is evolving toward complexity, differentiation, and non-balance. The theoretical research on the latter is the international hot spot in recent years. However, the traditional framework of Gestalt visual analysis makes no attempt to addressing the problems of complexity and non-balance of urban systems. The pathways for urban visual analysis introduced above grasp the dynamic foundation for the interaction between man and urban space system in the fundamental way. For the future study, it is necessary to construct an adaptive visual analysis tool under the framework of complex system. Microscopically, the empirical study of Gestalt should be freed from the mystical terms such as "situation," "notch," and "accidental awareness." The properties inside the urban visual system should be grasped fundamentally in order to expand the application in urban planning and design.

References

Batty M (1979) On planning processes. In: Goodall B, Kirby A (eds) Resources and planning. Pergamon, Oxford, pp 17–50

Campen C (1994) Gestalt from Goethe to Gibson theories on the vision of beauty and order. Utrecht University, Utrecht

Chen T (2007) Integrated principle of urban design. Tianjin University, Tianjin

Holland J (1996) Hidden order. HarperCollins UK, London

Koffka K (1935) Principles of gestalt psychology. Publ. Lund Humphries, London

Qian X, Jingyuan Y, Ruwei D (1990) A new discipline of science—the study of open complex giant system and its methodology. Nature 1:3–10

Taine (1983) Artistic philosophy (trans: Lei F). The People's Literature Publishing House, Beijing

Vernant J-P (1996) The origin of Greek thoughts (trans: Haiying L). SDX Joint Publishing Company, Beijing

Yang J (2004) Through the complex jungle. J Renmin Univ Chin 5:17–24

Yin D (1999) Introduction to design. Hunan Science & Technology Press, Changsha

Zhang T (2011) Urban construction in the 21st century under the guidance of 20th century planning theory. Urban Plann Forum 3:1–7

Recuperator Concept Design of Low-Emission 2-MW Gas Turbine

Qiao-ling Kong, Zhi-guo Dang, Guang-di Xu, Chuang Gao and Wei-guang Huang

Abstract Due to energy shortage and air pollution, the distributed energy supply system has been quickly developed. Because it can improve the effectiveness of natural gas utilization and save energy and reduce emission. The core equipment of the distributed energy supply system is the gas turbine which is rated up to 2 MWs in China. So a recuperated cycle gas turbine was presented in this paper. Three types of recuperator such as the tabular, plate fin, and primary surface recuperators were designed and analyzed. The result shows that the primary surface recuperator has more advantages than the other two. It has higher compactness and less weight. If the whole manufacture process is well developed the primary surface recuperator will have more competitiveness in the future recuperator market. What is more, the increasing of the effectiveness of the recuperator will result in a dramatic increase in the recuperator size and weight. So when a recuperator is designed it must be taken into consideration as an integrate part of the overall energy conversion system.

Keywords Compactness · Gas turbine · Recuperator · Primary surface recuperator

1 Introduction

Due to energy shortage and air pollution, small distributed energy supply system (DESS) has become the main trend of the DESS with the advantages of small investment, quick return, low emission, low-energy transmission loss, and high energy efficiency. As in China, the DESS is mostly demanded in the densely populated areas with stable heat load and high technical management level such as large venue area of 10,000 m^2, communities, airports, medical and other industrial

Q. Kong (✉) · Z. Dang · G. Xu · C. Gao · W. Huang
Center for Clean Energy Technology Research, Shanghai Advanced Research Institute, CAS, Shanghai 201210, People's Republic of China
e-mail: kongql@sari.ac.cn

S. Feng et al. (eds.), *Low-carbon City and New-type Urbanization*,
Environmental Science and Engineering, DOI 10.1007/978-3-662-45969-0_18

and civil facilities, and future home users. So the power of gas turbine for the DESS is about 1–2 MW. To improve the gas turbine efficiency and reduce the emission is the key technical problem in improving the economic performance and in large scale promoting the DESS. As we know the use of regenerative cycle can improve the gas turbine efficiency, however, the feature of the heat recuperator will affect the gas turbine. So the recuperator was simply described and analyzed in this paper.

2 Parameters of the Gas Turbine

The gas turbine applies low-pollution combustion technology of which NOx emission is lower than 15 ppm,the power is 2 MW, simple cycle efficiency is 26 %, and the exhaust temperature is as high as 56 °C. If a recuperator is used to recycle the exhaust heat to preheat the compressed air and improve its temperature, it will reduce the fuel consumption in the combustor to achieve a certain turbine inlet temperature which can improve the gas turbine efficiency. The principal of recuperated cycle is shown in Fig. 1. When the recuperator efficiency is 0.9, the exhaust temperature can reduce to 327 °C and the system efficiency can be improved to 37 %. The different gas turbine efficiency changing with the recuperator efficiency is shown in Table 1. It shows that the higher the recuperator efficiency, the higher the gas turbine efficiency. But the volume of the recuperator becomes bigger and the cost increased for the higher efficiency. So in order to select the recuperator design parameters, the whole-system economical performance should be taken into consideration.

Fig. 1 Regenerative cycle (recuperator efficiency = 0.9)

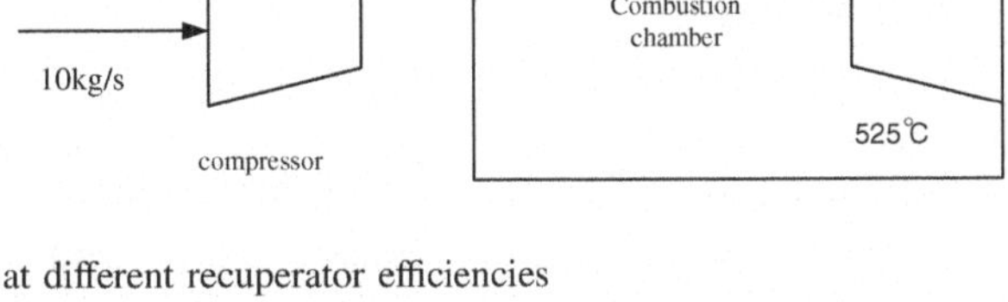

Table 1 Gas turbine parameters at different recuperator efficiencies

Recuperator efficiency	Efficiency of gas turbine (%)	Exhaust gas temperature (°C)	Compressed air temperature (°C)
0.95	38.05	316	514
0.9	37.2	327	503
0.85	36.32	339	491
0.8	35.43	350	478.7
0.75	34.74	361.5	468.6
None recuperator	26.6	560	288

3 Types of Recuperator

With the development of micro gas turbine DESS, the recuperator of micro gas turbine was widely studied by researchers from all over the world. McDonald (2003, 2000) summarized the status of the recuperator technology from structure, surface geometry, material, and challenges ahead. Fend et al. (2011) developed a ceramic high-temperature heat exchanger, and the performance of the heat exchanger was tested. Aquaro and Pieve (2007) described the technology status for high-temperature nuclear recuperator and compared different types of recuperator for certain design parameters. Early applications of recuperator are mostly tabular or plate fin heat exchangers. The tabular and shell heat exchanger has large volume, and the plate fin heat exchanger needs brazing–welding process which is of higher cost. In recent years, the primary surface recuperator (PSR) has been quickly developed. PSR is inherently resistant to cyclic thermal fatigue because it can flex to relieve thermal and mechanical strains. An additional benefit is the intrinsic natural damping characteristics derived from stacking cells together with multiple friction interfaces. These several contact points absorb displacements from vibration sources (Ward 1995). Printed circuit heat exchanger (PCHE) has not been used as gas turbine recuperator; the manufacturer simulated the gas turbine operation condition to test the PCHE transient response feature in order to promote PCHE to be applied as gas turbine recuperator (Grady and Dewson 2004). In China, several universities and research institutes (Liu and Cheng 2008) design and test the PSR recuperator of micro gas turbine, but those recuperators are only at laboratory test and have not been used in engineering application. So the PSR for 2-MW gas turbine was designed and compared with the tabular and plate fin recuperator in this paper.

4 Primary Surface Recuperator Design

4.1 Input Parameters for the Recuperator

According to the gas turbine recuperated cycle calculation, when the recuperator efficiency is 0.9, the input parameters are shown in Table 2.

4.2 Flow Passes Parameters

The cross section for cross-wave (CW) surface for PSR is shown in Fig. 2. The principal of selecting the pass parameters is to keep the heat transfer rate of both exhaust gas and air almost the same in order to let the comprehensive performance for heat transfer, flow and investment match with each other. The parameters for the corrugated board are $b_h = b_c = 0.65$ mm, $a_h = 0.5$ mm, and $a_c = 0.4$ mm. The equivalent diameter of exhaust gas side is 1.5 mm and the air side is 1.2 mm.

Table 2 Input parameters for the recuperator

Working fluid	Compressed air	Exhaust gas
Mass flow rate (kg/s)	10	10.11
Inlet temperature (°C)	296	525
Inlet pressure (kPa)	799	103
Allowable pressure drop (%)	1	3
Efficiency	0.9	

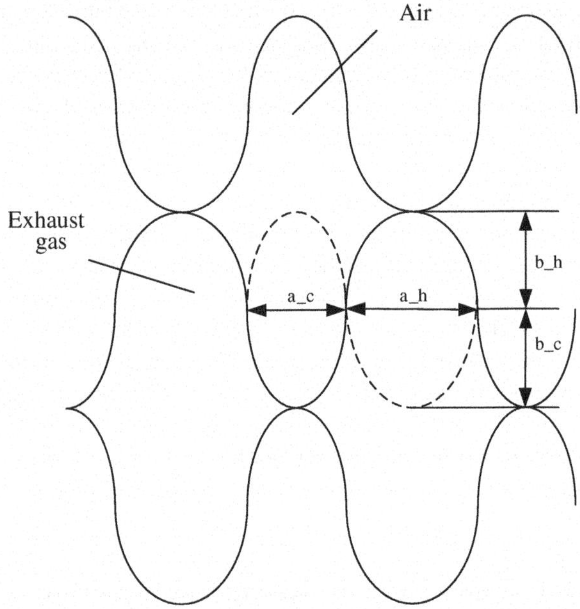

Fig. 2 Diagram of elliptical cross section for PSR

4.3 Thermal Hydraulic Calculation for Primary Surface Recuperator

Convective heat transfer coefficients of air and exhaust gas are calculated according to the convective heat transfer empirical formula in rectangular slot (Liu and Cheng 2008):

$$Nu = 0.0031\,Re^{1.18}\,\mathrm{Pr}_{0.4}(h/w_c)^{0.19} \tag{1}$$

$$\alpha = \frac{Nu\lambda}{de} \tag{2}$$

Equation (1) is suitable for $Re < 1{,}000$, $h/w_c = 1$–9, and the equivalent diameter is between 0.5 and 1.5 mm, where h and w_c are the height and width of mini rectangular slot, respectively. In this paper, they are elliptical long axis 4b and short axis 2a in Fig. 2. The property of fluid is calculated using the average temperature of the inlet and outlet.

Flow resistance is calculated according to the Eq. (3) (Yang et al 2003).

$$f = \frac{72}{Re} \tag{3}$$

where Re is calculated using the equivalent diameter. The total pressure drop Δp for the air and exhaust gas is composed of three parts: They are flow resistances of inlet Δp_i, counter flow section Δp_n, and outlet Δp_e. That is,

$$\Delta p = \Delta p_i + \Delta p_n + \Delta p_e \tag{4}$$

where

$$\Delta p_n = \frac{1}{2} f \rho u^2 \frac{L}{de} \tag{5}$$

where L is the length of the flow pass, de is the equivalent diameter, u is the average velocity of the fluid, and ρ is the average density of the fluid. In fact, the inlet and outlet flow resistance is far less than the counter flow part so they can be neglected.

4.4 Results Analysis

According to the input parameters for 2-MW gas turbine shown in Table 2, three types of recuperator such as PSR, plate fin, and tube and shell are designed. The results are shown in Table 3.

From the data shown in the table, the PSR has large compactness. And the total volume, heat transfer surface, and weight are less than the other two heat exchangers, and it has better economical performance. The pressure drop for the gas turbine recuperator is very strict, so the velocity of the plate fin and tabular heat exchanger cannot be very large; otherwise, the pressure drop will surpass the allowable value, and the plate fin heat exchanger size is also limited by the size of the brazing furnace so the flow pass cannot be as shorter as the PSR.

In the gas turbine recuperator design process, the overall investments and system efficiency of the gas turbine must be comprehensively considered while blindly pursuing high efficiency and small pressure drop of the recuperator. To some extent, a little improvement of the recuperator efficiency will cause large volume and cost increase. It shows in Fig. 3 that the weight and volume of the recuperator change with the efficiency. When the efficiency of the recuperator increases from 0.9 to 0.95, the weight of the recuperator increases from 756 to 2,379 kg. That is, the efficiency increases 5.5 %, but the weight increases 214 %, and the volume

Table 3 Design results for different types of recuperator

Items	PSR		Plate fin		Tube and shell	
	Air	Exhaust gas	Air	Exhaust gas	Air	Exhaust gas
Equivalent diameter (mm)	1.2	1.6	1.9	3.8	5	
Re	479	507	1,012	723	1,220	780
Convective heat transfer coefficient (w/m^2 K)	183	147	53		51	90
Overall heat transfer coefficient (w/m^2 K)	81		–		27	
Pressure drop (Pa)	2,851	7,197	1,310	3,050	6,900	2,900
Plate number	1,392	187	186	–	–	–
Size (length × width × height) (m)	1 × 0.342 × 1.84		1.8 × 1 × 1.5		Φ2.2 × 4.5	
Core volume (m^3)	0.612		2.7		14.1	
Core weight (kg)	770		3,383		31,058	
Core compactness (m^2 m^3)	1,399		470		352	

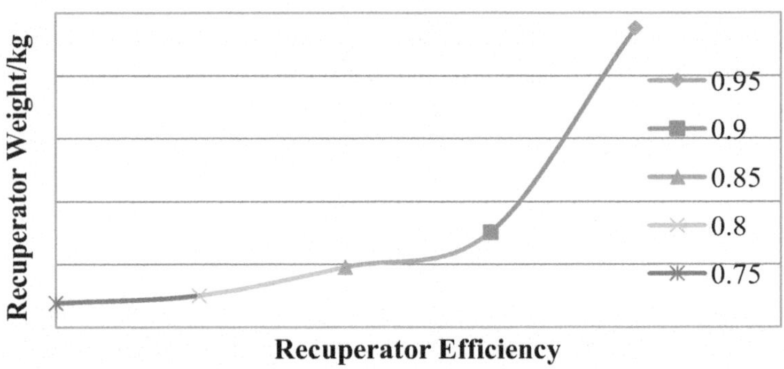

Fig. 3 Weight of the recuperator changes with the efficiency

increases too. While the recuperator efficiency increases, the logarithmic mean temperature difference decreases and the heat transfer surface increases. And with the limited pressure drop value, the velocity cannot be too high so the overall heat transfer coefficient is low.

5 Questions to Be Solved

There are still some questions to be solved to develop the PSR:

1. Build the high-temperature heat exchanger thermal–hydraulic test rigs and use numerical simulation method to obtain the empirical equation for calculating the heat transfer and flow resistance;

2. Build the manufacturing technology including punching, folding, pressing, and welding to reduce the cost;
3. Develop fatigue and creep analysis method for high-temperature recuperator and build the high-temperature cycle fatigue and vibration test rigs to improve the service life of the recuperator; and
4. Study the dynamic response characteristic of the recuperator under part load and its impact on the gas turbine systems.

6 Conclusion

The recuperator of the 2-MW gas turbine was designed and analyzed in this paper. The selecting of the recuperator parameters must be optimally analyzed taking the comprehensive performance and economical performance of the gas turbine into consideration. The development of the PSR must develop the test rigs, manufacturing technology, and design database; the more important thing is to keep up with the gas turbine application. Good marketing prospect will promote the technology development.

References

Aquaro D, Pieve M (2007) Nucleare high temperature heat exchangers for power plants: performance of advanced metallic recuperators. Appl Therm Eng 27:389–400

Fend T, Völker W, Miebach R et al (2011) Experimental investigation of compact silicon carbide heat exchangers for high temperatures. Int J Heat Mass Transf 54:4175–4181

Grady CJ, Dewson SJ (2004) Effects of HTGR core transient behavior upon the primary recuperator. In: Proceedings of ICAPP'04. Pittsburgh, pp 397–409

Liu Z, Cheng H (2008) Multi-objective optimization design analysis of primary surface recuperator for microturbines. Appl Therm Eng 28:601–610

McDonald CF (2000) Low-cost compact primary surface recuperator concept for microturbines. Appl Therm Eng 20:471–497

McDonald CF (2003) Recuperator considerations for future higher efficiency microturbines. Appl Therm Eng 23:1463–1487

Ward ME (1995) Primary surface recuperator durability and applications. In: Turbomachinery technology seminar paper TTS006/395. Solar Turbines, Inc., San Diego

Yang J et al (2003) Design analysis of primary surface recuperators for marine gas turbines. J Eng Therm Energy Power 18(4):365–368

Green Campus as a Pilot Site Towards Low-Carbon City: Enlightenment from Cornell Climate Action Plan

Tian-ren Yang

Abstract After green building initiative, much more attention has been paid to green campus transformation. Green campus serves as a pilot site for low-carbon city. This paper summarized the development process of green campus while analyzed the current situation of China's green campus initiative, including overemphasis on energy efficiency of single building, lack of integrated campus efficiency monitoring system, and insufficient financial support. Based on the experiences and enlightenment from Cornell University's Climate Action Plan, the sustainable development of green campus in China has been discussed, in the aspects of single building, community/campus, and city. On building level, campus building energy technology demonstration should be strengthened and popularized. On community/campus level, regulatory agencies should be established to take charge of low-carbon campus overall planning and monitoring. On city level, campus, as a regional experimental field of sustainable development, should promote low-carbon transformation of the city in specific ways.

Keywords Green campus · Low-carbon city · Climate action plan · Cornell University

1 Introduction

Currently, the necessity of developing green buildings has been widely recognized across the world. When green buildings are developing vertically and horizontally in all countries, the development of green campuses has become a key direction for

T. Yang (✉)
College of Architecture and Urban Planning, Tongji University,
1239 Siping Rd, Shanghai 200086, People's Republic of China
e-mail: tianren.yang@hotmail.com

T. Yang
College of Architecture, Georgia Institute of Technology, Atlanta 30332, USA

S. Feng et al. (eds.), *Low-carbon City and New-type Urbanization*,
Environmental Science and Engineering, DOI 10.1007/978-3-662-45969-0_19

transforming green buildings in a mature way. The campus, where it is equipped with a great deal of building facilities as well as living facilities and scientific research facilities, is a big energy consumer in the society. Meanwhile, the campus is undertaking the great responsibility of training talents for the society and leading sustainable development of the society as a key field for educating people and developing technologies. Therefore, in order to achieve a sustainable development of universities and even the whole society in future, it is significant to construct a resource-saving, environment-friendly, and green college campus and to incorporate the idea and practice of sustainable development into various areas, such as college teaching, scientific research, talent training, and social service (Tan et al. 2012).

As an experiment plot and demonstration area of ecological city construction and a complete system in the social ecological system, green campus is featured with certain independent ecological functions while being closely related to surrounding ecological, economic, and social systems. With the in-depth development of the construction of green campuses, however, the sustainable development of campuses has been stuck in bottleneck. Moreover, it is an important period for China to develop green buildings, ecological cities, conservation-oriented campuses, and green campuses (Tan 2013). This paper is aiming at exploring and discussing technical paths for achieving sustainable development of green campuses in three levels [building–community (campus)–city] through concluding successful experiences from the development process while referring to experiences and enlightenment of Cornell University from "Climate Action Plan" for solving the bottleneck problem in the process of constructing and developing green campuses in China.

2 Development Process of Green Campuses

So far, the process of developing international green campuses can be concluded into five stages: the initial stage for environmental protection education, the stage for developing sustainable education, the stage for practicing green campus, the stage for demonstrating green campus, and the stage for promoting green campus (Table 1).

1. Environmental protection education. The concept of green campus was originally proposed in The Stockholm Declaration published by the United Nations Human Settlement Assembly in 1972. The declaration has raised concerns about environmental protection education in the society.
2. Developing sustainable education. In 1990, John Mayer from France made The Talloire Declaration, advocating for a 10 action plan that integrates sustainable development with college teaching, scientific research, operation, and publicity. It is the first official commitment made by a university administrator about sustainable education in colleges and universities. In 1994, UNESCO proposed to incorporate the idea of sustainable development into school education and

Table 1 Development process of green campuses

Year	Event	Key topic	Stage
1972	The Stockholm Declaration	Education of general knowledge on environmental protection	Environmental protection education
1990	The Talloire Declaration	Sustainable education of colleges and universities	Developing sustainable education
1994	EPD program (UNESCO)	Incorporate environmental education into development education, population education, etc	
1997	Thessaloniki meeting in Greece (UNESCO)	Sustainable education	
1997	Pilot project of green university	Construction of the pilot project	Practicing green campus
2006	AASHE	A system for tracking, assessing, and commenting sustainable development	Demonstrating green campus
2007	ISCN	Sustainable campus	
2007	ACUPCC	Declared to construct the campus of colleges and universities into a "carbon neutrality" park	Promoting green campus
2010	The centre for green schools (USGBC)	Work with people in the area of education to promote the construction of "green campus" upon built environment	

launched the EPD program. In 1997, UNESCO convened a meeting in Thessaloniki (Greece); in the meeting, the idea of "sustainable education" was defined, which indicates that environmental education was no longer education on environmental problems, but that it began to combine with education on peace, development, and population to form a philosophy of "sustainable development education."

3. Practicing green campus. In 1997, some colleges and universities in the United States, out of a purpose to improve environmental protection consciousness and promote sustainable development education practice and campus construction, launched the Campus Consortium for Environmental Excellence (C2E2), a non-profitable organization or alliance. Supported by the United States Environmental Protection Agency (EPA), the alliance began to construct pilot projects vigorously. In the same year, the George Washington University (GWU) began to practice its pilot project for constructing a green university with a view to making the university the first green university in the world.
4. Demonstrating green campus. In 2006, colleges and universities in the United States and Canada co-founded the "Association of the Advancement of Sustainability in Higher Education" (AASHE), which developed "the Sustainability Tracking, Assessment and Rating System" (STARS). With this system, every college or university can self-evaluate its sustainability and make corresponding development policies by based on the evaluation report. In April of 2007,

Switzerland established an International Sustainable Campus Network (ISCN) with a view to constructing a platform for sharing experiences and information on how to construct sustainable development colleges and universities in a global range.

5. Promoting green campus. In June of 2007, the United States' practice of including sustainable development into operation and course practice in colleges and university yielded its most important result, that is, "the American College and University Presidents' Climate Commitment" (ACUPCC). ACUPCC was joined by 284 American college and university presidents who then jointly declared to construct college campus into a "carbon neutrality" park and try to reduce the direct or indirect emission of carbon dioxide through planting trees, saving energy, using renewable energy, and other ways. In 2010, USGBC founded the Centre for Green Schools, which was targeting at promoting the construction of "green campus" with people in the field of education upon the built environment.

After that, concepts such as "Eco-campus" and "Sustainable campus" began to appear one after another in the process of constructing college campus. From environmental education course to in-depth exploration on the idea of green colleges and universities, the content of constructing green colleges and universities was perfected constantly.

Under the background of new urbanization, colleges and universities are closely connecting to cities (Cornell University 2013). Therefore, the construction of green campus will enter a new stage with the construction of low-carbon cities. Green campus will not only cover content on planning, architecture, and landscape, but that comprehensive and integral consideration should be given to its urbanism and sociality. In future, the development of green campus will primarily be presented in the following areas: (1) The sustainable development idea should be included into the talent training systems of colleges and universities; (2) green and environmental protection technology innovation will become an important field of academic researches; school facilities, along with innovation of energy-saving and emission reduction technologies and their research, development, and promotion, will be featured with great potential; (3) colleges and universities will pay attention to the radiation influence of green campuses on the society and surrounding communities so as to actually turn green campus into a pilot place and demonstration area of ecological city construction.

3 Existent Problems on Practicing Green Campus in China

3.1 Focus on Single Building Mostly

Currently, the practice of green campus in China tends to regard single building as starting points, which has resulted in the absence of overall grasp over various spatial elements between buildings and open spaces. Meanwhile, too much attention to indoor environment of campus buildings is likely to lead to an overlooking over comfort of outdoor environment. Furthermore, green campus is often considered in an isolated way in the overall city ecological framework, with little emphasis on positive interaction between campuses and surrounding cities, society and ecology and a lack of systematic consideration to its matching with infrastructure construction in surrounding areas.

3.2 Low Level of Campus Management Efficiency

Since a campus has many buildings, the need for energy grows rigidly and the energy consumption and water consumption per student in the campus are higher than the average level of urban residents. Under the background when China is adjusting its social and economic industrial structures while transforming and developing its industrial production structure, the total energy consumption of the campus is increasingly taking a bigger and bigger proportion of the total energy consumption of the society. In China, however, colleges and universities are public institutes funded by the central or local governments. As a result, a severe shortage of energy-saving consciousness and energy cost-control consciousness has been existing in these places for a long time. Moreover, there is not a comprehensive and systematic sustainable development and construction standard for campus energy consumption management in levels from management to planning and from planning to operation. Hence, it is urgent to develop an idea of low-carbon development so as to promote the movement of green campus with a digital management model under the principles of saving energy, land, water, materials, and environment (Wu and Zisong 2013).

3.3 Inadequate Stamina for Construction

Since the operation and construction of green campuses is a time-consuming and huge project, every school should conduct a systematic and scientific theoretical study according to its characteristics of running before practicing. Nevertheless, inadequate knowledge on the importance of green campuses has directly resulted in a shortage of capital and necessary hardware and software facilities. As a result of that, there is a lack of continuity for developing green campuses when the work is done only by some staffs and part-time teachers with some public facilities.

4 "Climate Action Plan" by Cornell University

Under the framework of AUPCC, hundreds of college and university presidents in the United States committed to start it in colleges and universities and reduce the emission of greenhouse gases (Zhang and Zhang 2011). The content includes the

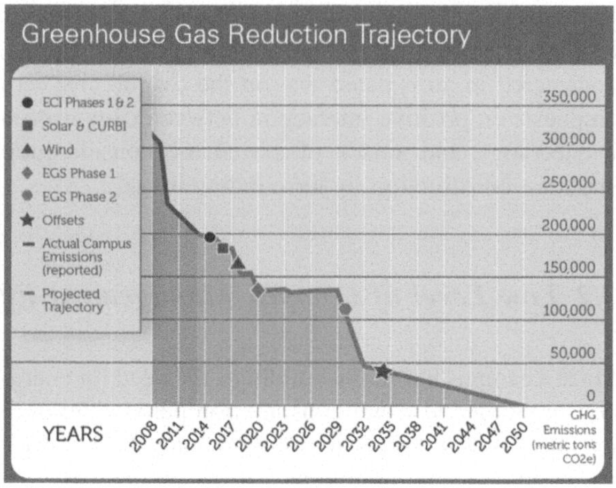

Fig. 1 Greenhouse gas reduction trajectory. *Source* Cornell climate action plan and roadmap (2014–2015)

determination of action plans and the objective of neutral carbon emission. In 2007, Cornell University made a "Climate Action Plan," which regards carbon neutrality as a strategic direction for promoting core tasks of the school, such as teaching, scientific research, and community service. According to the plan, the emission of greenhouse gases is expected to decrease to zero in 2050 (Fig. 1).

4.1 Quantitative Emission Reduction Target and Stage Plans

Seven neutrality key actions are put forward: (1) complete phase 1 of the energy conservation initiative (ECI) and initiate phase 2 to conserve energy in campus buildings; (2) integrate building energy standards and energy modeling into the building design, review, and approval process in order to maximize energy efficiency; (3) optimize the campus heat distribution system to increase efficiency and cost-effectiveness and to facilitate the integration of Cornell's future energy sources; (4) capitalize on more than 50 campus waste streams and other university-owned biomass resources to generate renewable energy through Cornell University Renewable Bioenergy Initiative (CURBI); (5) eliminate the combustion of fossil fuel for campus heating by developing an enhanced geothermal system (EGS) hybridized with biogas, prepare a preliminary design and phased implementation plan for a hybrid enhanced geothermal system, and build a demonstration project; (6) support the expansion of regional wind generation capacity and integrate wind power into Cornell's renewable energy portfolio; and (7) implement broad-based, mission-linked carbon management strategies such as forest management, carbon capture and sequestration, and community projects to offset unavoidable university emissions. Every aspect is calculated in its effort to carbon emission quantitatively.

4.2 Reconstruction of Single Green Building

When practicing the "Climate Action Plan," 96.8 % "green campuses" required new buildings to meet LEED or equal effect evaluation standards, 65.5 % proposed the proportion of renewable resources, and 81 % clarified the proportion of the using of clean energy meeting standard for green certificate and came up with countermeasures from various aspects, such as single-building energy, materials, and illumination (Gan et al. 2012).

The importance of reconstructing single green building lies in that it cannot only improve energy efficiency but also make use of renewable resources. In Cornell University, both new and reconstructed teaching and laboratory buildings are regarding attaining LEED certificate as an important goal of development.

4.3 Pilot Projects of Campus Energy Technologies and the Monitoring System

As with energy action, Cornell University has employed two certified energy managers and senior engineers for overall monitoring the supply–demand balance between hydraulic/thermal power plants (the supplier) and campus buildings (the demander) under the college energy management plan.

1. Application of renewable energy technologies
   From a macroperspective, optimization of energy distribution system in the campus has significantly improved the efficiency and cost-effectiveness of energy using and facilitated the integration of new energy technologies of Cornell University in future. In the projects, the necessity to update or maintain short-term and long-term pipelines is analyzed and proven through collecting real indoor data so as to guarantee the necessity and feasibility for using new energy technologies for different buildings. Meanwhile, Cornell University is planning to use its existing more than 50 technologies (such as enhanced geothermal system, gas-supplying energy, wind energy, and carbon management strategy) for developing renewable biological energy sources and establishing relevant demonstration projects (Fig. 2).
2. Real-time campus energy monitoring system
   To monitor energy data in a more effective and quantitative way, Cornell University has established a real-time campus energy consumption monitoring system (Cornell Energy Dashboard). With this system, the teachers and students of the school can get to know energy consumption of every building in the school through Internet (Fig. 3). Based on an induction system, Cornell Energy Dashboard is able to monitor heat supply, ventilation, air conditioning, refrigeration, illumination, fireproofing, smoke detectors, public facilities, and elevators and to record the data. The installation of a great deal of sensors has made

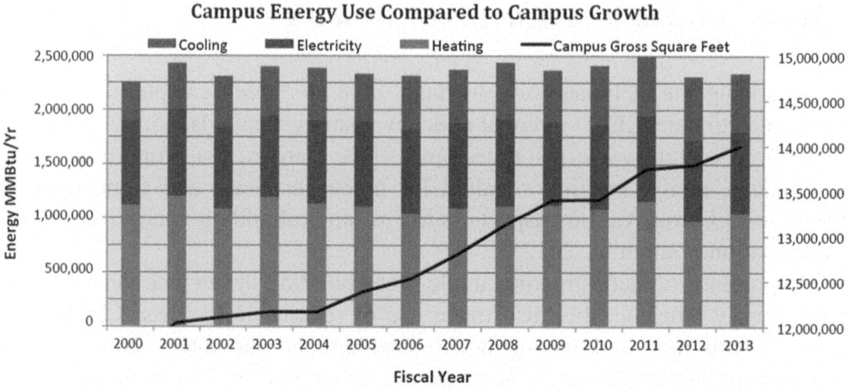

Fig. 2 Campus energy use compared to campus growth. *Source* Cornell climate action plan and roadmap (2014–2015)

Fig. 3 Cornell energy dashboard. *Source* building-dashboard.cornell.edu

sure the smooth operation of automatic control system, heat recovery system, and illumination system of Cornell University. Finally, the real-time campus energy monitoring system is used for planning all subsystems as a whole and integrating data.

4.4 Demonstration Effect as a Regional Low-Carbon Community

When constructing itself into a demonstration case of green campus, Cornell University has spread its influence and demonstrative technologies to surrounding communities. For example, its most famous "Energy Conservation Engagement Program," "Climate Literacy Outreach," and "STARS" have been applied in surrounding communities as mature experiences or demonstrative technologies.

The excellence of Cornell University in the field of public participation and its interdisciplinary cooperation have provided it with distinctive advantages on the issue of sustainable development. Since the issue of climate changes can only be solved by regional scales, the campus, as a pilot place for ecological and low-carbon technologies, can actually contribute its value only when its demonstration effect is spread to a wider regional network. In fact, many colleges and universities in the United States have come to realize the importance of regarding public/private sectors as their partners for speeding up the construction of green campuses; meanwhile, the colleges and universities are trying to practice in a bigger community by based on the campus.

4.5 Institutions for Guaranteeing Implementation of the Overall Plan

In order to facilitate the implementation of the pioneering plan of "Climate Action Plan," Cornell University has established a new international management mode: establish a sustainable campus committee and employ "climate action plan" managers, special office staffs, and special administrative staffs for seeking for solutions to climate changes and how to promote construction of campus sustainability. As a matter of fact, all colleges and universities in the United States have established corresponding leading institutes in the process of starting and promoting green campus construction, such as Environment Committee, Green Campus Office, Environmental Working Group, Environmental and Social Institute, Comprehensive Waste Management Center, and environment-related connecting Web sites and organizations. These institutes are to formulate policies on campus environment in colleges and universities while organizing, directing, and guiding colleges and universities to construct green campuses (Han 2011).

5 Enlightenment for China's Green Campus Construction

So far, China's colleges and universities have made remarkable achievements in constructing "conservation-oriented campus" launched in 2006. Meanwhile, the energy conservation and emission reduction in campuses have been developed from the initial spread of idea to demonstrative practice of conservation-oriented campus to the current comprehensive green campuses with depth. However, as commented by ACUPP on "Climate Action Plan," green campus is not a simple low-carbon plan but covers measurements on the setting of specific energy-saving goals, facilities of relevant scientific research projects, and determination of examination systems and future promotion. Therefore, green campuses in China are to be constructed and developed step by step, from points to areas and being innovated and improved constantly.

5.1 Formulate Scientific and Feasible Action Plans and Establish Management Institutes

From the experiences of the United States on constructing "green universities," most green campuses have set goals and action plans by based on their characteristics, which has significantly contributed to their long-term and lasting success. Setting a strategic goal of becoming a green university is an ideal anticipation of colleges and universities about constructing green universities from a macroperspective. When setting goals, therefore, it is required to base on current status while coming up with the development objectives and action plans for constructing green universities by based on an analysis on current status of colleges and universities and making plans and implementation measures in different stages.

Furthermore, construction of green campuses is a complicated project, requiring a powerful core institute to manage in a unified way from the planning of campus environmental protection projects to promotion of green scientific research results to the society. In management level, it requires to establish a mechanism for promoting planning and coordination between government top design and multiple departments so as to strengthen the bottom-to-up (government leading) and top-to-down (self-initiated by schools) complementary development and perfect management mechanisms of all areas. For example, the "sustainability campus committee" founded by Cornell University and other institutes (such as "green committee" and "green office") founded by other green campuses have played a critical role in management and provided abundant experiences for China with regard to construction and management of green campuses.

5.2 Demonstrative Application of Energy-Saving Technologies for Campus Single Building and Their Popularization and Promotion

The demonstrative value of green buildings never lies in the accumulation of advanced energy-saving technologies—but in the adaptable combination of them. Meanwhile, no matter whether it is newly constructed or reconstructed projects, consideration should be given to relationship between buildings and surrounding sites and natural landscapes as well as interaction with surrounding building groups.

In addition, tools for evaluating designs of green buildings should be put forward in the stage of popularization and promotion, including quantitative tools for evaluating costs of carbon emission in future and qualitative tools for society, environment, economy, institute priority, and other areas.

5.3 Energy Management of Campus Overall Planning and Energy-Saving Monitoring

Green campus is an important and strategic step from single green building to overall green communities and low-carbon cities. In the process of constructing a campus to reconstructing it, planners should actively implement environmental protection ideas such as saving energy, saving water, saving materials, and saving land while giving consideration to campus ecological network and energy network planning to create an ecological and effective campus environment (Zeng 2007). Moreover, guarantee circular flow of various types of physical energy in the campus ecological environment and exchange of all information through a network-type ecological structure so as to effectively bring all ecological factors of the campus ecological system into play.

From the Cornell Energy Dashboard, we can see that the project of constructing a campus energy-saving monitoring platform is a demonstrative project of colleges and universities. Purpose of the construction is to realize digitization of energy consumption, dynamic management, visual data, and indexation of energy saving. The practice of international college energy platform indicates that data of green campus should be collected step by step: collect energy consumption information of buildings in all campus areas to a transfer station and submit it to the energy source monitoring center. In fact, some schools (such as Tongji University) in China have developed systematic, real-time, and dynamic energy source monitoring platforms that are similar to "Cornell Energy Dashboard"; moreover, their platforms are featured with energy subindex measurement, which can provide beneficial guidance for energy saving.

5.4 The Campus Is Regarded as an Experimental Field of Regional Sustainable Development

Having integrated scientific research, education, planning, operation, citizen participation, and other fields, colleges and universities are playing as an experimental field of sustainable low-carbon development.

1. Promote green transformation of communities surrounding the campus by based on green campus

   A college or university is a smart hub of a city, while middle schools and elementary schools are distributed among urban communities according to a fixed radius. The practice of regarding green campus as a practice base and displaying vivid examples will give impetus to students, parents, and community residents, therefore facilitating the green transformation of surrounding communities with half efforts (Fig. 4).

   In the level of macroplanning, campus should correspond to the relationship between the development of nature, society, and historical culture of a city and

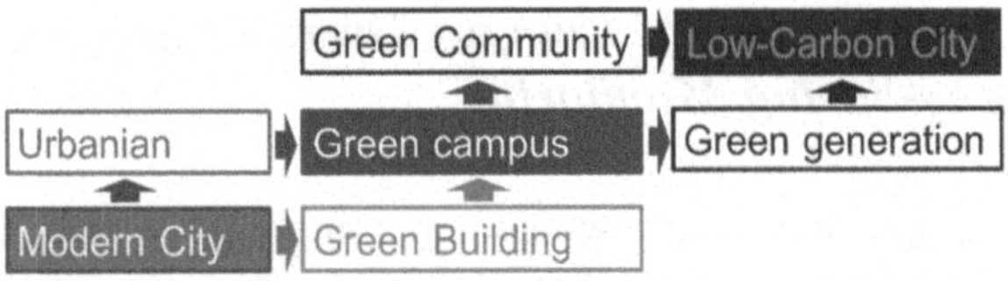

Fig. 4 Campus as the transitional scale between city and single building

integrated land use. Currently, newly constructed campuses of many colleges and universities have significantly driven the development of their surrounding areas. Furthermore, coordination between campus location and general urban planning goals is good for colleges and universities to rely on basic natural environment resources and social environment resources of the city; also, it is good for the city to achieve better development by relying on economic and social benefits from aggregation of colleges and universities. As with current campus reconstruction projects, however, studies should focus on means of using renewable resources and lowering energy consumption, as well as the interaction between the campus and surrounding communities.

2. Strengthen public participation and promote the interactive relationship among schools, enterprises, and society

   Multiple-subject and multiple-level cooperation and communication among schools, enterprises, and society should be strengthened. In the United States, an integrated and rational interactive relationship among schools, enterprises, and society has been established in green colleges and universities. Colleges and universities should cooperate actively with the government and seek for policy and financial support from the government; also, they should cooperate with enterprises so as to know current market demands, transform the latest green scientific research results of colleges and universities into social productive forces, and better serve the development of enterprises and society. Contemporary colleges and universities should be viewed as part of the society, serving the society and being responsible for social environment. However, China's traditional education philosophy has separated schools from enterprises and society, which in fact is wasting educational resources. Therefore, the unified and interactive idea learning among schools, enterprises, and society in American green campuses is something worth learning and imitating by China for constructing green campuses.

6 Conclusion

From perspective of ecology and low-carbon emission, campus ecological structure is functioning as a connecting link between the proceeding [urban (suburb) ecological system] and the following (campus ecological nodes). Hence, considering the three levels [building–community (campus)–city] of green campus planning from a general and systematic perspective is significant for both the campus and the whole urban ecological system.

It can be concluded from the experience of Cornell University on constructing green campus that green campus has become an experimental field and demonstrative area of low-carbon city construction. When creating a natural, comfortable, and safe space for teachers and students to study, live, and contact, green campus has realized sustainable development of colleges and universities. Meanwhile, it has promoted green transformation of communities around the campus and finally applied the results of green campus in a new round of developing and studying low-carbon cities.

References

Cornell University (2013) Climate action plan update and roadmap 2014–2015, 2014

Gan J, Hu Y, Wang M, Liu Z (2012) Enlightenment form green campus construction in U.S. In: Proceedings of the 8th international conference on green and energy-efficient building, vol 7

Han Y (2011) Experience and enlightenment of the construction of "Green University" in the United States. Univ Acad 03:67–72, 66

Tan H, Chen S, Zhang N (2012) Chinese green university report (6) 2011. CGUN, Shanghai

Tan H (2013) Development and thoughts of China green campus. Constr Sci Technol 12:25–29

Wu Z, Zisong W (2013) Introduction of green campus evaluation criteria. Constr Sci Technol 12:20–21, 24

Zeng Q (2007) Ecological network construction on new campus planning. Huaqiao University, China

Zhang J, Zhang Y (2011) Energy conservation actions and measures in American University Campus. Meitan High Educ 01:97–100

The SWOT Analysis and Path Selection of the Construction of Low-Carbon City in Our Country

Ting-xiu Liu, Jian-ying Niu and Jin-sheng Zhou

Abstract City is the important carrier for creating material wealth. However, it is the major consumer of energy and major emitters of greenhouse gases. So low-carbon cities and ecological civilization have become the strategic choices for sustainable development in our economic society. Starting from the concept of low-carbon city, this text uses SWOT model to analyze the low-carbon city and puts forward the new path for the low-carbon city construction with Chinese characteristics.

Keywords Constructional path · Low-carbon city · Sustainable development · SWOT analysis

1 Introduction

In 2005,the hurricane Katrina made landfall in Louisiana and Mississippi which caused tens of thousands of houses submerged and hundreds of thousands of homes powered off. In 2009, the heat wave swept India, and the temperatures in the capital New Delhi are as high as 48 °C which killed at least 100 people. In 2012, the lowest temperature for over 100 years and the biggest snowfall for decades occurred in

T. Liu (✉) · J. Niu
School of Humanities and Economic Management, China University of Geosciences, Beijing 100083, People's Republic of China
e-mail: 609024096@qq.com

J. Zhou
School of Land Science and Technology, China University of Geosciences, Beijing 100083, People's Republic of China
e-mail: 656525822@qq.com

S. Feng et al. (eds.), *Low-carbon City and New-type Urbanization*,
Environmental Science and Engineering, DOI 10.1007/978-3-662-45969-0_20

parts of the Europe which caused more than 500 people dead. The frequent occurrence of climatic anomaly draws the extensive concern of various countries around the world. In 2005, the Kyoto Protocol took effect. In 2007, the United Nations Framework Convention on Climate Change (UNCCC) drew up the "Bali Roadmap" to respond the climate change. In September 2013, IPCC announced that the average temperature of 2003–2012 increased by 0.78 °C than 1850–1990; sea level rises faster. During 1901–2010, the global average sea level rose by 0.19 m, 1.7 mm per year on average. The average rate reached to 2.0 mm per year during 1971–2010; the average rate reached to 3.2 mm during 1993–2010; gas concentrations of greenhouse gases in the atmosphere rose. In 2011, the gas concentration of CO_2, CH_4, and N_2O reached to 391 ppm, 1,803, and 324 ppb, respectively, 40, 150, and 20 % higher than pre-industrialization.

As a regional economic center, transportation center, and information center, more than half of the population in the world gather in the cities which create 85 % of global GDP. However, at the same time, city consumes 75 % of global energy, discharges 80 % of greenhouse gas, and causes waste of resources, deterioration of air quality, unreasonable layout, traffic congestion, and other issues. Therefore, solving the urban problem, constructing the low-carbon city, and taking the road of low-carbon development have become an international trend. As a great power in global energy production and consumption, China should respond to the international call, take the responsibility, make use of the international opportunity, and construct low-carbon cities to realize the sustainable development in cities (Liu 2009).

2 The Concept and Characteristics of Low-Carbon City

The concept of "low carbon" proposed under the background of accelerated urbanization process responses to global climate change and serious resource waste. In 2003, Britain officially puts forward the concept of "low-carbon economy" for the first time in energy white paper *the future energy—create a low-carbon economy*. Then, Japan, the USA, Sweden, Denmark, and other countries begin commit to the construction of low-carbon society. Scholars from our country also put forward their views from different aspects on "low-carbon city" (Li 2010).

The first director of the representative office in UNEP, Xia Kunbao, believes that low-carbon city means to develop urban low-carbon economy, build up a resource-conserving and environment-friendly society, and construct benign and sustainable energy ecological system through the low-carbon production and low carbon consumption. Scholars such as Fu Yun and Wang Yunlin believe that the low-carbon city can be developed by changing the social economic operation mode of mass production, mass consumption, and mass wastes, optimizing the energy structure, conserving energy, reducing emissions, recycling, reducing greenhouse gas emissions as possible as can, and finally realizing clean, efficient, low carbon, and sustainable development of the city (Fu et al. 2008). Zhuang Guiyang considers that low-carbon city is a city that the government regards the construction of low-carbon society as the blueprint, the enterprises take the development of low-carbon industries and low carbon production as the leading pattern, and citizens insist on concept and behavior characteristics of low-carbon lifestyle. Huang Tao and Feng Hui think that low-carbon city is the important spatial carrier for developing circular economy which benefits to carbon emission reduction and is the major means and important way to construct a low-carbon city. Scholars such as Song Ding think that low-carbon city is to change the original mode that regards human as the single center, single subject, and single value; to set up a mode of double centers, double subjects, and double value between human beings and nature; and to realize the harmonious development of man and nature and a new urban form of sustainable development. WWF defined that low-carbon city refers to the city which can maintain energy consumption and carbon dioxide release at a lower level under the premise of rapid economic development.

Based on the views of experts and my own reflection on low-carbon city, this paper puts forward that without sacrificing economic development, low-carbon city is a new developing mode for city to develop low-carbon economy as the strategy; to take economy, safety, systematicness, dynamic, and regional as the characteristics; to innovate low-carbon technology as the support; to change the mode of production and lifestyle as the means; to construct low carbon energy, low-carbon buildings, and low-carbon transport as content; to minimize urban greenhouse gas emissions; and to make economy, society, and environment to realize synchronous development. Economy refers to saving costs and producing huge economic benefit. Safety refers to reduce emissions of greenhouse gases such as CO_2, to reduce the pressure of ecosystem, to protect human's survival environment, and to produce good social benefits and ecological benefits by developing the industry of low consumption and pollution. Systematicness means that to construct a low-carbon city is a complex systematic project which needs the participation of government, enterprises, and citizens. Dynamic means the construction of low-carbon cities is not just limited to the script for urban planning. It needs a good practice and perfect planning in the implementation process. Regional means low-carbon city is influenced by the geographic position, natural resources, and regional culture of a city. In that case, in the process of construction should we consider the local actual situation, respect natural law, give play to our own advantages, adhere to the urban character, rather than blindly follow others and lose our own characteristics (Dai 2009).

3 SWOT Analysis of the Construction of the Low-Carbon City

Currently, our country adheres to the resource conservation and environmental protection as the basic national policies, exerting to promote green development, cyclic development, and low-carbon development. Constructing low-carbon cities has become a major reform of the urban outlook on development and is an important orientation of the future urban development. This article analyzes the construction of low-carbon city in our country from the four angles of the SWOT theory.

3.1 Strengths

3.1.1 Superior Natural Environment

With a vast territory of our country, the climate, hydrology, geology, soil, and vegetation present different characteristics in various regions and produce different urban landscapes, architectural forms, and architectural culture. Such as the architectures in Anhui Province, the streets distribute properly. Row upon row of houses and country yards reflect the Anhui's landscapes. The natural landscapes and human landscapes are in harmony, and the Huizhou culture is reflected incisively and vividly. For another example, Hainan Province's climate is rainforest climate which is mild and rainy. This climate fits for growing forests or bamboo. Therefore, the architectures are usually constructed by wooden supports which are beneficial to aeration cooling, rainproof, and water draining. Besides, the charming architectural style attracts the tourists from all over the world to visit. In a word, the rich natural environment in our country makes the building space and architectural style different. Adjusting measures to local conditions, constructing houses according to the land and the unity of nature and human, and displaying the unique style of the city personality are so important.

3.1.2 Rich Natural Resources

Our country is an integral part of the world's energy market, and sustained growth of energy offers important support for the urban construction and development. First of all, the broad geographical area in our country and all kinds of rich land resources provide a broad space for city construction. Secondly, from the climatic resources, most areas in China are located in the medium and low latitude area. Even if the residents live in the most north latitude or the lowest floor can also meet the basic needs of sunlight by adjusting the height of the building and the separation distance between the buildings. Thirdly, the gross quantity of mineral resources is large in our country. Raw coal production reached to 3.65 billion tons in 2012, and among which, the output of crude oil reached to 207.48 million tons and the output of natural gas reached to 107.2 billion cubic meters. The 12th Five-Year Plan on Natural Gas Development proposed clearly that by 2015, our country's natural gas supply capacity will reach to 2015 cubic meters or so, among which, the conventional natural gas will reach to about 138.5 billion cubic meters, coal gas will reach to about 15–18 billion cubic meters, the development and production of coalbed methane will reach to about 16 billion cubic meters, the population who use the natural gas in the city and county will account for about 18 % of the total population.

3.2 Weaknesses

3.2.1 Unreasonable Spatial Arrangement

Urban spatial layout can reflect the use of land of a place, and a reasonable urban spatial layout can not only save the land but also benefit for carbon emission. In the procession of the urban construction, some cities choose the spreading structure to build the dwelling district, recreational areas, and working area dispersedly which increase urban population flow and bring the city transportation with huge pressures. Some cities' industrial parks occupy the areas with large population which not only wastes land resources but also produces wastewater, waste gas, and solid waste during the production. It pollutes the urban environment seriously and is not beneficial for the construction of low-carbon city (Chen 2009).

3.2.2 Unadvanced Technological Level

Low-carbon technology is an important way for cities to abandon the high-carbon technology pattern and to achieve the leap development of the economy. Besides, it is also the key to improve the urban core competence. There are some disadvantages in our country's technology area; the development of clean coal combustion technology, solar photovoltaic power generation, large wind turbine, and so on are lack of independent research and development and innovation and mainly depend on the introduction of foreign advanced technology and equipment. The governmental investment in science and technology is not enough. Nowadays, the proportion of research and development spending in GDP is less than 1.5 %, and there is a big gap compared to the 3 % of the world's leading country's level (Jenny and Will 2008). Enterprises pursue the short-term interests and economic interests and weigh the relationship between the input and output. Technology research, development, and the update means the increase of investment and cost, and enterprises will still choose the old production technology under the condition of the blind result which will limit enterprise's independent research level of technology. In addition to, as a research personnel training base, advanced technology research base, colleges and universities failed to play its core role, and the result of the research cannot be effectively applied to practice. And the integration of industry, university, and research has no good implementation.

3.2.3 Imperfect Policy of the Law

At present, our country enacts a series of policies of the law on environmental protection (Cai and Wang 2007). Circular economy promotion law of the People's Republic of China establishes the framework of law and policy for constructing low-carbon cities. And it provides legal protection for promoting the transformation of our country's mode of economic development model. National assessment report of climate change points out that our country should gradually establish systems and mechanisms of slowing down the climate change to reduce carbon emissions and to provide a good legal and policy environment for our country's construction of low-carbon city. It is very important for the legal protection of low-carbon city construction in our country. But the legal system about promoting the construction of low-carbon city is not perfect in our country and mainly presents the following: Firstly, no law is directly linked to the construction of low-carbon city and no law can make sure the goal of development and basic measure of low-carbon city fundamentally and broadly. Secondly, the law about the areas such as oil, natural gas, nuclear power, and wind power is still blank. Thirdly, the operability of some laws is bad. Among the laws and regulations about environmental protection in our country, only water pollution prevention and control law and Atmospheric Pollution Prevention Law have some clauses about the total quantity control of pollutant emission which will not be able to effectively support the work of monitoring and controlling the total quantity of pollutant emission.

3.3 Opportunities

3.3.1 The Powerful Drive by International

Internationally, many countries have reached a consensus to set up a low-carbon concept, construct low-carbon cities, and develop a series of action laws and development strategies. In 2008, eight countries in G8 summit indicated that they would seek the cooperation of other signatory of UNCCC and together regard a 50 % reduction of the global greenhouse gas emissions as a long-term goal. Britain develops clean energy technology market and encourages generating power by renewable energy; Sweden regards the disposal and utilization of waste as one of the core of low-carbon city construction (Cai 2009). They use sludge to produce marsh gas, use wastewater to irrigate the nursery garden, and use rubbish to supply power and heat. They focus on the development of renewable energy, vigorously develop and utilize hydropower, solar energy, biomass energy, and other new energy, and make every effort to get rid of the dependence on fossil fuels thoroughly by 2020. Those experiences and measures are worthy for our reference.

3.3.2 Enough Attention Paid by Our Country

In 2008, a pilot development project about low-carbon city was launched in Shanghai and Baoding. It puts low-carbon economy and low-carbon society as the development direction. Shanghai selects the pilot construction from office buildings, hotels, shopping malls, and other large commercial buildings, publicizing the energy consumption, carrying out the energy audit, and searching for ways to improve the energy efficiency in large buildings. Baoding carries out six key projects such as the China Power Valley construction project, solar city construction project, urban ecological environment construction project, low-carbon operation in office building demonstration project, low-carbon community demonstration project, and the low-carbon city traffic system integration to promote low-carbon city construction. Afterward, the practice of low-carbon city enters into a rapid development stage. Guiyang, Hangzhou, Wuxi, Dezhou, Nanchang, Jilin, Zhuhai, Xiamen, and other cities have put forward the conception of low-carbon city. The obvious results they have achieved will be more conducive to promote the construction of low-carbon city carried out in the whole country (Bi 2009).

3.4 Threats

3.4.1 The Difficulties in the Transformation of Traditional Industries

Viewed from the industrial structure, the secondary industry acts as the second pillar industry for economic growth, whose output has accounted for about 45 % of GDP. Even if the total value of the tertiary industry in 2013 was slightly higher than the secondary industrial production for the first time, it still cannot change the dominance of the secondary industry (Table 1). In the aspect of the internal industrial structure, the six high-energy-consuming industries have more than 9 % growth over the previous year from 2009 to 2013. Even though the growth rate of them has slowed down in recent two years, the proportion of high energy consumption industry is still relatively large, compared with the low proportion of high technological content, high added value, but low energy consumption industries (Fig. 1). China has focused on the development of labor-intensive and energy-intensive industries for a long time and has become the world's largest

Table 1 The proportion of three main industrial output values accounted for GDP in China over the years

Proportion / Year / Industry	1990	2000	2010	2011	2012	2013
primary industry	27.2	15	10.2	10	10.1	10
The secondary industry	41.3	45.9	46.8	46.6	45.3	43.9
The third industry	31.5	39.1	43	43.4	44.6	46.1

Source 2010–2013 statistical bullets for national economic and social development

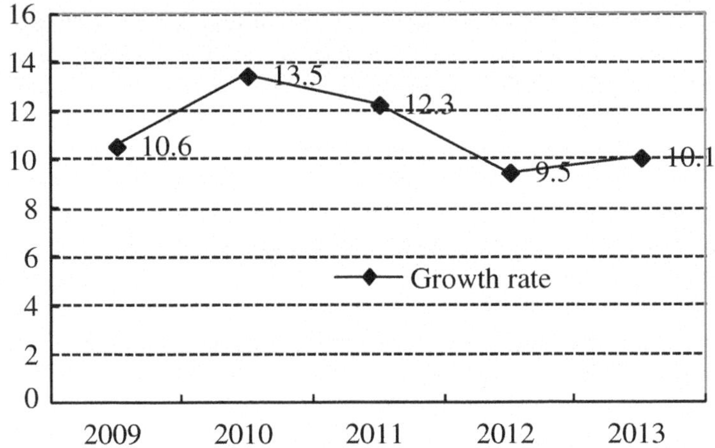

Fig. 1 Growth rate of energy-intensive industries in China over the previous year. *Source* 2010–2013 statistical bullet in for national economic and social development

manufacturing base, but it has also leaved the high consumption, high emission, and high pollution for itself in the course of export products. Industrial structure is an important determinant of economic growth mode, because it can adjust and optimize the industrial structure, as well as realize the low-carbon development which has become a major challenge for building low-carbon cities.

3.4.2 The Unreasonable Energy Consumption

Firstly, in terms of energy consumption structure, the total energy consumption in China was 37.5 tons of standard coal in 2013, an increase of 3.7 % over 2012, consumption of crude oil increased by 3.4 %, natural gas consumption increased by 13.0 %, and electricity consumption increased by 7.5 %. In the electric power industry, water and electricity have accounted for only about 20 %, and thermal power has accounted for over 77 %. Energy consumption depends on coal for more than 70 %, and high carbon became the absolute dominance (Wu 2010). As for the hydropower, nuclear power, wind power, solar energy, and other new energies with cleanness and persistence, these are lack of attention utilization and promotion. Secondly, in terms of energy consumption, as the biggest nation of steel, coal, and cement consumption in the world, our country's unit energy consumption is 3.1 times of the world average, 7.6 times of Japan, 4.2 times of the USA, and 1.5 times of India (Li and Lv 2011). The seriously unreasonable energy consumption is another major challenge in the construction of low-carbon cities.

3.4.3 The Complicated Balancing Act

Building a low-carbon city is a long-term process. A low-carbon urban construction planning with high objectivity and operability takes 20 years to be significantly effective, far beyond the tenure of normal local officials. On the basis of the balance between long-term interests and the immediate interests, officials are more insistent upon image and vanity projects, as well as promotion concept, unwilling to focus on works which cannot see the benefits in short term. So the construction of low-carbon city is always shelved. Besides, our country has been taken economic situation as a principal element to measure the local strength and competitiveness, focusing on the dominant position of the secondary industry, at the expense of the environment for the growth of GDP, neglecting industrial transformation and environmental protection. Even if some places are committed to the development of new industries to change the past mode of high-carbon development, they still pay more attention on building the low-carbon industrial park such of the vanity projects, which cannot achieve the city's low-carbon industrial development fundamentally. Therefore, immediate and long-term interests, economic growth, and environmental protection, quotas among various industries have become the serious challenge to the construction of low-carbon city (Chen and Chu 2009).

4 Path Selection of Low-Carbon City Construction

Faced with the current situation of excessive resource consumption, serious environmental pollution, and ecosystem degradation in our country, building low-carbon cities and taking the low-carbon development path becomes an inevitable choice for China. Strengthening the construction of ecological civilization and building a Wild China, exploring a unique low-carbon urban construction path with multi-stakeholder involvement is the reliable guarantee to enhance the international image and to enhance the comprehensive national strength.

4.1 The Government Leading

4.1.1 To Be Based on Urban Planning and to Grasp the Overall Layout

Urban planning emphasizes balance among the economic, social and ecological, including urban industry, construction, traffic, and forest planning. In the respect of urban industrial planning, it aims at reducing the pace of development of high-carbon industries, accelerating economic restructuring. In respect of urban architecture planning, it aims to save land resources and improve land use efficiency (Yao et al. 2006). In the respect of urban transport planning, the rapid growth of car ownership in China increased from 24 to 137 million in the decade of 2003–2013, which brought enormous pressure to the urban transport and the environment. National implementation is a low-carbon transportation planning led by public transportation, with bicycles and walking as avocations, focusing on the development of rail transportation (see Fig. 2). In the respect of urban forest planning, it is good to build more ecological landscape rather than cement squares, to set limits on building the high-energy artificial waterfalls and fountains, to develop vertical greening, and to mitigate urban heat island effect and increase the capacity of carbon absorption (Pan et al. 2008).

4.1.2 To Perfect the Legal System and to Strengthen the Financial Support

In respect of legal construction, although many laws and policies have carried already, there still are existing defects as legal gaps which need us to improve the construction of laws and regulations. At present, the National Development and

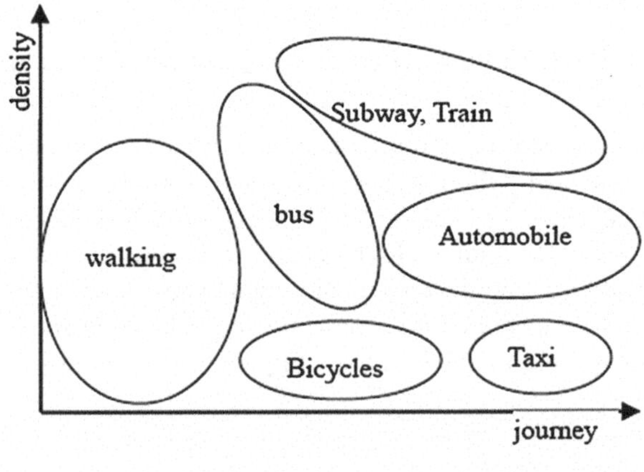

Fig. 2 The planning chart of low-carbon cities' sustainable transport development

Reform Commission, the NPC Environmental and Resources Protection Committee, the State Council Legislative Affairs Office, and other departments jointly set up a leading group to complete the drafting on climate change in order to further promote low-carbon urban construction.

In the respect of system construction, firstly, rebuild cadre's performance appraisal system. A lot of places in our country assess cadres focusing on economic growth, ignoring energy conservation, environmental protection, etc. This requires the optimization of assessment indicators and the improvement of the supervision mechanism, by refining the indicators, setting clear targets, as well as responsibilities for each unit (Cai and Wang 2007). Secondly, formulate government procurement details of low-carbon products and list enterprises of low energy consumption and less pollution onto government's prior procurement catalog. Thirdly, establish a strict industry access system and open up green channels for projects of new technology, high value added, low consumption, and low emissions, and strictly set limits on investments of high pollution and high-energy projects.

In the respect of financial support, use revenue, price regulation, fiscal subsidies, and other methods to guide enterprises' development direction and public consumption direction. Take the following event for instance: direct state subsidy for those high-grade green building which have more than 60 % energy saving makes extensive use of renewable energy in the construction industry. Another example is funding more than RMB 30 billion to promote nearly more than 90 million sets of energy-saving appliances, and more than 3.5 million fuel-efficient cars drives the public to the low-carbon consumption.

4.1.3 To Pay Attention on Self-Practice and to Broaden Communication Channels

The government should carry out e-government affairs and use modern information technology such as computer networks and communications to gradually realize the paperless office and form an efficient, honest, concise, and fair government operation mode. As for the others, government broadcasts low-carbon concepts, develops low-carbon education, and guides the low carbon consumption through television, radio, newspapers, network, and other media. Schools guide teenagers to cultivate good living habits and establish a healthy lifestyle.

4.2 Enterprise Leading

As the emitters of CO_2, the main factors causing environmental pollution, enterprises are the backbone of low-carbon urban construction, which should be improved from the following aspects: firstly, transfer enterprises managing goals.

Most enterprises take profit maximization as the goal, pursuing immediate not long-term interests, pursuing economic interests, and ignoring the environment. Companies should realize that protecting the environment is their disposable corporate duty, corporate responsibility, and social responsibility, and they should combine environmental protection with profits creating. Secondly, transfer enterprises managing models. The past business model of enterprises is "high pollution, high energy consumption but low output and low efficiency" which not only is the high cost of production, but also causes great damage to the environment. Now through the development of clean technologies, reducing dependence on traditional fossil energy sources, improving the "carbon productivity" (Ma 2009), and using modern technology and modern management methods promote enterprise transformation spanning development. Thirdly, transfer enterprise market and products. Enterprises should seize opportunities bought by low-carbon economy to exploit low-carbon commodity market, for example, the production of energy-saving lamps, night-lights, chopsticks, LED, TV, handkerchiefs, rechargeable batteries, bags, and other low-carbon goods. It was estimated that here would be nearly 8.5 million people engaging in wind energy, solar energy, and other areas of work in 2030. The engagement of low-carbon industries and the production of low-carbon products become a new driving force to promote enterprise development and urban development.

4.3 Organizational Assistance

Society organizations play an important cooperative role in building a low-carbon city. For example, financial institutions establish "energy saving, environmental protection, green finance," and other concepts and further open green credit projects to those industries and fields meeting low-carbon requirements, which not only enhance the image of the organization and gain commercial interests, but also achieve the goal of reduction of CO_2 emission 213.731 million tons. All of them can be said to serve multiple purposes. Another example is China Low Carbon Economic Union (CLEA), a national non-profit cooperative organization integrating technicality, association, speciality, and platform. Over the years, the organization has set their aims at spreading the concept of low-carbon, advocating low-carbon life, building low-carbon cities, and developing low-carbon economy. It provides a reference for the government in enacting policies, regulations, and industrial standards of the development of the low-carbon economy regulations. It assists the government's guidance and supervision in low-carbon urban construction and development of low-carbon industries. In addition, universities and research institutes attach great importance to research, scientific, and technological talents training and the improvement of independent innovation capability. With major scientific projects as a link, they make the priority of development of clean energy, development, and utilization of renewable energy and other core technologies to realize the effective sharing of information of the low-carbon technology research and development, equipment, platforms, and enterprises in order to provide technical support and personnel support for building the low-carbon cities.

4.4 Public Participation

The public acts as a social subject, and building a low-carbon city is not only the government's responsibility, but also the social responsibility to individual, who should do the following things. First, fulfill the obligations of low-carbon construction. For example, everyone should contribute to the greening of our country as a citizen in the annual March 12 which is called "voluntary tree planting day." Another example is the Car-Free Day on September 22 each year. The citizens should actively advocate Car-Free Day and take public transportation, changing the car-based transport modes. (Cui 2009). Second, establish a concept of low-carbon consumption. Acting as a type of value and culture, carbon consumption concept promotes to the spirit of frugality and opposes "face consumption," "luxury consumption," and "one-off consumption" (Liu 2012). Citizens should establish a correct concept of consumption, in order to guide their consumption patterns and lifestyles. Third, cultivate a low-carbon lifestyle. According to statistics, if nationwide household widely adopt saving lamp, there would save more than 70 billion KWH a year; if the existing more than 100 million domestic refrigerators replaced all energy efficient, it would save more than 40 billion KWH a year. The two combined could save more than the amount of electricity produced by the Three Gorges Power Station (Zou 2008). This shows that even the subtle lifestyle such as purchase of low-carbon electrical appliances, low-rise stairs climbing, taking along with shopping bags, a multi-purpose water and so on will have a huge impact on society. Every society is a separate entity, as long as every citizen can start small, start now, and start bit by bit, and the power will be condensed into large torrent to promote the building process of China's low-carbon society and low-carbon city.

5 Conclusion

Building low-carbon cities is the important ways of city orderly and efficient operation. In the process of building the low-carbon city, each region should follow the foundation of regional and national development strategies, giving full play to the government's predominant role, enterprise's leading role, organization's assistance, the roles of the participation of citizens, as well as the international reference, combining with the characteristics of itself in order to blaze a new path which takes building a low-carbon society as its blueprint, makes companies with low-carbon economy as its goal, and makes citizens with low-carbon concept as a guide. Only learning from foreign advanced experience and seeking the cooperation opportunity can we better and faster realize the economic growth, energy security, environmental protection, can the "3E" target have synchronous development, coordinated development and sustainable development.

References

Bi J (2009) The post-crisis era, the path of low-carbon city construction in our country. Nanjing Soc Sci 11

Cai L (2009) Low carbon economy: the green revolution and global innovation competition pattern. Economic Science Press, Beijing

Cai W, Wang S (2007) A tentative study on the predicament and countermeasure of our nation's environmental enforcement. J Yunnan Adm Coll 2

Chen Q (2009) The planning ideas on our country's construction of low carbon city. Mod City Res 11

Chen F, Chu D (2009) Confirmation of connotation, model and targeting strategies to the low carbon city research. Urban Plan Forum 4

Dai Y (2009) The concept evolution and first exploration of measure about low carbon city development. Mod City Res 11

Fu Y, Wang Y, Li (2008) The research on development path of Low carbon city. Impact Sci Soc 2

Jenny C, Will F (2008) A low-carbon future: spatial planning s role in enhancing technological innovation in the built environment. Energy Policy 12:23–25

Li X (2010) The global climate change rules and its effects for the world economy. Truth Seeking 4

Li F, Lv Y (2011) Economic transition and the rise of low carbon economy. National School of Administration Press, Beijing

Liu Z (2009) The necessity and policy suggestions of low carbon economy development in China. J Party Sch CPC Cent Committee 6

Liu B (2012) Study on the urban low-carbon economic assessment based on subjective and objective factors. China Economic Publishing House, Beijing, p 45

Ma X (2009) The energy strategies of low-carbon city construction. Urban Build 12:74–75

Pan H, Tang Y, Wu J, Lu Y, Zhang Y (2008) Spatial planning strategy for 'low carbon cities' in China. Urban Plan J 6:57–64

Wu H (2010) Path exploration of the industrial policy in Chinese high-carbon industry transferring into low-carbon industry. Soc Sci 10:27–34

Yao R, Koen S, Li B (2006) Sustainable urban and architecture design. China Building Industry Press, Beijing, p 165

Cui Y (2009) Study on the low-carbon city development mode based on fast public traffic. China Popul 19:11–16

Zou W (2008) Enormous potential in household energy conservation and emission reduction. Contemp Guangxi 7:59

Regional Power Production and Consumption Forecast Solutions

Ting-yu Gao, Xiao-fei Yang, Cheng-gang Cui, Zhi-cong Zhu, Wei-jun Liu, Shi-feng Wang and Dan Liu

Abstract There are three energy modes in Shanghai Advanced Research Institute-CAS: power production systems (solar power, wind power, combined cooling, heating and power, solid oxide fuel cell), power storage systems, and power consumption unit (electric vehicles, laboratory equipment, computer room, canteen, elevators, air conditioning systems, etc.). Power forecast for production and consumption accurately can improve operational safety and efficiency of power networks and equipment and reducing gas and electricity consumption, is a foundation for power production and storage, and used to optimize scheduling decisions and to ensure the safety of energy systems and economic and rational operation. Solution in the park combined energy production, storage, and consumption of the entire supply chain. Power production and consumption forecast system architecture, functional division, predicted core processes, and other aspects are given in regional power production and consumption forecast solutions. It has integrated into the energy management system in the Shanghai Zhangjiang Hi-Tech Park in Pudong through demonstration and application, verifying its effectiveness.

Keywords Regional power · Power production and consumption forecast · Forecast solution · Power load forecast

1 Introduction

Economic development of the energy consumption is increasing contradiction between dwindling energy reserves which are becoming increasingly serious energy crisis threatening the world economy. China is in the full period of industrialization,

T. Gao (✉) · X. Yang · C. Cui · Z. Zhu · W. Liu · S. Wang · D. Liu
Cultural Heritage Science Conservation Laboratory, Shanghai Advanced Research Institute-CAS, No. 99 Haike Road, Pudong District, Shanghai 201210, People's Republic of China
e-mail: gaoty@sari.ac.cn

S. Feng et al. (eds.), *Low-carbon City and New-type Urbanization*,
Environmental Science and Engineering, DOI 10.1007/978-3-662-45969-0_21

rapid economic development, and low-energy efficiency in recent years, and appearing in many areas with varying degrees of "three shortages" (electricity shortage, coal shortage, and oil shortage) is a manifestation of this contradiction (Sun et al. 2005). However, in the global warming trend, Western countries have introduced low-carbon economic development strategies and policies. These policy measures are mainly given as follows: the transformation of traditional high-carbon industries, strengthening the low-carbon technology innovation, positive development of renewable energy and new clean energy, application of market mechanisms and economic leverage to promote corporate carbon reduction, and strengthening the international carbon reduction collaboration (Ren 2009).

In recent years, Smart Grid has become the hot topic on the development trend of the future grid at home and abroad (Hu 2009). Institute for Advanced Study in the Shanghai park has formed a certain scale micro-grid: wind and solar power generation systems, distributed energy supply systems, fuel cell power generation systems, energy storage systems, ground source heat pump ice storage system, electricity, and other buildings and laboratory unit, and in order to better support the park energy, energy efficiency, and the use of thermal energy production and consumption, it is necessary to build an efficient energy forecasting solution, and the program based on BP neural networks, support vector machines, gray system, and other predictive models seamlessly combines historical data acquisition, historical data identified abandoned pseudo-real deal, and curve statistical analysis of historical data to assess the forecasting model to predict the results of the correction into a single package. The subsequent solution referred to as the energy forecasting system software package.

2 System Architecture

The system architecture of the energy forecasting system software package consists of logic architecture and physical architecture. It is described as follows.

2.1 Logic Architecture

Logic architecture describes functions, lays relation, modules interfaces, etc. The details of the above references are presented in Fig. 1.

(1) Basis data layer
The energy forecasting system software package involves various types of data such as load power, meteorology, economic, social policy, and so on. So the database needs to create, use, organize, and manage variety of data, and support to query variety data. The database provided the real-time data simultaneously for online adaptive rolling forecast on forecast model.

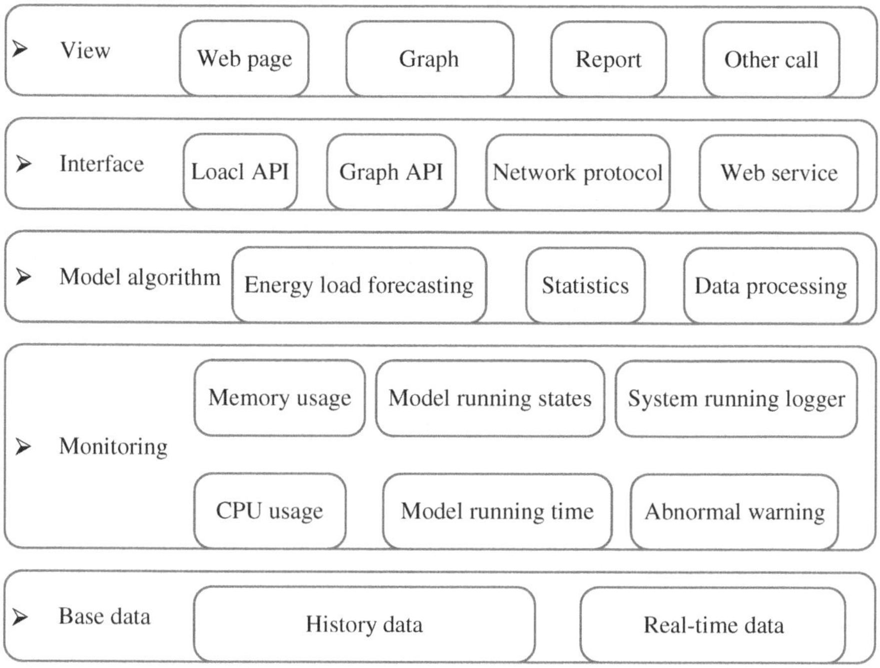

Fig. 1 Energy forecasting system logic architecture

(2) System monitor layer
Variety of models and algorithms have been used in the energy forecasting system software package. It is more important to surveillance the system status, for example, model run state, modules logger, run time, abnormal values, memory usage, and CPU usage.

(3) Model and algorithm layer
In order to adapt to power load forecast for each time period (super short term, short term, medium, and long term), more forecast models have been supported in software. And some algorithm has been designed on statistics to the period history data and its vacancy to fill and identify exception.

(4) Development interface layer
The energy forecasting system software package has been more flexible and scalable, and some interfaces include local application programmer interface (API) (save, printer, report forms), graph API (tables, curve points chart, bar graph), network protocol interface file transfer protocol (FTP), remote method invocation (RMI), and Web service interface.

(5) Application view layer
The energy forecasting system software package will provide graphical user interface, forecast result chart, graph, report form, and third-party call interface.

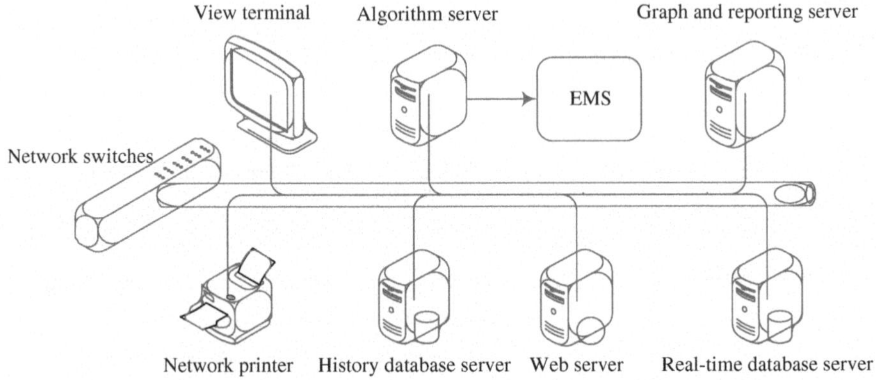

Fig. 2 Energy forecasting system physical architecture

2.2 Physical Architecture

The energy forecasting system software package has been deployment on browser system. Figure 2 is a description of networks' topological structure on necessary devices on local networks.

3 System Function

3.1 Data Configuration

Data configuration means the parameter configuration between the energy forecasting system software package and energy history data sets. The parameters have been supported by the types of data source, class, time period, and so on. By these parameters, the energy history data sets can be imported easily into the energy forecasting system software package.

(1) Data source file
The energy forecasting system software package has been supported to import to flat file (suffix names txt, csv, excel) and database (Microsoft SQL Server 2008, Oracle 2010).

(2) Data source types
The energy forecasting system software package has been supported by types of data, such as types of devices power load, cool and heat load, meteorology, economic (electricity prices) and social policy (brownouts), and so on.

(3) Time range of data sources
Time units such as year, month, day, hour, and minutes are respectively corresponding to the forecast time stage such as the medium or long term, short term and super short term. Time range includes start date and end date.

3.2 Data Processing

As usual, the energy forecasting system software package has been provided two operations to history data on power load, cool/heat load, and power consumption on the following detailed description:

(1) Basic operations
The energy forecasting system software package has been supported basic data processing function, such as view, edit, find, save, create, delete, and order by date–time.

(2) Advanced operations
The energy forecasting system software package has been supported advanced data processing function, such as data validation and vacancy data recovery. The data validation will filter each energy data by some conditions and view in different colors or category display. Usually, abnormal data were validated by the boundary conditions such as greater than, equal to, greater than or equal, less than, less than or equal, and not equal.

3.3 Energy Load Forecast

Power load production and consumption forecast is the core function on the energy forecasting system software package. It is described as follows:

(1) Forecast aim
There are two functions: the energy production forecast and energy consumption forecast. Energy production prediction included producing electricity and heat production. Power generation equipment involves solar and wind. The heat production is the heat dissipation such as the heat consumption envelope structure (including walls, windows, roof, and floor), the new air cooling loads, lighting or body, and so on. Energy consumption prediction is divided into two types: power consumption and heat consumption forecast, where the main power consumption equipment has power sockets, lighting, and air conditioning. The heat consumption equipment mainly has building envelope, lighting equipment, and cold air infiltration amount.

(2) Short-term forecast model
The energy forecasting system software package includes some short-time forecast models, such as BP (back-propagation), neural network model (Fang 2011), SVM (support vector machine) regress model (Peng 2011), time series model (George et al. 2011), such as auto-regressive (AR), moving average (MA), auto-regressive and moving average model (ARMA), and chaos time series regress model (Bu et al. 2009; Zhang et al. 2004; Ren et al. 2010).

(3) Medium- and long-term forecast model
Gray system regress model (Liao et al. 2011) and fuzzy line regress model (Yu and Yang 1995) as medium- and long-term forecast model have been designed based on the energy forecasting system software package.
(4) Holidays forecast model
Linear interpolation forecast model (Kaotanzad et al. 1998) as holidays forecast model has been designed based on the energy forecasting system software package.

3.4 Index Access

The energy forecasting system software package has been supported to assert to the forecasting result. The detailed description has been provided on the following:

(1) Result content
There are displayed some result, such as absolute error, the total error, statistics, and so on.
(2) Output form
Output to support: local save, local printer, network printer, FTP publishing, intelligent mobile terminal publish, and e-mail release.
(3) Show form
The form of forecast results shows the user by tables and curve point map. The form of error results shows the user by curve points chart, bar graphs, and tables. The result of the statistical analysis shows the user by the column chart and form. Time series result shows the user by the autocorrelation coefficient and partial correlation coefficients. The result of the significant test supports the curve point map and the chart.

3.5 Statistical Analysis

The energy forecasting system software package has been supported by basic statistical analysis of energy production and energy consumption.

(1) Energy production statistics
The building heating load is statistic and analyzed according to the total value of building heating load in accordance with the year, month, week, day, hour, and peak, average, and supporting reports and charts. And it will be support reports and charts. The power consumption of the equipment (lighting, sockets, air conditioners, heat pumps, other) is statistic and analyzed according to the total power load value with the year, month, week, day, hour, and peak, average, supporting reports and charts.

(2) Energy consumption statistics
The building heating load statistics according to the total value of building heating load in accordance with the year, month, week, day, hour, and peak, average. And it will be support reports and charts. The power consumption of the equipment (lighting, sockets, air conditioners, heat pumps, other) according to the total power load value with the year, month, week, day, hour, and peak, average, and it will be support for reports and charts.

3.6 Information Support

The energy forecasting system software package has been supported by the assessment of the various functional modules such as performance prediction model.

(1) Performance consumption
The energy forecasting system software package will provide the occupancy of the memory and CPU in 24 h. And the time consuming of the module running in the system will be statistic, such as the configuration data, the data processing, the model training, and the prediction model, and the results show the data output.
(2) Model states
Some statistic values are recorded including the number of calling model, the largest number of cycles, and so on. And it will display the output such as saving, printing, publishing FTP, and e-mail functions.
(3) Exception
Abnormal operation generating on each module of the system has been monitored. And the result of it has been statistically reported by log files. Abnormal value has been output by saving, printing, publishing FTP, and e-mail functions.

4 Energy Load Forecast

The core functionality of the energy forecasting system software package presents on the energy load forecasting process (Fig. 3), including database interface, data processing methods, configuration prediction model, offline forecasting, real-time data interface, error analysis, online adaptive prediction scrolling function.

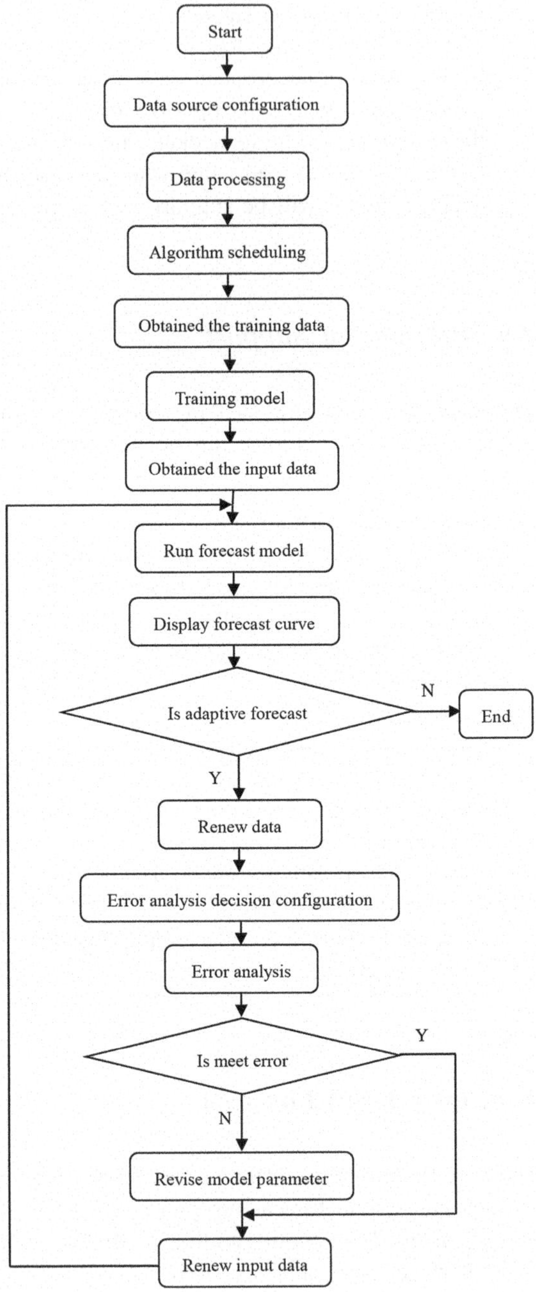

Fig. 3 Energy load forecasting software package flowchart

4.1 Data Source Configuration

This module requires the user to interact with the system through the interface and mainly configure the following parameters:

(1) Data source files (txt, csv, excel, various databases files).
(2) Types of data source (power, heat, weather and the economy).
(3) Time type of data sources (year, month, day, hour, minute), the time interval, and time ranges.

4.2 Data Processing

Data from data source need to be cleaned and corrected, so that the user required interacts with the software to configure the methods that process the data. There are two main functions of software.

(1) Basic functions such as data processing, data display, edit, modify, delete, add, sort, and search.
(2) Advanced data processing such as exception data validation and data processing vacancy.

4.3 Algorithm Scheduling

The parameters can be configuration for multi-user mode. Different users on the system can create multiple sets of parameters, and the same user on the system can create multiple sets of configuration parameters.

This module requires the user to interact with the system through the interface, and there are two functions: offline and online prediction scheduling parameters.

(1) Offline prediction scheduling parameters (Table 1).
(2) Online prediction scheduling parameters (Table 2).

4.4 Obtain the Train Data

The training data sets are defined in the offline scheduling algorithm configuration parameters.

Table 1 Offline scheduling algorithm configuration

Parameters name	Parameters description
Forecast type	Energy production, energy consume, metro
Forecast index	Power, heat, temperature
Forecast time type	Short term, mid-term, long term
Forecast models	BP neural network model, time series models, support vector machine model, linear regression models, time series, gray time series regression, fuzzy linear regression, linear correlation
Model parameters	Model parameters recode the structure of the model
Forecast steps	Forecast time window, short-term forecast range generally within 0–3 h in 15 min intervals, short-term forecasting at intervals of 0–3 days to hours, and long-term forecasts for weeks or months and years interval
Train data configuration	Extract the training data
Input data configuration	Extracting the input data
Max error	The maximum allowable training error

Table 2 Online scheduling algorithm configuration

Parameters name	Parameters description
Start numbers	Setting the starts numbers on each day
Start time	Setting the start times on each day
Error configuration	Setting the methods computing error of online forecasting
Maximum error	Setting the methods maximum error of online forecasting

4.5 Training Model

There are two steps to train model, one of it is to choose train data and the other is to learn by input and output on model.

4.6 Obtain the Input Data

The input data sets are defined in the offline scheduling algorithm configuration parameters.

4.7 Running Forecast Model

The forecast model will run according to the input data and the forecast model parameters. So the forecast result will be computed.

4.8 Index Assert

(1) The forecast curve shows the user by Web page.
(2) Some result shows the user such as the maximum error of certain period of time, the error curve, and the average error etc.
(3) Some results are output to the user by methods such as save, print, curves, charts, email, and publish to FTP servers.

4.9 Is Adaptive Forecast

If you start the adaptive prediction function, then you want to configure online algorithm scheduling parameters. Otherwise, nothing does it.

4.10 Renew Data

The new data will be got from the database and renew it to the energy forecasting system software package.

4.11 Error Analysis

Some mathematical formulas are given as follows:

A. Absolute error

$$\mathrm{AE} = \left|\hat{Y} - Y_i\right| \tag{1}$$

B. Relative error

$$\mathrm{RE} = \frac{\hat{Y} - Y_i}{Y_i} \times 100\,\% \tag{2}$$

C. Mean absolute error

$$\mathrm{MAE} = \frac{1}{M}\sum_{i=1}^{M}\left|\hat{Y} - Y_i\right| \tag{3}$$

D. Mean absolute percentage error

$$\mathrm{MAPE} = \left(\frac{1}{M}\sum_{i=1}^{M}\frac{\left|\hat{Y} - Y_i\right|}{Y_i}\right) \times 100\,\% \tag{4}$$

On the above formula, $\hat{Y}$ represents the predicted load, Y_i represents the actual load, and M represents the number of the prediction point.

4.12 Revise Model Parameter

The model revised according to error analysis result. Different model used different strategies.

4.13 Renew Input Data

The data will be renewed from database on the time windows, and it will be the input for next forecasting.

5 Conclusion

(1) Various types of energy within the Shanghai Institute for Advanced Studies Park need a solution. On the above, from the perspective of software engineering to optimize energy consumption to provide a predictive scheduling solution.
(2) The solution from system architecture, system function modules, and core processes in detail provided the process of forecasting energy generation and consumption in different timescales.
(3) The scheduling process with a predictive model is the robust predictions for different types of scalable and can be implemented, and the prediction model can be a arbitrary, flexible configuration.

Acknowledgments This work was supported in part by the Chinese Ministry of Housing and Urban-Rural Development under Grant 2013-K8-25 the Strategic Priority Research Program of the Chinese Academy of Sciences, Grant No. XDA06000000 and the Chinese Academy of Sciences Innovation Engineering under Grant KGCX2-EW-321.

References

Box GEP, Jenkins GM, Reinsel GC (2011) Times series analysis: forecasting and control. In: Wang C, You M, Hao Y (eds) Times series analysis: forecasting and control, 4th edn. China Machine Press, China

Bu Y, Wen G, Li H (2009) Nonlinear adaptive predictor for chaotic time series. J Comput Appl 29 (11):3158–3160

Fang F (2011) Research on power load forecasting based on improved BP neural network. Harbin Institute of technology

Hu X (2009) Smart grid—a development trend of future power grid. Power Syst Technol 33 (14):1–5

Jiuru Y, Zehua Y (1995) A fuzzy linear regression model and its application. J Syst Sci Information(4):32–37

Kaotanzad A, Af'khami-Rohani R, Maratukulam D (1998) ANNSTLF—artificial neural network short-term load forecaster-generation three. IEEE Trans Power Syst 13(4):1413–1422

Liao F, Liu Q, He H, Cheng Y, Xu C, Yao J (2011) Bus load forecasting based on improved gray model and meteorological elements. Power Syst Technol 35(10):183–188

Peng G, Si H, Yu J, Yang Y, Li S, Tan K (2011) Modification and application of SVM algorithm. Comput Eng Appl 47(18):218–221

Ren L (2009) Enlightenment from foreign policies of developing low carbon economy. Dev Res (2):23–27

Ren H, Zhang X, Sun C, Wen J (2010) Regulation regression local forecasting method of multivariable chaotic time series in short-term electrical load. Comput Sci 37(7):220–224

Sun P, Gu X, Liu J, Wang Q (2005) Decomposition analysis of energy consumption in China. Res Sci 27(5):15–19

Zhang B, Liu X, Wan J, Liu P, Cheng S (2004) Load forecasting based on chaotic time series and analysis of its key factors. Power Syst Technol 28(13):32–36

Internal Resistance of Lead-Acid Battery and Application in SOC Estimation

Wei-wei Li, Li Cheng and Wei-ming Ding

Abstract In order to improve the performance of electric vehicle, some battery life tests have been carried out to acquire the relevant conclusions about battery internal resistance during charging and discharging and establish the resistance equivalent model for the real-time identification, all of which above are set up as the evaluation index of state of charge (SOC). The results demonstrated that internal resistance can reduce the inaccuracies of SOC estimation and car mileage.

Keywords Lead-acid battery · Internal resistance state of charge (SOC) · Low carbon

1 Introduction

In recent years, electric vehicles gain competitive advantage in new global carbon economy, while battery maintenance becomes very important in electric vehicles as UPS. Much research on battery internal resistance has been carried out to improve the accuracy of battery SOC estimation and the reliability of battery. As we know, lead-acid battery resistance is divided into three parts: ohmic resistance, electrochemical resistance, and concentration polarization resistance. Ohmic resistance consists plate resistance, electrolyte resistance, separator resistance; electrochemical polarization resistance is decided by current density; concentration polarization resistance is influenced by electrolyte's diffusion speed.

At present, many equivalent circuit models for battery have been built. The typical one of them is called PNGV which is shown in Fig. 1 (Enthaler and Gauterin 2013). In this model, two kinds of polarization resistance are combined to capacitors C_p and resistors R_p in parallel; E represents open-circuit voltage of the battery; R_o is the battery internal ohmic resistance. Obviously, the increase of battery internal resistance would lower battery discharge voltage U, which has a direct impact on

W. Li (✉) · L. Cheng · W. Ding
School of Energy and Environment, Southeast University, Nanjing 210096
People's Republic of China
e-mail: vivlee1990@163.com

S. Feng et al. (eds.), *Low-carbon City and New-type Urbanization*,
Environmental Science and Engineering, DOI 10.1007/978-3-662-45969-0_22

Fig. 1 Resistance equivalent circuit model

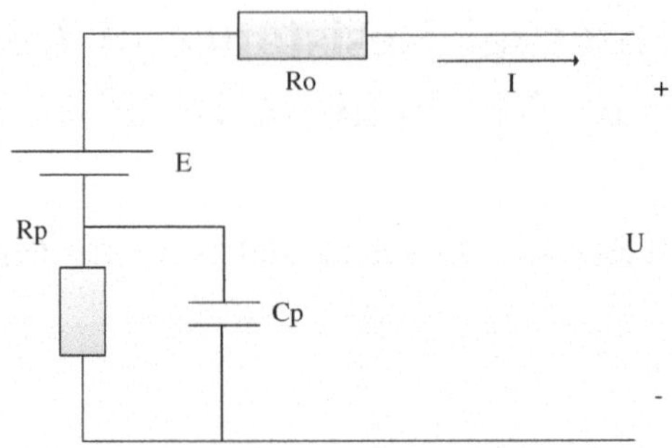

battery SOC and discharge time. Therefore, the internal resistance, especially the ohmic resistance, will be an important parameter to estimate the SOC of battery.

In practice, the complexity of battery internal resistance results in the measurement to be difficult. It only can be measured by external voltage and current. DC discharge method is a common one (Kurisawa 1997). Applying a load current to the battery, the ohmic resistance is equal to the voltage drop value divided by the current: $R = \Delta U/I$. Another typical one is AC injection method. Read out the voltage and current with the same frequency after importing a certain frequency AC signal into the battery, then $R = U_{av}/I_{av}$. The two methods indicated above are simple and accurate for the measurement of the internal resistance; however, both of them cannot meet the requirement of online measurement and practical application. To realize the real-time measurement of resistance, this chapter proposed the resistance identification after establishing the internal resistance equivalent model. The internal resistance obtained is applied in the estimation for SOC.

2 The Resistance Characteristics

Firstly, taking some 12 Ah lead-acid batteries as test subject, several experiments of battery discharge and charge were carried out to get the test dates, such as cell voltage, current, and internal resistance during charging and discharging.

As shown in Fig. 2, the internal resistance rises as an exponential foundation during the discharge process due to the formation of lead sulfate crystals ($PbSO_4$) (Hu et al. 1999). As the state of charge SOC exceeds 60 %, the internal resistance decreases slowly with the slow decreases of terminal voltage. After SOC is below 40 %, the internal resistance rises sharply. At the same time, the terminal voltage is also plummeting. As a result, the internal resistance must be a major reason for the decline of voltage and SOC.

In order to recognize the resistance characteristic during charging, the battery discharged before is charged. The difference between charging and discharging is shown in Fig. 3. Obviously, the internal resistance during charging is also decreasing as an exponential foundation with the decline of the charging current. By contrast, the resistance during charging process is less than discharging in the same SOC.

At the same time, battery lifetime experiment indicated that discharge current also has influence on internal resistance. Taking three full charging lead-acid batteries with a similar performance to discharge, as shown in Fig. 4, the change of internal resistance under different current for discharging has the same trend. Obviously, the battery internal resistance increases faster along with the enhancement of discharging current.

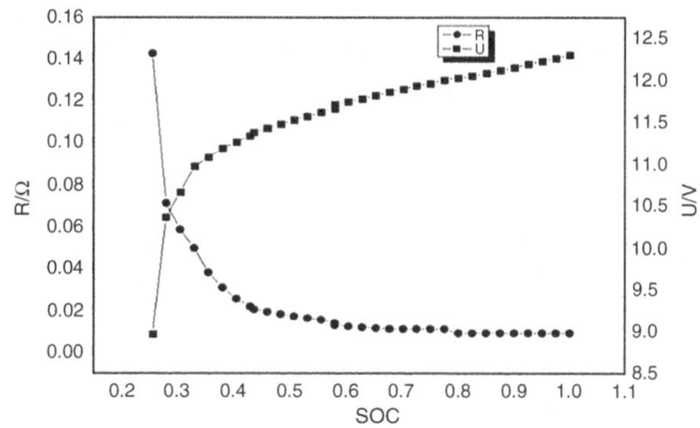

Fig. 2 The discharging internal resistance

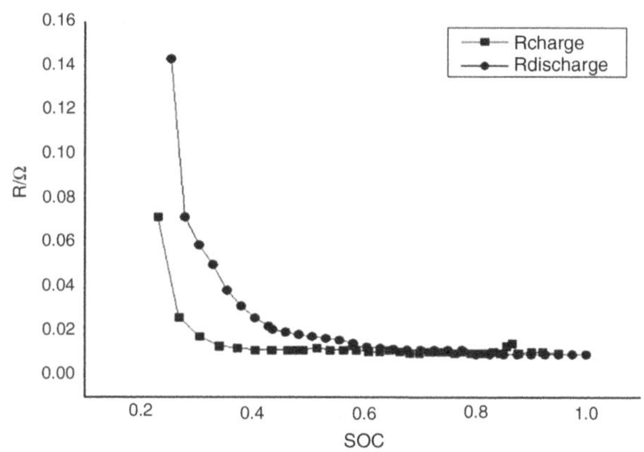

Fig. 3 The charging and discharging resistance

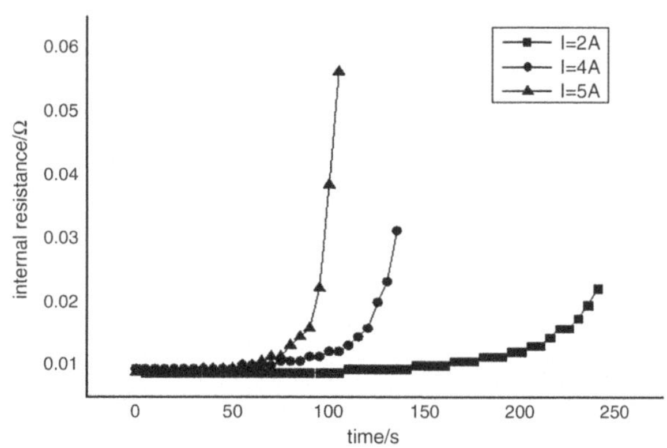

Fig. 4 The internal resistance under different discharging currents

Figure 5 shows the resistance of two 12 Ah lead-acid batteries with different health conditions. Apparently, deteriorated cell's resistance which is higher than the healthy one increases sharply in case that SOC is less than 75 %.

Figure 6 is the internal resistance of a 200 Ah lead-acid battery in good condition. Comparing with Fig. 2, it can be obtained that the greater the rated capacity is, the smaller the internal resistance will be. Figure 7 is the resistance change of another 200 Ah battery in discharging process which is inferior and irreversible. The resistance is 10 times larger than the healthy one when it is full charged.

The lead-acid battery internal resistance model established by PNGV are all simulated in (Wei et al. 2009); the real-time identification can be carried out by the BP algorithm. The flow chart is in Fig. 8 (Ling et al. 2013). Taking the actual values of current excitation and the gathering voltage as input, the applicability of this simplified model to identify internal resistance can be confirmed. The sum of squares due to error (SSE) is equal to 0.003 (Fig. 9).

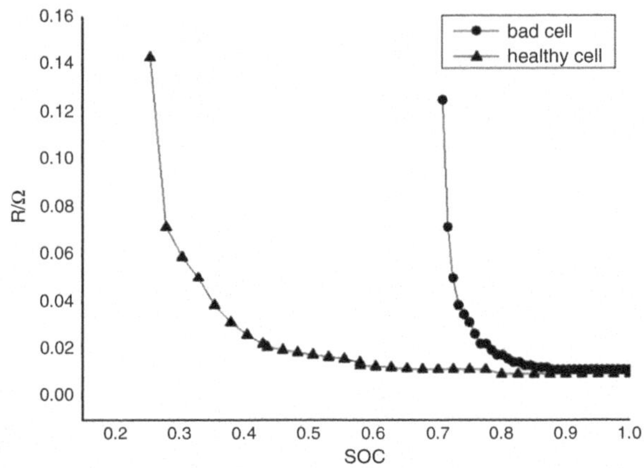

Fig. 5 The internal resistance under different conditions

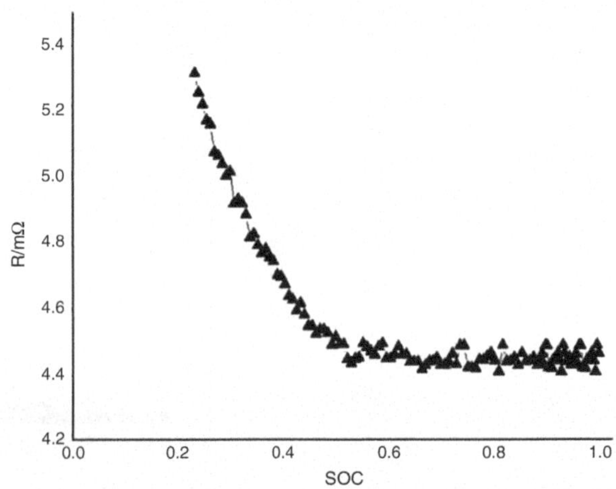

Fig. 6 Healthy battery of 200 Ah

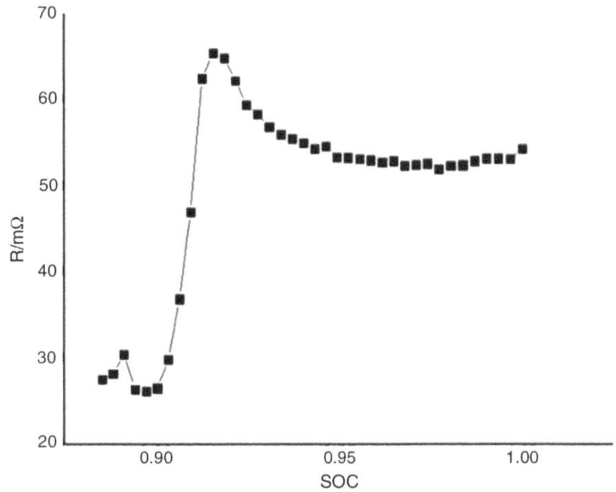

Fig. 7 Deteriorated battery of 200 Ah

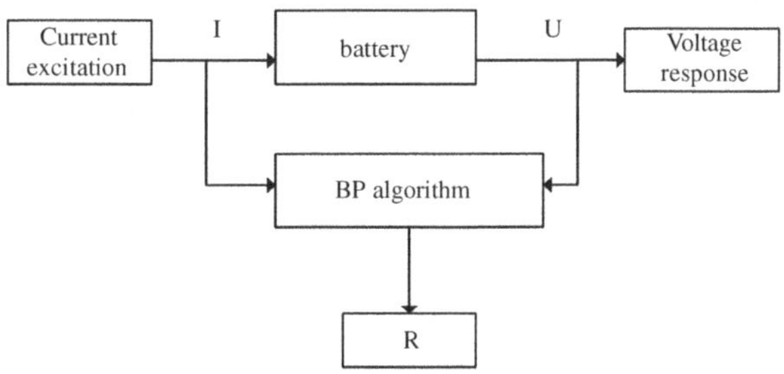

Fig. 8 Internal resistance identification model

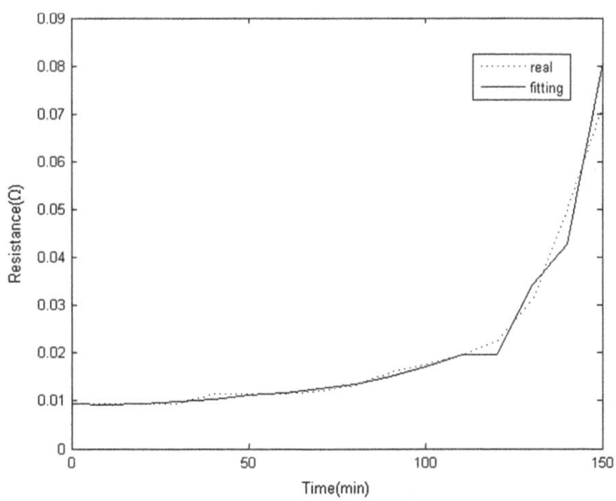

Fig. 9 Identification result of internal resistance

3 Estimation for SOC

Figure 1 shows the equivalent circuit model of lead-acid battery model; the relationship between SOC and internal resistance can be inferred as follow:

Firstly, according to the model's voltage relationship, the response to the DC current can be obtained (taking the discharging process as an example):

$$V = E - I(R_p + R_o)a \tag{1}$$

According to Yang et al. (2010), the SOC is linear with respect to the electromotive force:

$$\text{SOC} = aE + b \tag{2}$$

Based on Fig. 1 and Eq. (1), the identification model for SOC estimation with external parameters is obtained:

$$\text{SOC} = a(V - I(R_p + R_o)) + b \tag{3}$$

Besides, SOC is directly related to internal resistance from Fig. 2. A Para-curve instead of exponential function can be applied to express the relationship:

$$\text{SOC} = mR^2 + nR + l \tag{4}$$

After a synthetic analysis of the above, the new identification model for SOC is presented, based on electromotive force and internal resistance

$$\begin{aligned} \text{SOC} = aE + bR^2 + cR + d &= a(V + I(R_p + R_o)) \\ &+ b(R_p + R_o)^2 + c(R_p + R_o) + d \end{aligned} \tag{5}$$

During the constant current discharge, terminal voltages and current are recorded for the same time interval. After, the resistance is solved by Fig. 8:

$$\begin{aligned} \text{SOC}_0 &= a(V_0 + I(R_{p0} + R_{o0})) + b(R_{p0} + R_{o0})^2 \\ &\quad + c(R_{p0} + R_{o0}) + d \\ \text{SOC}_1 &= a(V_1 + I(R_{p1} + R_{o1})) + b(R_{p1} + R_{o1})^2 \\ &\quad + c(R_{p1} + R_{o1}) + d \ldots \\ \text{SOC}_n &= a(V_n + I(R_{pn} + R_{on})) + b(R_{pn} + R_{on})^2 + \\ &\quad + c(R_{pn} + R_{on}) + d \ldots \end{aligned} \tag{6}$$

Fig. 10 SOC estimation

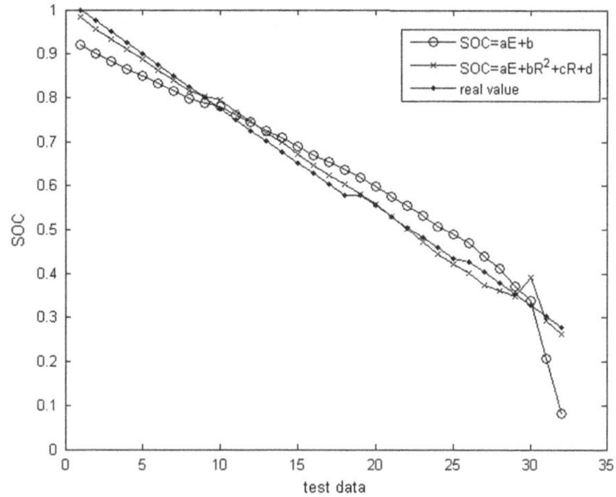

Turning the formulas above into matrix form:

$$Y = [\mathrm{SOC}_1, \mathrm{SOC}_2, \mathrm{SOC}_3 \ldots \mathrm{SOC}_n]^T, \theta = [a, b, c, d]^T$$

$$\phi = \begin{bmatrix} V_0 + I(R_{\mathrm{p}0} + R_{\mathrm{o}0}) & (R_{\mathrm{p}0} + R_{\mathrm{o}0})^2 & (R_{\mathrm{p}0} + R_{\mathrm{o}0}) & 1 \\ V_1 + I(R_{\mathrm{p}1} + R_{\mathrm{o}1}) & (R_{\mathrm{p}1} + R_{\mathrm{o}1})^2 & (R_{\mathrm{p}1} + R_{\mathrm{o}1}) & 1 \\ \ldots & \ldots & \ldots & \ldots \\ V_{\mathrm{n}} + I(R_{\mathrm{pn}} + R_{\mathrm{on}}) & (R_{\mathrm{pn}} + R_{\mathrm{on}})^2 & (R_{\mathrm{pn}} + R_{\mathrm{on}}) & 1 \end{bmatrix} \tag{7}$$

$$Y = \phi\theta$$

According to least square method, this matrix equation is solved:

$$\hat{\theta} = [a, b, c, d]^T = (\theta^T\theta)^{-1}\theta^T Y \tag{8}$$

Compared the SOC obtained by parameter identification of this equivalent model with the SOC estimation value by electromotive force (E) just based on Eq. (2) and the actual value of SOC by Ah method, the partial result can be seen in Fig. 10. The error of this model can be controlled in 0.06 and has a higher precision than the model (2).

4 Conclusions

Based on several charging and discharging experiments, the characteristic of lead-acid battery internal resistance is analyzed in this chapter. Meanwhile, the simplified equivalent circuit model is obtained and established at last. Through the calculation according to the equivalent model and the least square method and the comparison with the experiment result, this method effectively improves the accuracy of SOC estimation without a deep discharge and off-line operation, which confirm the serviceability of this model for SOC estimation well. This kind of accurate online monitoring is very important for batteries to improve the reliability of UPS and prolong the life of batteries, all of which will promote the development of electric vehicles and realize the low-carbon city life gradually.

References

Enthaler A, Gauterin F (2013) Significance of internal battery resistance on the remaining range estimation of electric vehicles. In: Proceedings of the international conference on connected vehicles and Expo (ICCVE) 2013, (67), pp 94–99

Hu X, He Q, Tong Y (1999) Studies on the internal resistance of valve-regulated lead acid battery. Chinese J Power Source (23):302–303

Kurisawa I (1997) Internal resistance and deterioration of VRLA battery-analysis of internal resistance obtained by direct current measurement and its application to VRLA battery monitoring technique. In: Proceedings of the international telecommunications energy conference, pp 687–694

Ling R, Wei X, Dai H (2013) A simulation of lithium-ion battery ohmic resistance identification. In: Proceedings of the 9th IEEE vehicle power and propulsion conference, pp 41–44

Wei X, Zhu B, Xu W (2009) Internal resistance identification in vehicle power lithium-ion battery and application in lifetime evaluation. In: Proceedings of the international conference on measuring technology and mechatronics automation, 468:388–392

Yang Y, Li K, Cheng T (2010) Introducing the model of electromotive force-impedance estimating the SOC of battery on line. In: Proceedings of the electrical measurement and instrumentation, vol 37(3), pp 16–19

An Intelligent Collision Avoidance System for Metro Railway Vehicles Based on the Synthesis of Multiple Acoustic Methods

Wen-jin Zhou, Qiao-lian Zhou, Qi Deng and Xiao-dong Li

Abstract As one of the most energy-efficient mass transit systems, metro railway has increasingly make use of modern wireless communication technologies to improve the efficiency and safety. In this chapter, we discuss some acoustic distance measurement and proximity detection methods, which may be integrated into an intelligent onboard proximity warning and collision avoidance system in order to increase the metro operation efficiency by safely improving the headways when the automatic train control (ATC) system is switched into a degraded mode. Those acoustic methods are based on the reception, processing, and identification of acoustic signals that are either excited by the moving vehicles or artificially generated by loudspeakers. Example system is introduced and partially validated by online tests. It is shown that the acoustic methods, when implemented independent of the railway signal and communication systems, have potential to improve the transit efficiency and safety without any wayside equipment.

Keywords Acoustic ranging · Collision avoidance system · Mass transit · Intelligent vehicle · Proximity detection

1 Introduction

The automatic train control (ATC) systems have been widely used in urban railway transit, which has increasingly integrated modern wireless communication technologies to further improve the efficiency and safety, such as the recent radio frequency (RF) communication-based train control (CBTC) (Morar 2010).

W. Zhou (✉) · X. Li
Shanghai Advanced Research Institute, Chinese Academy of Sciences, Pudong, Shanghai 200120, People's Republic of China

Q. Zhou · Q. Deng
Technical Center of Shanghai Shentong Metro Group, Minhang, Shanghai 200031, People's Republic of China

S. Feng et al. (eds.), *Low-carbon City and New-type Urbanization*, Environmental Science and Engineering, DOI 10.1007/978-3-662-45969-0_23

A typical ATC system consists of the wayside, onboard, and data communications equipment which meet the stringent availability and reliability criteria. However, system degrading or even halt due to various external or inherent causes is sometimes inevitable in essence, which not only influences the transit capacity, but also makes the system actually unable to guarantee the operation safety as well. There is therefore an increasing demand on the auxiliary proximity warning equipment with low failure rate and less maintenance effort in order to ensure the operation efficiency and safety for metro railway vehicles, in particular when the ATC system has to be switched to the degraded mode.

One of the solutions is the development of the collision avoidance system (CAS). From the cost and maintenance point of view, a CAS for metro railway needs to be cost-effectiveness and totally onboard as it is designed to complement current vehicle control and signaling systems that are failsafe and have taken into account the redundancy of equipment. The desired ranging distance of CAS needs to exceed the stopping distance when emergency brake applied plus the worst-case positional uncertainty. During the degraded mode of ATC, the minimum warning distance is desired to be at least 200–300 m depending on the speed of metro railway vehicles. Since the metro railway environment has limited the usage of satellite navigation or vision-based methods (Erick 2002; Strang et al. 2006; Maire 2007), one of the solutions is the development of the CAS based on the time of flight methods, such as the radio-based distance measurement, which has shown the potential of application not only for metro railway vehicles but also for the high-speed train (Lienard et al. 2004; Elhillali 2010; Wang et al. 2009).

Another wireless distance measurement method is to use acoustic waves. For vehicular CAS, the application of acoustic methods is not new. Commercial solution such as the ultrasonic radar is currently used for automobile parking distance control. However, due to high attenuation of ultrasound in air, the ranging distance of ultrasound may be no more than dozen of meters, which is too short to offer the desired performance for metro railway vehicles even ignoring the influence of ambient noise. Moreover, the application of conventional ultrasonic pulse-echo method may suffer unacceptable false alarms in metro railway where many wayside objects cause lots of suspicious echoes. In the indoor scenario of metro railway, the acoustic reverberation may further constrain the application of ultrasonic pulse-echo method since the echoes from the approached vehicle might be confused with the reverberation echoes and therefore inexplicable. Indeed, if the audio frequency acoustic wave rather than the traditional ultrasound is used, lots of effective ranging modes can be applied and the ranging distance may be significantly increased.

For wireless distance measurement in metro railway environments where usually exists various disturbances in either mechanical or electromagnetic spectrum, simply using the redundant CAS equipment to improve the system availability and reliability may have both cost and common cause failure issues (Flammini 2012). In fact, the metro railway vehicle, when is regarded to be dangerous to the others, can produce some physical signs in the form of mechanical waves that may be used for proximity detection. Due to this fact, the redundancy can be realized by the active acoustic ranging method in combination with the passive acoustic method.

In this paper, we investigate some active acoustic ranging methods and passive acoustic proximity detection methods that may be used for metro railway vehicles, especially in an onboard way, which has never been used for the vehicular CAS application.

2 Active Acoustic Ranging

2.1 Pulse-Echo

In order to minimize the influence of the acoustic reverberation that may confusing the received echo signal in the railway tunnel, the low-frequency zero-order wave mode might be used since it is the only non-dispersive wave that can be readily generated without producing the unwanted high-order dispersive waves, as shown in Fig. 1. Owing to the low attenuation, the acoustic echo from the leading vehicle, if dominated by the zero-order mode, can propagate a long distance and be received and identified for location purpose.

The echo signal from a vehicle is mainly relative to the cross-sectional geometry of tunnel and vehicle, although the axial characteristics and length of the vehicle also contribute to the time history pattern of echoes. Typical reflection coefficient of the low-frequency zero-order mode wave from a vehicle in a modern circular tunnel excavated by boring machines is about −18 to −10 dB, which can be estimated from experimental and numerical analysis.

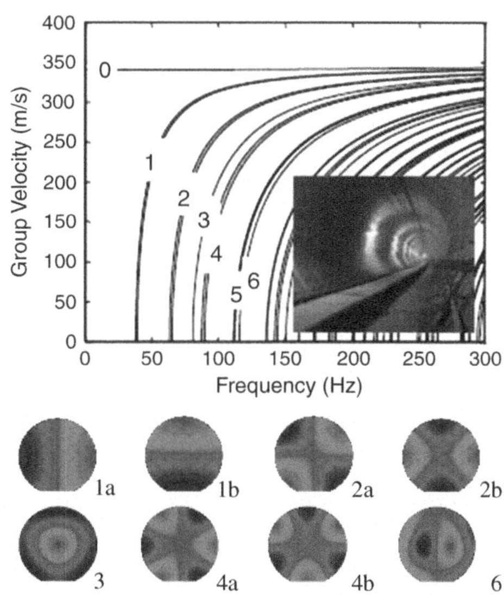

Fig. 1 Group velocity dispersion and mode shapes of guided waves in a typical metro railway tunnel (ID = 5.2 m)

2.2 Single-Way Pitch-Catch

The single-way pitch-catch method estimates the vehicle spacing by the analysis of received acoustic signal strength as the signal transmitted is known and the basic relationship between the transmission loss and the travelled distance can be readily evaluated in the metro railway environments. The relative velocity that further indicates the severity of a collision hazard can be calculated by the Doppler frequency shift of the received signal. In practice, the frequency that is suitable to single-way pitch-catch vehicle spacing estimation should be in several kilohertz at least, below which the acoustic signal strength may fluctuate significantly along the transmission path due to the wave dispersion in lots of metro railway tunnels.

Although the vehicle spacing measured may not be as precise as the time of flight approaches, the single-way pitch-catch method is convenient to implement, and apt to combine with other methods such as the passive proximity detection methods or the two-way pitch-catch active ranging method.

2.3 Two-Way Pitch-Catch

If the leading vehicle is equipped with an acoustic transponder, the vehicle spacing can be computed in a two-way pitch-catch manner. First, the following vehicle transmits a coded acoustic interrogation, which is received and identified by the transponder of the leading vehicle; the transponder then sends back an acoustic response to the receiver of the following vehicle. In this way, the real-time distance L from following vehicle to the leading vehicle can be computed after the accurate identification of the response signal, which leads to

$$L = (c_0 - v_1 + v_2)\Delta t/2 \tag{1}$$

where v_1 is the velocity of the following vehicle that is computed from the Doppler frequency shift, v_2 is the velocity of the leading vehicle that is calculated through the frequency shift Δf_2 of the received responding signal which has a known transmitting frequency f_2.

$$v_2 = c_0 - \frac{c_0^2(f_2 + \Delta f_2)}{f_2(c_0 + v_1)} \tag{2}$$

The two-way pitch-catch method is able to offer more precise ranging results than the energy-based single-way pitch-catch method.

2.4 *Hybrid Acoustic/Radio Ranging*

The radio waves can propagate very well in tunnels, although they may suffer severely from the multipath fading in some preferred frequency band (Didascalou et al. 2001). The transponder in the leading vehicle can emit the acoustic response according to the radio interrogation from the following vehicle in which the radio serves as the timing gauge to calculate the time of flight of the acoustic response from the leading vehicle. The radio may be in the ISM band, or some specially allocated band, depending on the local radio regulation in the metro railway environment.

3 Vehicle-Induced Acoustic Wave and Proximity Detection

Most railway collision accidents happened in the way that one stopped or slowly running vehicle being rear-ended by another vehicle with higher speed. When the railway vehicles are running, they usually generate various acoustic waves. Some types of the acoustic waves can propagate a long distance and at a speed much larger than the metro vehicle itself, and thus be measurable far ahead of the vehicle as the early sign of approaching.

Two types of acoustic response are generated by moving vehicle. One is the airborne sound, which propagates in air inside of the tunnel, and another is the structure-borne-sound traveling in either the tunnel or highroad structure. Both of airborne sound and structure-borne sound can propagate in a very long distance if in form of guided waves, especially in the metro railway tunnels, as their transmission paths are shown in Fig. 2. Normally, if the vehicle speed is higher, the stronger acoustic signal can be received by the leading vehicle. This allows the passive methods to be used either independently for early warning of proximity, or complement for the active acoustic ranging.

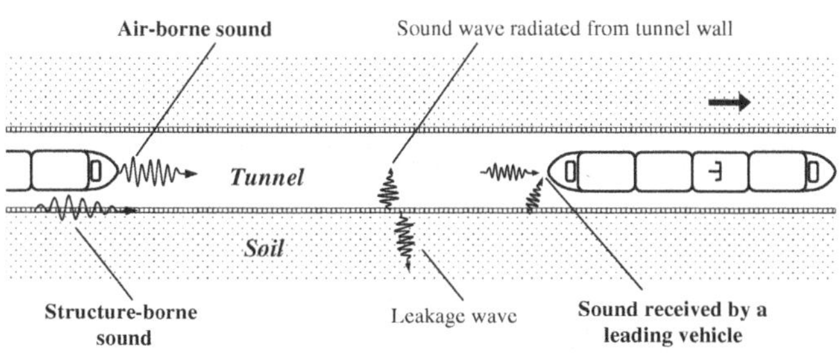

Fig. 2 Acoustic wave transmission paths from a following vehicle to a leading vehicle in railway tunnel

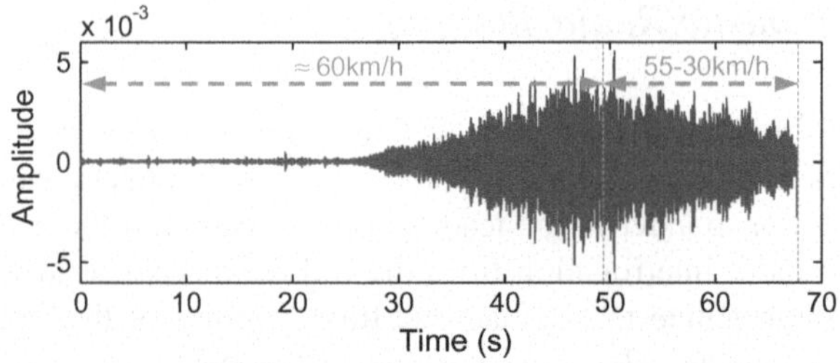

Fig. 3 Narrowband acoustic signal recorded before a metro vehicle passing by the microphone at the 67th second (measured in Shanghai Metro Line 2)

3.1 Structure-Borne Acoustic Waves

Many structures in the railway can be regarded as structural waveguides, including the tunnel wall, highroad, rail, and overhead power line. Those structures can carry structure-borne acoustic waves that are excited by the moving vehicles, and the radiated sound can be readily measured by the receivers far away from the vehicles.

Figure 3 displays a typical time history acoustic signal measured in Shanghai Metro Line 2 during normal operation. Through the spectrum analysis of long time monitoring in multiple locations, such kind of signal cannot be found in the background noise when there are no approaching vehicles, which means that the proximity can be easily identified by a stopped or slowly running vehicle, either by pattern analysis or simply setting a warning threshold.

3.2 Airborne Acoustic Waves

A typical type of airborne acoustic wave that may be used for approaching warning is the pressure wave. The pressure wave in the railway tunnel is mainly in the form of the low-frequency acoustic pulses which are excited by the rapidly moving vehicle passing by the non-homogeneous air cavities in a homogeneous tunnel (Yoon et al. 2001). They can travel a very long distance in the tunnel until being attenuated by the multiple cavity non-homogeneities or the obstacles such as vehicles in the tunnel.

Another type of pressure wave is caused by the vehicle moving-induced air flow in the railway tunnel. In case that the following vehicle running evenly toward the leading vehicle, the pressure wave may simply exhibited as the gradual pressure augment rather than the acoustic signal although the pressure fluctuation is usually observed. The time history signal of the pressure variation captured in the rear of the leading vehicle indicates how the following vehicle is approaching and can be used as early warning signals since the typical warning time may exceed 30 s.

The low- and mid-frequency airborne noise mainly resulting from the wheel–rail interaction can also be used as approaching signal in the tunnel scenario, as they are well guided and travel a long distance, especially when the leading vehicle is stopped with a low ambient noise level.

3.3 Distance Estimation

When the leading vehicle has detected the suspicious proximity, it has to inform of the approaching vehicle. The alarm information had better be transmitted by radio though acoustic signal is also feasible. Once the following vehicle receives the radio wave modulated with the raw acoustic signal or the extracted signature, it can calculate the spacing through the correlation analysis of the approaching signal from the leading vehicle and the acoustic signal recorded by the following vehicle itself.

The transmitting of the alarm signal can be performed in the single-way pitch-catch mode, and the following vehicle can then estimate the distance in two independent ways, since the reception of warning signal that has specific attenuation characteristics has already indicated that the leading vehicle may be not far away.

As the passive method can detect the proximity hazard by monitoring the early acoustic signal generated by the following vehicle in a long distance that is usually exceeds the typical dangerous proximity, in order to reduce false alarms, sensitivity of the proximity detectors can be tuned well below the level that can avert most of undesired alarms.

4 System Description and Test Results

4.1 System Setup

The setup of the acoustic CAS combining the active ranging and passive proximity detection is shown in Fig. 4. The system is powered by the vehicle battery since the traction power loss used to significantly increase the hazard of vehicle collision.

The active acoustic ranging module consists of the loudspeakers, microphone arrays, power amplifiers, audio data acquisition and processing unit, and a radio communication unit if the hybrid acoustic/radio ranging mode is integrated. The acoustic proximity detection module simply comprises some proximity sensors, a low-power processing unit, and a communication unit. Most of the acoustic transducers need be installed outside the cabin, as shown in Fig. 5.

The active acoustic ranging module and acoustic proximity detection module are each associated with an independent vehicle motion detector. The output of two modules is collected by a logic voter. The logic can be voted 1oo2 (one-out-of-two) with or without decision fusion, depending on the receiver operating characteristic of the system.

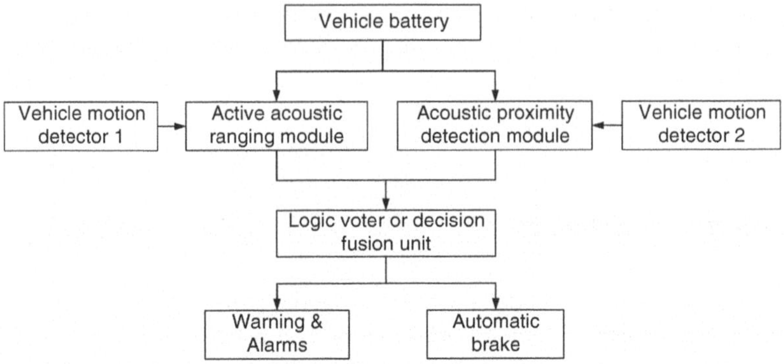

Fig. 4 Example setup of a CAS combining the active acoustic ranging and passive acoustic proximity detection

Fig. 5 Onboard transducers of acoustic CAS installed on a vehicle in Shanghai Metro Line 11

Fig. 6 A snap of the warning display of the acoustic CAS that is switched to the active ranging mode, where the vehicle keeps following a leading one at the speed of 49 ± 5 km/h in a metro railway tunnel

The output of the acoustic CAS can be indicated by a warning display equipped in the driver cabin, as shown in Fig. 6. The output can also be sent to the automatic brake system through transistor to transistor logic if necessary.

4.2 Vehicle Motion Detection

In the CAS, the detection of vehicle motion is desired since the system is aimed to be totally onboard for maintenance concern and independent as necessary for reliability concern. The system had better to be aware of the vehicle's basic status including the speed and direction in order to intelligently switch into the corresponding working state. In such way, the onboard system can perform more effectively with the speed adaption treatment and prevent the signal interference between vehicles in different directions. The vehicle direction can be identified through vehicle velocity measurement by the acoustic ranging module. The speed of the vehicle can also be estimated by recording and analysis of the ambient noise, since the vehicle running-induced noise level depends highly on its speed.

Although the vehicle motion can be detected by different acoustic methods, the auxiliary vehicle motion detectors such as the inertial device may further improve the system intelligence and performance without affecting the reliability if integrated into the system in a failsafe way.

4.3 Test Results

Evaluation of the performance of the acoustic CAS is performed on 55 vehicles of Shanghai Metro Line 11. As the vehicle rarely approaches each other during normal operation, accelerated online test can be made through continuous tracing between vehicles during maintenance time.

Figure 7 shows the continuous active ranging results recorded when the vehicle running after another one at the speed from 30 to 55 km/h and spacing from 200 to 300 m. It is shown that the system can trace the leading vehicle in satisfactory assurance. The real-time distance between two vehicles can be continuously

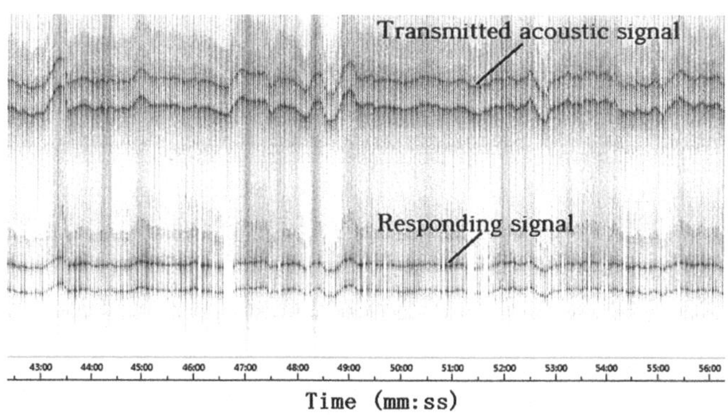

Fig. 7 Time-frequency representation of the continuously active ranging

computed and updated except when the relative velocity is below the predefined threshold, e.g., v1-v2 < −6 km/h in this case, which implies that the leading vehicle will run away quickly and the frequent responding to a transmitted interrogation is therefore considered to be unnecessary.

5 Conclusion

This study investigates some active acoustic ranging methods and passive acoustic proximity detection methods for metro vehicle collision avoidance purpose. As the passive proximity detection can be realized with much less effort and cost, it is combined with the active ranging in order to reduce the total functional failure rate of CAS with relative low cost of hardware.

An example system has been designed to illustrate the feasibility and performance of the synthesis of multiple acoustic methods in a real metro railway environment, which can be operated in multiple modes to improve the capability of resistance from noise and jamming. Based on the acoustic methods proposed, the system can provide the following functions depending on the way the system output being adopted:

Automatic vehicle motion detection;
Automatic forward distance measurement and approaching warning up to 300 m;
Backward alarm to the probably approaching vehicle;
Automatic emergency brake (combining with brake system).

The proposed acoustic CAS allows the vehicle to be capable of listening and talking to each other and has potential to significantly improve the operational efficiency and safety of vehicles when the ATC is in the degraded mode.

Acknowledgments Work supported by the Science and Technology Commission of Shanghai Municipality (STCSM) under grant No. 12231200102.

References

Didascalou D, Maurer J, Wiesbeck W (2001) Subway tunnel guided electromagnetic wave propagation at mobile communications frequencies. IEEE Trans Antennas Propag 49:1590–1596

Elhillali Y (2010) Enhanced high data rate communication system using embedded cooperative radar for intelligent transports systems. Transp Res Part C 18:429–439

Erick MJ (2002) Collision avoidance using GPS device and train proximity detector. US Patent 6,345,233

Flammini F (2012) Railway safety, reliability, and security: technologies and systems engineering. IGI Global, Hershey

Lienard M, Degauque P, Laly P (2004) Long range radar sensor for application in railway tunnels. IEEE Trans Veh Technol 53:705–715

Maire F (2007) Vision based anti-collision system for rail track maintenance vehicles. In: IEEE conference on advanced video and signal based surveillance, AVSS, IEEE, pp 170–175

Morar S (2010) Evolution of communication based train control worldwide. In: Institute of engineering and technology (IET) Professional Development Course on Railway Signaling and Control Systems 2010 (RSCS 2010). doi10.1049/ic.2010.0106

Strang T, Meyer M, Hoerst MZ (2006) A railway collision avoidance system exploiting ad-hoc inter-vehicle communications and galileo. In: Proceedings of the 13th ITS world congress, London, pp 8–12

Wang C, Cheng X, Laurenson D (2009) Vehicle-to-vehicle channel modeling and measurements: recent advances and future challenges. IEEE Commun Mag 47(11):96–103

Yoon TS, Lee S, Hwang JH, Lee DH (2001) Prediction and validation on the sonic boom by a high-speed train entering a tunnel. J Sound Vib 247(2):195–211

Developing Tailor-Made Urban Environmental Policies for China's Low-Carbon Cities—Implications from Japan and Germany

Xiao-long Zou and Yan Li

Abstract China's urban conditions are facing dire challenges due to the incremental development paces and deteriorating eco- and environmental situations. To cope and pursue urbanization in a sustainable manner, the Chinese government has strenuously enacted an array of national urban environmental policies. This paper reviews these policies from "garden city," "eco-city" to "low-carbon city" to determine the foci of the environment-related urban environmental policy developments under different periods in China. Two case studies of the renewable energy-exporting town of Rhein-Hunsrück district in Germany and the eco-town of Kitakyushu city in Japan are examined and analyzed to obtain enlightening factors in terms of their urban environmental policies as well as the references and implications from their successes and limits. Major findings indicate that the local autonomy and flexibility in policymaking and civic participation profoundly contributed to the successful switch to renewable energy in Germany's case. For Japan's case, government policy and financial support played significant role in transforming the industrialized city of Kitakyushu into sustainable practices center of eco-city. These experiences could offer insightful references to China's low-carbon urban developments from different perspectives.

Keywords China · Low-carbon city · Eco-city · Urban environmental policy · Renewable energy · Waste recycling

1 Introduction

With the largest population in the world, over half of which is living in urban areas (United Nations and World Population Prospects 2012). China has been enjoying the dividends led by phenomenal economic development and rapid urbanization processes, while suffering various problems that come hand in hand in the urban areas:

X. Zou (✉) · Y. Li
Graduate School of Ritsumeikan Asia Pacific University, Jumonjibaru1-1, Beppu 874-0011, Japan
e-mail: zoufuchen@gmail.com

S. Feng et al. (eds.), *Low-carbon City and New-type Urbanization*,
Environmental Science and Engineering, DOI 10.1007/978-3-662-45969-0_24

shortage of housing, traffic congestions, environmental pollutions, food and soil security, public health crises, and so on. To tackle these challenges, a series of countermeasures have been taken, featured in the symptoms of different periods, trying to cure these ills. From the 1986's first local government's attempt in developing "eco-city" project in Yichun city of Jiangxi Province to the recently released grand strategy of "National New-type Urbanisation Plan (2014–2020)" to redirect the nation's urbanization development toward a human-centered and environmentally friendly path, the Chinese government has put relentless and continuous efforts into reconciling urbanization and economic development ventures with the need for seeking balance between nature and development in a sustainable manner.

A variety of frameworks, indicator systems, initiatives, and programs of "eco" and "low-carbon" cities have been seen in China. Several major national frameworks or programs have been proposed different governmental entities, such as "the Eco-Garden City" program by the Ministry of Housing, Urban–Rural Development,[1] "the Eco-County, Eco-City and Eco-Province Indices" by the Ministry of Environmental Protection,[2] and "the Low-Carbon City" programs by the National Development and Reform Committee (NDRC).[3] These programs have attracted a great number of Chinese cities to participate. However, unlike the MEP or MoHURD that offers specific indicator systems, NDRC's low-carbon city program is still in the trial phase, with selected piloting provinces and cities to come up with their own development master plans. From the trajectories of China's national urban environmental policies, it can be observed that the focus has shifted from that of dealing with pollution and setting up greenery in the urban areas led by eco-garden cities, to the pursuit of eco-cities where urban planning and management are integrated based on ecosystem merits, and then to the development of low-carbon cities that features in reducing carbon emissions by decoupling of fossil fuel uses and improving energy efficiencies. As per the new norm, carbon reduction has been brought up to the top national policies in China, as in most parts of the world. In 2009, the State Council announced the target of reducing its GDP carbon intensity by 40–45 % by 2020 compared to the 2005 level (2020); later in the 12th Five-Year Plan, a binding target of 17 % CO_2 reduction per unit GDP from 2011 to 2015 was set as a national goal (2011), which greatly incentivizes the low-carbon city development in China.

There are many literatures reviewing on the eco-city concepts, frameworks, and indicator systems (Zou and Li 2014; Joss and Tomozeiu 2013; Li and Gang 2012). Another school of literatures studied low-carbon cities in terms of concepts (Yang and Li 2013), developments (Li et al. 2012) and promotion of individual low-carbon cities (Feng and Zhang 2012; Bi et al. 2011; Lehmann 2013). However there are few studies conducted to evaluate the policy-driven low-carbon cities implemented in China

[1] Ministry of Housing Urban and Rural Development (MoHURD), Eco-Garden City Standards and Selection Criteria. MoHURD: Official Website.

[2] Ministry of Environment Protection (MEP), Eco-County, Eco-City, Eco-Province Index, C.M.o. E. Protection, Editor., MEP: Offical website.

[3] National Development and Reform Commission (NDRC), Low Carbon City Initiatives. NDRC: Offical website.

(Khanna et al. 2014). This paper aims to comb through the major national urban environmental policies to summarize the characteristics in a conceptual sense and focus on the current status quo of Chinese low-carbon cities, to offer readers a holistic view of China's urban environmental policy transformations. Under the current tide of low-carbon urbanization, what could be the lessons learned from the leading countries such as Japan and Germany for the low-carbon city developments in China? The German case of Rhein-Hunsrück district demonstrated how rather flexible energy policies have helped to transform that region from an energy importer to an exporter. And the Japanese eco-town project in Kitakyushu indicated that the technological innovations have significantly contributed to the city's low-carbon target. Both of them would offer suitable references to China's low-carbon city development.

2 Development of China's Major National Urban Environmental Policies

2.1 International Background

Modern urban environmental policies can be dated back to the post-industrial revolution UK, when England's Ebenezer Howard proposed the concept of "garden city" in the late nineteenth century, as a means to mitigate the concentrated and polluted urban working, living space, and environment with smaller resident settlements of 30,000–50,000 people that are surrounded by green spaces such as parks and trees. This conceptual framework has laid the foundation for future urban environmental policies across the world. After the Second World War, specifically between the 1960s and 1970s, many industrialized countries like the United Kingdom, Germany, France, and Japan recovered and prospered significantly due to the rapid economic growth coupled with fast urbanization, meanwhile beginning to suffer the problems of housing shortages, deterioration of living conditions, and national environments, growing pollution-related problems like traffic, and diminishing green space, which gave birth to the first wave of sustainable urban environmental policies like the "ecopolis" in Germany and "amenity town" in Japan. These are the early concepts of today's "eco-cities" (Imura 2010).

2.2 China's Urban Environmental Policy Development

With the founding of the new nation in 1949, China during this time was still entrenched in previously implicated wounds of imperial and civil wars, as well as severe civil unrest. There were virtually no viable environmental legislations until the Chinese delegates attended the 1972 United Nations Conference on the Human Environment in Stockholm, after which the environmental awareness began to emerge in China. Since Deng Xiaoping's "Reform and Opening Up" policy sprang across in 1978 onward, China has achieved social and economic developments in astonishing paces at the expenses of shocking eco- and environmental costs. However, the fast economic growth and urbanization also provided the groundwork for subsequent national urban environmental policies in order to solve the problems that occurred (Bao 2012).

In the year 1986, Yichun city brought up the proposal of developing the first "eco-city" project in China, which was an indication of the local government's effort in tackling the local environmental and ecological challenges that arose during the period of rapid urbanization. In the early 1990s, the former Chinese Ministry of Construction (MoC) enacted the national framework of the "Garden City," focusing on the landscape and green space urban developments. This was the first national scale attempt to right the urban development into a more sustainable trajectory. In 2004, MoC's successor, the MoHURD, upgraded this framework into "Eco-Garden City," addressing with more pressing and comprehensive indicators for the related urban development. In the same year, the MEP came up with a national framework of eco-cities (containing three levels, county, city, and province), with more standardized concepts and indicators for urban development. In 2010, the NDRC initiated a nationwide call for "Low-Carbon City" developments, in an attempt to realize the CO_2 emission reduction target set in the 11th Five-Year Plan (refer to Fig. 1). According to a survey by the Chinese Society for Urban Studies (2012), 280 cities (out of the 287 cities that have been surveyed) have developed goals or plans for low-carbon city or eco-city projects, accounting for over 97 % of Chinese cities of municipality level and above.

The abundance in quantity of such projects does not necessarily guarantee the successful realization of urban sustainability or low-carbonization for cities. But it indeed displays the significant effort that the Chinese government has put into the transforming urban development strategies in a more sustainable manner. However, as of the present day, there has not yet been any well-established eco-city or low-carbon city in China that could offer scaling-up practices for other cities.

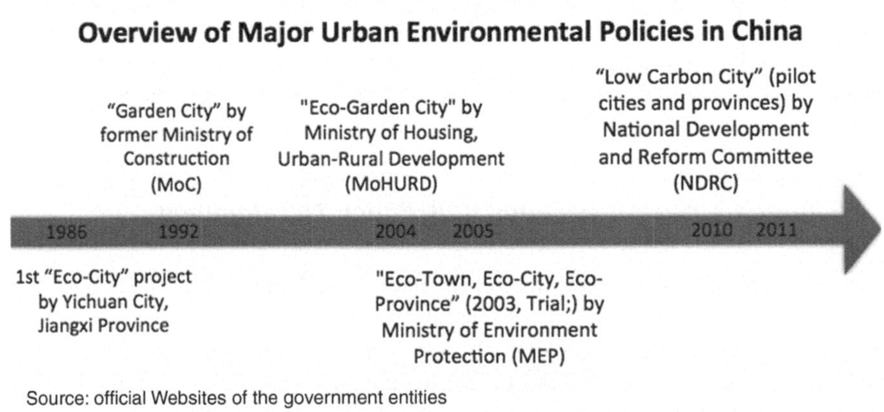

Fig. 1 Overview of China's major national urban environmental policies

Despite the existence of renowned cases such as Sino-Singapore Tianjin eco-city, these Sino-foreign projects are far from full completion and daily operations.

2.3 China's Low-Carbon Era

China took the lead among the developing countries to formulate and implement National Plan for Climate Change. NDRC (2007) issued the National Climate Change Program in accordance with UNFCCC provisions for addressing climate change. In 2009, the State Council announced a carbon intensity reduction target of 40–45 % by 2020 compared to the 2005 level. China took a further step of incorporating this target into the 12th Five-Year Plan (FYP), with a legally binding target of 17 % of CO_2 reduction per GDP unit from 2011 to 2015. Hence, low-carbon development has become the new norm in China's urban and economic developments. A series of policies, regulations, and frameworks have been put in efforts to support the related low-carbon projects, initiatives, and programs nationwide.

Since 2005, China's Ministry of Housing and Urban–Rural Development have formulated several policy measures to promote low-carbon public transportation and low-carbon constructions. In 2010, the NDRC officially put forward the initiative of experimental demonstration projects of low-carbon urban development in "5 provinces and 8 municipalities," which received active and extensive support from local governments. The following year, a second batch of trial provinces and cities was announced, totaled the trial places to eight provinces (including Beijing and Shanghai) and 34 cities [see (NDRC 2011)]. In 2011, State Council issued the Notice on Issuance of National Plan of Main Functional Areas, proposing "the development of low-carbon cities and reducing the intensity of GHG". In 2011, State Council issued the Opinions on Implementation of Division and Specialization of Key Departments in the Government Work Report, proposing to "advocate the experimental works of low-carbon cities" (Sun 2014). The Ministry of Finance (MoF) and MoHURD jointly issued the Notice on Implementation of Low-carbon Demonstration Town Pilot Program, promoting sustainable low-carbon developments in small cities and towns (2011).

After reviewing these major national urban policies in terms of their guidelines, frameworks, and indicator systems (where applicable) through substantial amount of desk research and comparisons, we have noticed several distinguishable features as well as shortcomings:

- China's urban environmental policies have mirrored the world's sustainable development trends especially since the 1980s onward, which normally reflected and address the major problems that occurred during that particular time under the global context. This suggests China's quick adaptation and responses to both domestic problems and international responsibilities.
- China's national urban plans are well accepted by local governments particularly in terms of the low-carbon eco-city projects and initiatives; however,

despite the large quantity and zealous participation, the successful cases that would offer larger-scale implementation are still insufficient. This is particularly due to China's vast geographic and geopolitical features, demographic characteristics as well as the uneven developments among cities, provinces, and regions. Developing local or regional appropriate adaptations of urban development models seems a rather reasonable principle.
- A shift from top-down implementation of policy to bottom-up policy development began to take shape, suggested by evolution of "Garden City," "Eco-Garden City," and "Eco-City" when the frameworks and indicator systems are developed by the national government body to be implemented by the local governments, to the development of "Low-carbon City," when local governments are required to develop their own comprehensive plans based on regional features. This is indeed a big advancement in the urban environmental policy developments.
- However, due to the lack of explicit definitions or clear distinctions of related concepts, such as "eco-cities," "low-carbon cities," or "low-carbon eco-cities," severe duplicities and overlaps are seen in the implementation of local urban actions. Many cities have participated in multiple pilot or trial programs, which causes complexity in implementations and more pressures on the local administrations and less efficiencies in program completions.

When it comes to the current status of low-carbon pilot cities in China, after reviewing their development master plans, Khanna et al. (2014) found that many of the low-carbon cities have too broad scopes and might not sufficiently tackle the essence of carbon mitigation (e.g., energy efficiency). Another key founding is that given the infancy status of low-carbon development, local city planners are still lacking knowledge related to low-carbon policies and practices. This suggests that China's low-carbon cities still have a long way to go; besides the eager pursuit of developing domestic low-carbon cities, actively referring to the international experiences, expertise of their best practices would produce valuable inputs to China's current endeavors.

3 International Practices: Lessons from Germany and Japan

From a global perspective, the low-carbon urban strategies vary in different regions and defined by particular local contest and country geopolitical settings. In Asia, countries like Japan and Singapore pay more attention carrying out the top-down management style to implement the national urban environmental policies into the local levels. China, despite its ambitious targets of GHG reduction and enthusiasm for developing low-carbon eco-cities, still needs to refer to the best available practices for inspiration and know-how. In the following part of this paper, we enlisted two case studies from Germany and Japan, the two leading countries in low-carbon sustainable urban development worldwide, to seek activities that would benefit China's local practices.

3.1 Case Study of Germany: "Rhein-Hunsrück District—from Energy Importer to Energy Exporter"

The district of Rhein-Hunsrück has a territory of 963 km^2 in the state of Rhineland-Palatinate in the southwest of Germany. There are approximately 103,000 residents in the district with 134 settlements, and 75 % of these settlements have less than 500 inhabitants. This district has the goal of converting the 290 million euros that spent on importing energy from outside of the region into regional community added value and jobs through improving energy efficiency and introducing renewable energy and switching its energy system into a highly efficient, local renewable sourced system by 2020. In 2009, the total electricity demand in the district was 473 million kWh, and in 2012, the share of electricity from renewable sources had reached 149 % and is expected to reach 286 % by 2015 (Fleck 2013).

The transformation process of Rhein-Hunsrück district started in 1999, when the local authority started to control the energy usage in the district-owned properties. In 2003, the district started to optimize the local building and improve energy efficiencies. In 2006, the district decided to develop their own comprehensive energy concept for the district. They commissioned a local research institute to integrate climate protection management schemes based on the already established concepts and aim to encourage more use of renewable energy and public participation, to achieve "zero emission" by 2020 Fig. 2.

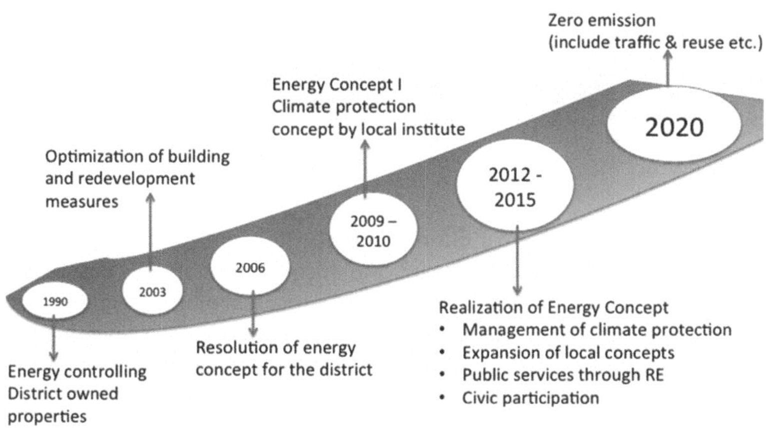

Fig. 2 Conceptual developments for Rhein-Hunsrück district's urban environmental policies (Fleck 2013)

From 1999 to 2009, these efforts had led to nearly 25 % reduction in heating demand, 5 % reduction in electricity demand, and 26 % reduction in water demand, approximately equivalent to 5,400 tons of CO_2 emission reductions and 1.13 million euros of cost savings. By 2012, their heating demand was further reduced by 26 % and water demand by 34 %, saving 9,500 tons of CO_2 emissions as well as costs up to 2 million euros. As a result, despite a doubling number of personal computers, installing of air-conditioned server and starting of catering and all-day school, their electricity demand increased only 1%. Without the additional energy control measures, the increase would have amounted to 30 % by 2012 (Fleck 2013).

A series of technologies have been implemented along the way, such as a district heating grid, heat pumps, photovoltaic systems, passive houses, biomass for heating, and an environmental education center and extracurricular learning center for environmental education. And if we set the total electricity demand of the district 473 million kWh in 2009 as 100 %, in 2012, the percent grew to 149 % for renewable energy (while the Germany national level was 22.9 %) with hydropower 3.6 %, wind energy 130.98 % (national 7.7 %), photovoltaic 12.10 % (national 4.7 %), and biomass 6.17 % (national 6.9 %). What is really notable is that privately owned wind and solar parks were established in municipal areas: 14 wind power plants with a permanent rent of ca. 300,000 euro per year for a duration of 20 years plus a percentage share and solar power plant with an output of 2 MW: The plant will pass into the ownership of the municipality after 25 years.

Source (Fleck 2013)

The district has devoted great efforts not only in adopting the technologies but also in transforming the education for the local communities, particularly the younger generations. For example, there are extracurricular educational facilities where kids can learn about renewable energy and ecological stewardship. Additionally, a public relation campaign focuses on improving early age children's eco-awareness and offering comprehensive information to the public with special marketing opportunities for citizen to contribute to the energy transition, community solidarity, and electricity cost savings.

3.2 Lessons Learned

In sum, the Rhein-Hunsrück district has made a comprehensive urban development plan with a focus on energy efficiency and application of renewable energy sources and implementation of eco-friendly technologies to maximize the local resources and conditions and greatly encourage the local stakeholder's participation in realizing this master plan. The results not only benefit their own residents, but also offer good references for China's low-carbon model town development.

3.3 Case Study of Japan: Kitakyushu Eco-Town Project

Japan is a leading country in terms of establishing low-carbon development strategies coupled with its technological advantages and financial power. In 2008, the Japanese government announced an ambitious goal of "Low-carbon Society" (LCS) by reducing 60–80 % of its CO_2 emissions (compared to 1990's level) by 2050. Later in 2009, then Prime Minister Hatoyama complemented this "ambitious pursuit" with a mid-term plan of reducing 25 % of CO_2 emission by 2020 (compared to 1990's level).

The city of Kitakyushu is located in the north of Kyushu Island of Japan, with a territory of 485 km^2. From the 1950s to 1970s, this heavy industrial city of iron manufacturing was severely polluted, especially air and water. The Dokai Bay of the city was so contaminated that it was gained the nickname of "Sea of Death," and public health suffered profoundly due to heavy pollution (see Fig. 3). Through decades of continuous and strenuous efforts of environmental protection and sustainable development, the city has returned to blue sky and clear water and has been awarded frequently with its efforts in sustainable urban transformation internationally and has been awarded or selected as a model city by the national government and international organizations including the UN and OECD (2013).

The eco-town project in Kitakyushu city encompassing the entire eastern section of the Hibiki Landfill Area was first approved by Japan's then Ministry of Industrial Trade and Industry (MITI) in 1997, which later became the Ministry of Economy, Trade and Industry (METI) in 2001. METI greatly promoted this project and

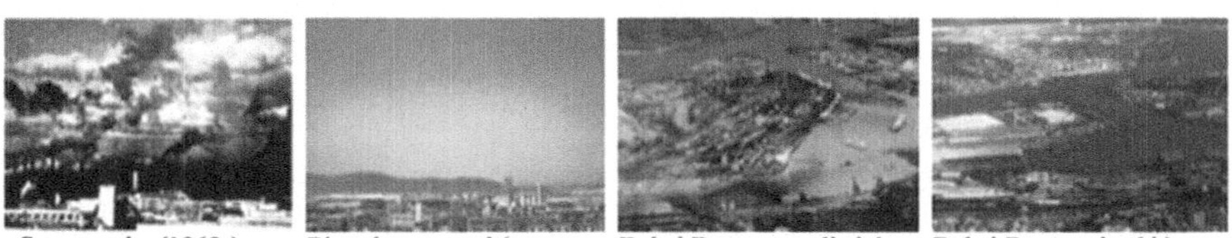

Smoggy sky (1960s) | **Blue sky restored (present day)** | **Dokai Bay once called the 'Sea of Death' (1960s)** | **Dokai Bay regained it's original beauty (present day)**

Fig. 3 Kitakyushu in its past and present [*Source* (Institute for global environmental strategies 2005)]

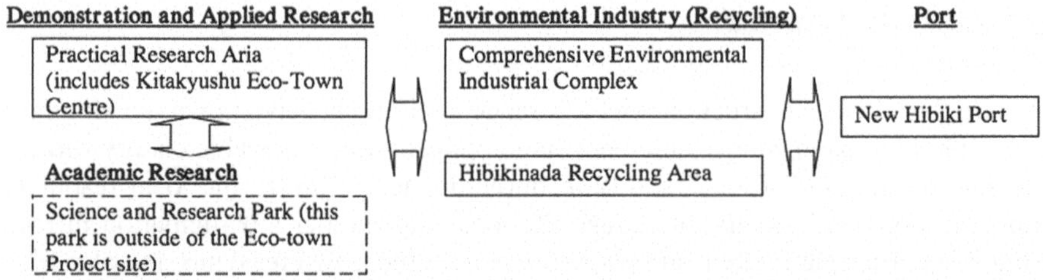

Fig. 4 Implementation strategy framework of Kitakyushu's eco-town project

offered subsidies for the constructions of infrastructure and marketing. The overall aim of this project is to promote zero emissions though reutilizing the wastes of local industries and to contribute to the Japanese 3R (Reduce, Reuse, and Recycle) Society Model. There are two stages designed for the project, the first stage (1997–2002) focused on "Recycle" and the second stage (2002–2010) focused on "Reuse". The implementation strategy is to combine academic research resources available in and outside of the city, the demonstration and applied research resources within the eco-town, and the nearby environmental industries, to develop a new recycling "green" site and ultimately a new port (Fig. 4).

The successful outcomes of this eco-town project of Kitakyushu city have made an international brand for Japan's local practices of eco-city initiatives; moreover, in terms of financial value, a total 50,200 million Japanese yen was invested, of which 7 % came from the private sector; over 1000 jobs were created. An accumulated 109,300 million yen was invested from 1998 to 2003.

3.4 Lessons Learned

Through the promotion and financial support of the government for the eco-town project, Kitakyushu city has transformed itself from a heavily polluted industrial city into an environmental industry city featuring in R&D in environmental technology and demonstration centers for resource recycling of the local industry with tangible economic gains. Through proper design, means of financing from both public and private sectors, integrated with the academic research and sciences, Kitakyushu city has become a good indication of how to redevelop an industrial city in a more sustainable, low-carbon manner.

4 Conclusion and Discussions

The previously enlisted two cases have demonstrated two different approaches to develop urban area into low-carbon eco-friendly cities of different population sizes, geographic features, and political systems. Germany's Rhein-Hunsrück district had transformed itself from a pure energy importer to an energy exporter by fully utilizing the local potentials for renewable energy production coupled with improvement of energy efficiencies and civic participations. For a city with the size of China's township level, Rhein-Hunsrück's approach could offer some reflecting insights to China's small-sized cities or townships' low-carbon sustainable developments.

The city of Kitakyushu in Japan indicates that both local and national governments would become the dominant factor for success or failure in a local region's transformation, especially for a traditional industrial city with population over one million. Though the government's political and financial support, more of the local resources and stakeholders from the industries, academic and general public could be integrated to serve the overall goal of eco-city/town development. A project at such a scale would only be possible with a functional government role well throughout the process.

China, as the world's second largest economy, is facing promises as well as challenges unprecedented to any other countries in the world. Blindly following other countries' methods or approaches for urban environmental development might not lead to the most desirable or applicable outcomes. However, continuing the "business as usual" development model is by no means sustainable with the current status quo. In an era when "low carbon" becomes the new norm, actively seeking best-practice examples and experiences would help China move in the right direction. More importantly, China needs to reinvent these ideas, concepts, and approaches based on its own geopolitical conditions and boldly pursue the sustainable urbanization practices with Chinese parameters.

References

Bao M (2012) Environmentalism and environmental movement in China since 1949. In: McNeill JR, Mauldin ES (eds) A companion to global environmental history, 2012. Blackwell Publishing Ltd, London

Bi J et al (2011) The benchmarks of carbon emissions and policy implications for China's cities: Case of Nanjing. Energy Policy 39(9):4785–4794

Chinese society for urban studies (2012) China low-carbon eco-city development report 2012. China Building Industry Press, Beijing

Feng Y, Zhang L (2012) Scenario analysis of urban energy saving and carbon abatement policies: a case study of Beijing city, China. Procedia Environ Sci 13:632–644

Fleck B (2013) Rhein-Hunsrück district—from energy importer to energy exporter. In: 4th International Conference on Circular Economy. Birkenfeld, Germany

Imura H (2010) Eco-cities: re-examining concepts and approaches. In: Liang FL, Gang C (eds) Towards a livable and sustainable urban environment: eco-cities in East Asia, 2010. World Scientific Publishing Company, Singapore

Institute for global environmental strategies (2005) Asia-Pacific Environmental Innovation Strategies (APEIS) Research on Innovative and Strategic Policy Options (RISPO) good practices inventory: KItakyushu eco-town project: online

Joss S, Tomozeiu D (2013) 'Eco-city' frameworks—a global review. University of Westminster: Online

Khanna N, Fridley D, Hong L (2014) China's pilot low-carbon city initiative: a comparative assessment of national goals and local plans. Sustain Cities Soc 12:110–121
Lehmann S (2013) Low-to-no carbon city: lessons from western urban projects for the rapid transformation of Shanghai. Habitat Int 37:61–69
Li Y, Gang X (2012) A report on some urban sustainability indicators practiced in the world, in reports of the city planning institute of Japan, No. 11, May 2012. http://www.cpij.or.jp/com/ac/reports/11-1_20.pdf
Li Z et al (2012) The development of low-carbon towns in China: concepts and practices. Energy 47(1):590–599
Ministry of Finance of the People's Republic of China (2011) Notice on implementation of low-carbon demonstration town pilot program.: online
National Development and Reform Commission's (NDRC) (2010) Notice on the initiation of trial low-carbon province and low-carbon city. NDRC: online
National Development and Reform Commission's (NDRC) (2011) Notice on the Second Batch of Trial Low-carbon Province and Low-carbon City, NDRC: online
National People's Congress (2011) The twelfth five-year plan on national economy and social development: Online
OECD (2013) Green growth in Kitakyushu, Japan. In: OECD green growth studies
State Council of the People's Republic of China (2009) China to cut 40 to 45 % GDP unit carbon by 2020: Online
Sun Y (2014) APEC Low-carbon model town project development in China. Department of International Cooperation, National Energy Administration, China. Presentation retrieved on Sep 2014, http://apecenergy.tier.org.tw/database/db/ewg43/ewg43/agenda8_chinaapec_low_carbon_town_demonstration_project_0229.ppt
United Nations, World Population Prospects (2012) United Nations: Online
Yang L, Li Y (2013) Low-carbon city in China. Sustain Cities Soc 9:62–66
Zou X, Li Y (2014) How eco are China's eco-cities? An international perspective. Int Rev Spat Plan Sustain Dev 2(3):18–30

Evaluation of Ventilation Effectiveness of Microscale and Middle-Scale Urban Green Belt Based on Computer Simulation

Xiao-ming Kuang, Jun Chen and Chang-feng Sun

Abstract As climate problems deteriorate nowadays, urban air path is playing a positive role in promoting urban air circulation, mitigating heat island effect, and reducing air pollution. Despite certain domestic and international researches on the planning and layout of urban air path, few are about the building factors of the air path itself. In this chapter, the authors performed computer quantitative simulation for microscale and middle-scale urban green belt, which is the object of study. By making single factor simulation and comparative analysis of the three factors: the air path width, the angle between air path direction and prevailing wind direction, and the height of obstacles in air path, the authors evaluated the ventilation effectiveness of air path, which can provide a reference for the planning and construction of air path during the rapid urbanization period in China.

Keywords Urban air path · Computer simulation · Evaluation of ventilation effectiveness

1 Introduction

In recent years, lots of researches and construction of regions and cities at home and abroad have shown that: urban air path developed for urban weak wind or static steady wind environment plays a decisive role in the promotion of urban air circulation, mitigation of the heat island effect, and reduction of air pollution.

X. Kuang
College of Architecture and Urban Planning, Tongji University, Shanghai 200092, China

J. Chen (✉) · C. Sun
Shanghai Tongji Urban Planning Design Institute, 8F, No.38 Guokang Rd, Shanghai 200092, China
e-mail: 450787313@qq.com

S. Feng et al. (eds.), *Low-carbon City and New-type Urbanization*,
Environmental Science and Engineering, DOI 10.1007/978-3-662-45969-0_25

As early as in the 1970s, Germany began the applied research of climate spatial planning and in the 1990s developed the assessment guidelines for urban climate environment VD13787 and technical standards (VDI 1997; Liu 2012). Japanese started this research in the late 1990s and published the Environmental Assessment Tool for Integrated Performance of Buildings—Heat Island Effect (CASBEE-HI) in 2008 (Association of architecture environment and energy conservation of Japan 2008). Hong Kong special administrative region also launched the Urban Climatic Map and Standards for Wind Environment—Feasibility Study in the year of 2003 and gradually incorporated the research results into the statutory plans for the planning and development of all regions of Hong Kong in 2007 (HKSAR Government Planning Department 2013).

At present, special plans for urban air path are ongoing in a number of cities in China. It has been proposed in the A Study of master plan of "Ecological Fuzhou" compiled in 2014 to plan ten urban air paths relying on the river network in Fuzhou (Fuzhou Urban and Rural Planning Bureau 2014); and in the 2012–2020 Master Plan for Wuhan City promulgated in 2009 to establish a "natural breathing" system of Wuhan by constructing ten urban air paths at different levels in the downtown and linking them with six identified ecological green wedges. The list goes on with Changsha, Hangzhou, Langfang, Nanjing, etc. It is clear that urban air path has become one of the academic issues of concern.

2 Current Researches of Urban Air Path

Currently, lots of studies focus on overall assessment of macroscale wind environment, arrangement way of air path planning as well as layout strategies of microscale block buildings, so we do not go further with the description of these research statuses quo here (Liu and Shen 2010; Chao et al. 2014; Yang 2012). However, researches on width, direction, underlying surface, and the other aspects of urban air path are few, and whose research status quo is as follows.

German scholar Kress believes that an air path with effective ventilation has the ground roughness below 50 cm; the length no less than 1,000 m; the width above 5 m; the height of internal obstacles less than 10 % of the air path width and below 10 m; and buildings or tall trees inside as few as possible. In the 2012–2020 Master Plan for Wuhan City promulgated in 2009, six green wedges are planned as first-grade urban air path, with their width varying from several hundred meters to one kilometer; in addition, the plan specifies the width of second-grade urban air path to be between 100 and 300 m. The Technical Guidelines for Urban Ventilation Planning *of Changsha City* prepared in 2010 specifies that the primary air path inside a city group should have a width greater than 50 m, and preferably greater than 100 m, and the secondary air path, with a width greater than 30 m, should echo with the primary air path (Changsha Urban and Rural Planning Bureau 2010).

The *Research of Evaluation Standard for Green Ecological Demonstration Zone in Beijing* developed by Beijing Municipal Planning Commission in 2013 also regards the direction and the width of air path as rigid control targets.

However, in these air path plans above, the determination of the relevant value yet needs scientific evidence and its reliability is doubtful. Take width for example, should it be the greater the better, or is there a certain threshold, that, an air path with a width equal to or greater than it will bring the maximum ventilation effectiveness? In addition, which is more conducive to the overall wind environment, is it a parallel between air path direction and prevailing wind direction, or a certain angle there between? And what limit on the height of obstacles (buildings or facilities) inside air path should be set to ensure its ventilation effectiveness? It needs take simulation one by one for each of the relevant factors before scientific conclusions can be reached for the questions above.

3 Research Assumption

The existing researches indicate that the ventilation effectiveness of air path depends on multiple factors, including the air path width, the angle between air path direction and prevailing wind direction, and the height of obstacles in air path.

In this study, a piece of urban micro-climate simulation software has been used to make quantitative simulation of air path at the level of block (Fig. 1), construct an ideal model, and assign different values to these factors, then verify them one by one. This study is mainly targets at single factors and does not cover the interaction among multiple factors.

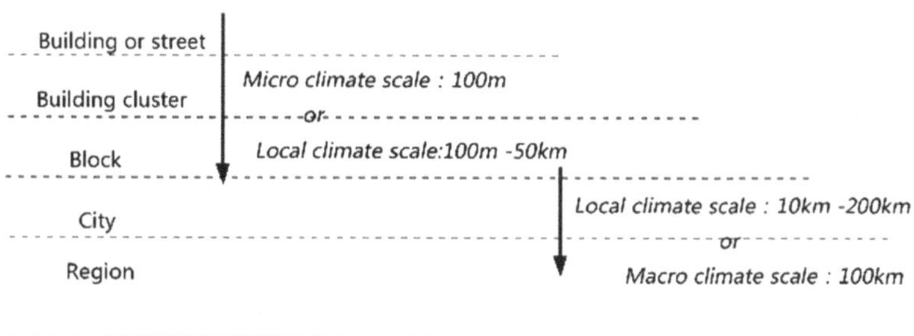

Fig. 1 The relationship between climate scale and urban scale

4 Ventilation Effectiveness Analysis of Single Factors of Microscale and Middle-Scale Air Path

4.1 Ideal Model

The ideal model for this study is set as follows: Assuming there is a research region dimensioning approximately 500 m × 500 m, with a total area of 25 ha, divided into four plots; the main orientation of the buildings in the plots is perpendicular to air path; buildings have a height of 40 m. Take the north–south air path in the middle of the plots as the main object of study (Fig. 2). The initial settings of the model are given as: temperature 27 °C, wind speed 3 m/s at a height of 10 m, and relative humidity 50 %.

4.2 Width Factor

Width is one of the factors important to the performance of the ventilation effectiveness of air path, as certain requirements are imposed on the width of air path by many plans or studies. In this study, five sets of values, 10, 20, 30, 50, and 100 m, were selected for analysis. The simulation results are shown in Figs. 3 and 4 (for comparison, the left and right sides of the simulation results of Fig. 3 have been cropped, similarly below).

The simulation results from Fig. 3 indicate that when the air path has a width of 10 and 20 m, the wind speed is 1.5 m/s or less, respectively, in as many as 57.22 and 47.59 % cases, and area with a wind speed ratio of 0.8 and above is only 3.09 and 5.09 % (Table 1), indicating poor ventilation. In fact, according to the fire

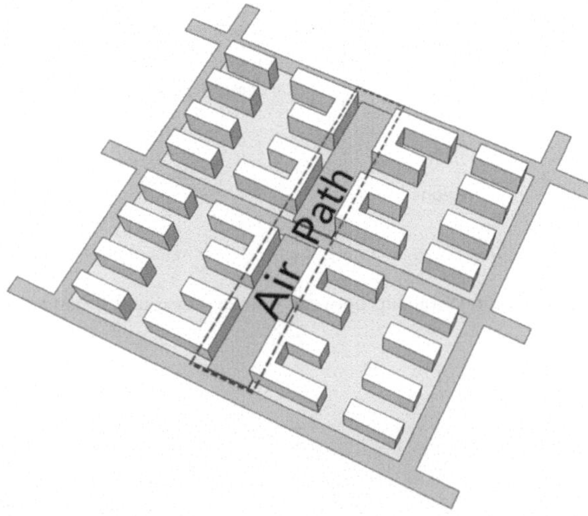

Fig. 2 Ideal model

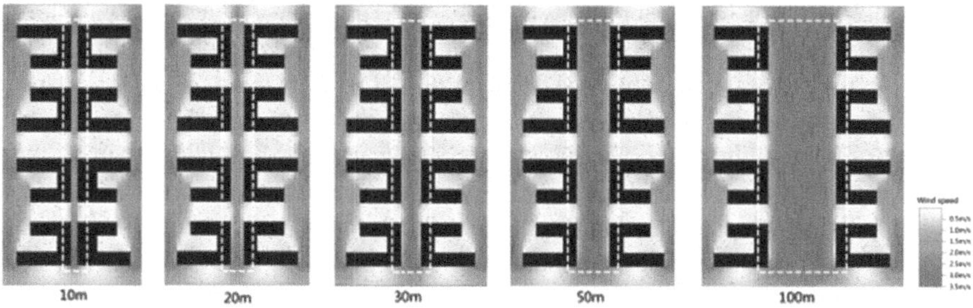

Fig. 3 Simulation results of wind field at the height of 1.5 m with different widths

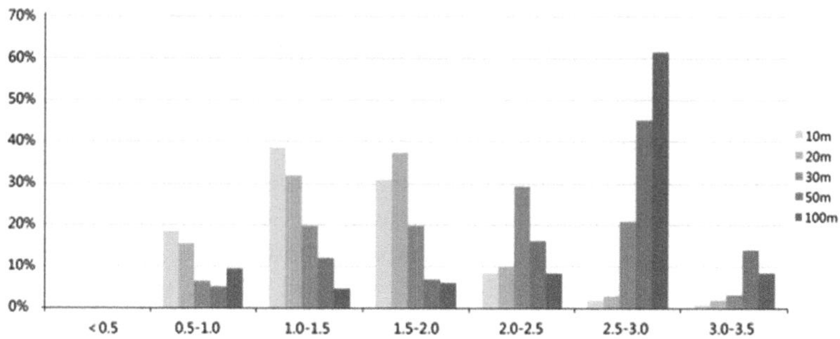

Fig. 4 Statistics on the simulation results of wind field at the height of 1.5 m with different widths

Table 1 Statistics on the simulation results of wind field at the height of 1.5 m with different widths

Wind speed (m/s)	10 m (%)	20 m (%)	30 m (%)	50 m (%)	100 m (%)
<0.5	0.00	0.00	0.00	0.00	0.09
0.5–1.0	18.59	15.58	6.54	5.11	9.56
1.0–1.5	38.63	32.00	19.82	12.21	4.78
1.5–2.0	31.08	37.23	19.90	7.04	6.62
2.0–2.5	8.61	10.09	29.48	16.27	8.58
2.5–3.0	1.99	2.94	20.91	45.26	61.62
3.0–3.5	1.10	2.15	3.34	14.10	8.75

design specification for buildings, multi-story buildings shall have a minimum spacing of 6 m with each other, and a multi-story building and a high-rise building shall have at least a spacing of 9 m, while high-rise buildings, no less than 13 m. Since the spacing among buildings can partially serve as an air path, a separate air path narrower than 20 m is of little significance. When an air path has a width up to 30 m, the proportion of static wind area is greatly reduced, with the area having a wind speed greater than 1.5 m/s amounting to 73.64 % and the area having a wind speed ratio of 0.8 or more increasing to 24.25 % (Table 1). When an air path has a width of 50 and 100 m, the flow inside the air path extremely smooth, with the area having a wind speed ratio above 0.8 accounting for 59.36 and 70.37 %, respectively, indicating good ventilation.

Therefore, the authors believe that in microscale and middle-scale, urban air path shall at least have a width of 30 m for them to function well; and 100 m to a threshold, above which air path will have extremely good ventilation effectiveness.

4.3 Direction Factor

The relationship between air path direction and prevailing wind direction is essential to ventilation effectiveness. In the Research of Evaluation Standard for Green Ecological Demonstration Zone in Beijing, air path direction is set as an important indicator and required to be parallel to prevailing wind direction. Is it most conducive to ventilation when they are parallel? In this study, we took an air path having a width of 50 m for example and simulated the five cases in which its direction was parallel to or formed an angle of 15°, 30°, 45°, and 60° with prevailing wind direction. The simulation results are shown in Fig. 5 (for convenience of comparison, the four figures of 15°, 30°, 45°, and 60° are set right).

The simulation results show that when air path direction is parallel to or forms an angle of 15° and 30° with prevailing wind direction, the ventilation effectiveness is quite good (Figs. 5 and 6): Area having a wind speed greater than 1.5 m/s, respectively, is 82.63, 83.42, and 81.02 %, and area with a wind speed ratio above 0.8, respectively, accounts for 59.34, 51.27, and 34.48 % (Table 2). When air path direction is parallel to prevailing wind direction, the wind quickly goes through the air path, resulting good ventilation. But the authors believe that while the angle between air path direction and prevailing wind direction is within 30°, not only the ventilation effectiveness of air path is good, but also the vortex area in the leeward side of buildings is reduced, creating a uniform outdoor wind field. When angle between air path direction and prevailing wind direction is 60° and 45°, the area having a wind speed ratio above 0.8 is significantly reduces to 2.04 and 0.36 %, and

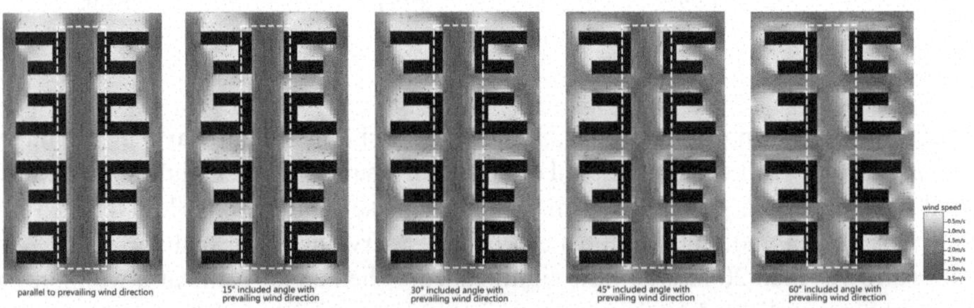

Fig. 5 Simulation results of wind field at the height of 1.5 m with different directions

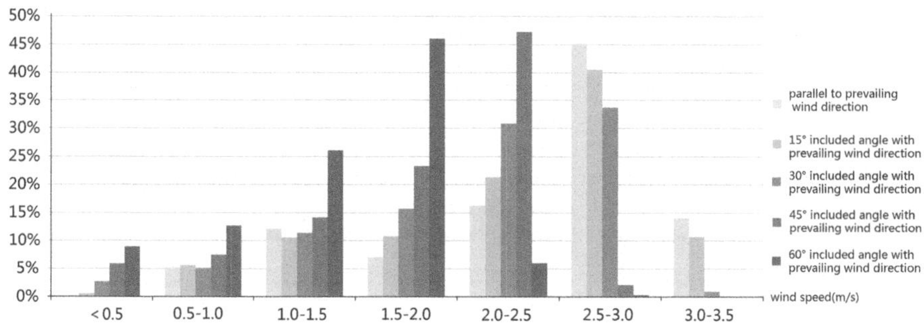

Fig. 6 Statistics on simulation results of wind field at the height of 1.5 m with different directions

Table 2 Statistics on simulation results of wind field at the height of 1.5 m with different directions

Wind speed (m/s)	0° (%)	15° (%)	30° (%)	45° (%)	60° (%)
< 0.5	0.00	0.51	2.65	5.80	8.87
0.5–1.0	5.11	5.54	4.93	7.35	12.67
1.0–1.5	12.21	10.53	11.40	14.10	25.99
1.5–2.0	7.04	10.72	15.76	23.31	46.16
2.0–2.5	16.27	21.43	30.78	47.40	5.95
2.5–3.0	45.26	40.75	33.64	2.04	0.36
3.0–3.5	14.10	10.52	0.84	0.00	0.00

the area having a wind speed below 1.5 m/s also increases to 27.25 and 47.53 %, resulting significant decrease in the ventilation effectiveness of air path.

As far as the evaluation of ventilation effectiveness is concerned, it is needless for air path direction to be absolutely parallel to prevailing wind direction. An angle is 30° or below has a small impact on the ventilation effectiveness of air path and is more conducive to the overall wind environment. In addition, due to various factors such as terrain, rivers, trend of road, and other restrictions, air path direction is not always parallel with prevailing wind direction. Therefore, allowing a certain angle between them can construct more air path and create good urban microclimate.

4.4 Obstacle Height Factor

Mainly in the form of urban green belt designed with the function of outdoor recreation, air path is interlaced with a certain number of service buildings. Such service buildings, as the obstacles in air path, have a great impact on the ventilation effectiveness of air path. What is more, the taller the buildings are, the greater the impact is. On the premise that air path direction is parallel to prevailing wind

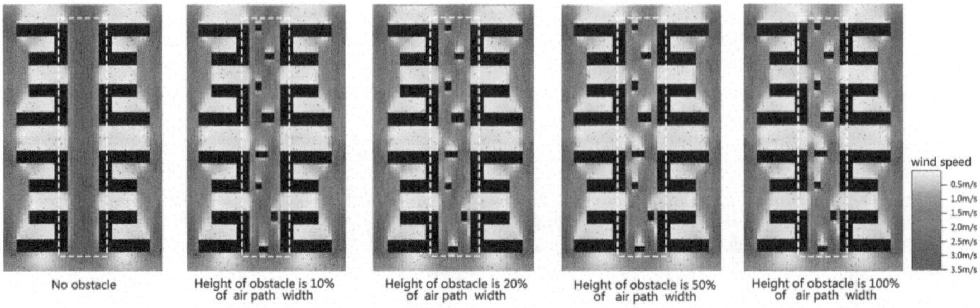

Fig. 7 Simulation results of wind field at the height of 1.5 m with different obstacle heights

direction, we took an air path having a width of 50 m for example, and simulated the five cases of no obstacles, obstacles having a height that 10, 20, 50, and 100 % of the air path width. The simulation results are shown in Fig. 7.

The simulation results show that the air path obstacles act as obvious barriers to wind. Regardless the height of obstacles, there is surely a certain shadow area of low wind speed in the leeward side of the obstacles (Figs. 7 and 8). When the obstacle height is 10 and 20 % of air path width, the overall ventilation effectiveness of air path is good, as the area having a wind speed greater than 1.5 m/s is, respectively, 64.42 and 58.52 % (Table 3), but still a significant drop from 82.63 % in the case of no obstacles. When the obstacle height is 50 and 100 % of air path width, the wind speed in the air path obviously slows down, as the area having a wind speed greater than 1.5 m/s is less than 40 % (Table 3), and the ratio of static wind area below 0.5 m/s increases significantly.

As can be seen from Fig. 9, while the height of the obstacles increases, shadow area of the low speed wind in the leeward side of obstacles gradually rises. When the height of obstacles reaches 50 % or more of the air path width, the area having a

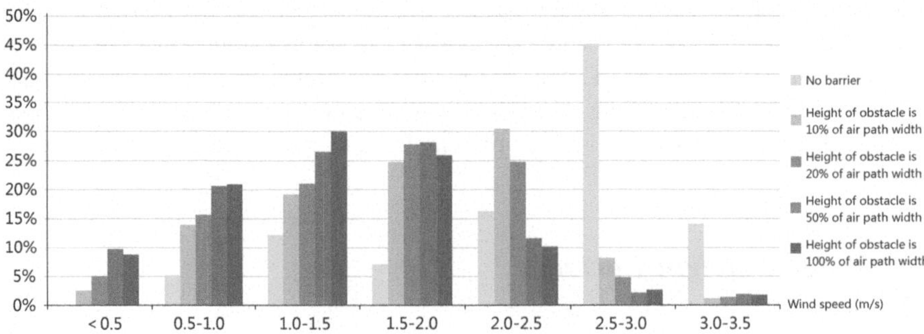

Fig. 8 Statistics on simulation results of wind field at the height of 1.5 m with different obstacle heights

Table 3 Statistics on simulation results of wind field at the height of 1.5 m with different obstacle heights

wind speed (m/s)	0 (%)	10 (%)	20 (%)	50 (%)	100 (%)
< 0.5	0.00	2.50	5.03	9.65	8.62
0.5–1.0	5.11	13.90	15.50	20.51	20.85
1.0–1.5	12.21	19.18	20.94	26.42	30.13
1.5–2.0	7.04	24.74	27.68	28.04	25.82
2.0–2.5	16.27	30.40	24.70	11.49	10.15
2.5–3.0	45.26	8.12	4.81	2.12	2.64
3.0–3.5	14.10	1.16	1.33	1.78	1.79

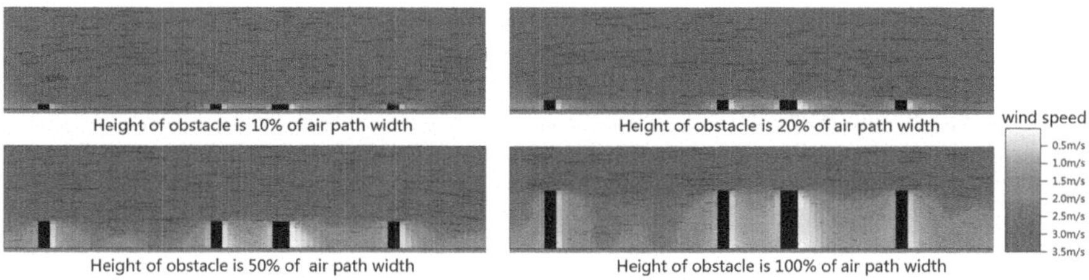

Fig. 9 Simulation results of wind field in Z direction with the height of 1.5 m with different obstacle heights

wind speed less than 1.5 m/s increases significantly. Therefore, the study suggests that the height of obstacles in air path should not exceed 20 % of air path width, and the shorter the better.

5 Conclusions and Prospects

This study focused on the ventilation effectiveness of microscale and middle-scale urban green belt and performed computer simulations on the air path width, the angle between air path direction and prevailing wind direction, and the height of obstacles in air path. The following conclusions are drawn:

1. In microscale and middle-scale, urban air paths shall have a width not less than 30 m, less than which the air path will be poorly ventilation. 100 m is the threshold for microscale and middle-scale urban air path, above which air path has a lower potential for promoting ventilation effectiveness.
2. When urban air path direction is parallel to prevailing wind direction, the air path has the maximum ventilation effectiveness; but an angle between the urban air path direction and prevailing wind direction less than 30° is more conducive to the overall wind environment.

3. The height of obstacles (buildings or facilities) in air path is negatively correlated to the ventilation effectiveness of air path, and the height of obstacle should be less than 20 % of air path width.

Subsequent studies will continue to focus on the evaluation of the superimposed impact of multiple factors on the ventilation effectiveness of microscale and middle-scale urban air path, as well as the evaluation of the ventilation effectiveness of macroscale urban air path.

References

Changsha Urban and Rural Planning Bureau (2010) Technical guidelines for urban ventilation planning of Changsha City (Revision for Approval)

Chao R, Yuan C, He Z, Wu E (2014) A study of air path and its application in urban planning. Urban Plan Forum 03:52–60

Fuzhou Urban and Rural Planning Bureau (2014) A study of master plan of "Ecological Fuzhou"

Liu S (2012) Comparative research on integration approaches of urban climate studies in urban planning between China and Germany. The Dissertation for Doctor Degree of Zhejiang University, Hangzhou

Liu S, Shen J (2010) Urban ventilation channel planning method based on local circulation: a case study of Stuttgart, Germany. J Zhejiang Univ (Eng Sci) 44(10):1985–1991

The Chinese University Hong Kong (2013) Urban climatic map and standards for wind environment—feasibility study. HKSAR Government Planning Department, Hong Kong

The Chinese University of Hong Kong (2006) Feasibility study for establishment of air ventilation assessment system. HKSAR Government Planning Department, Hong Kong

VDI (1997) VDI-guideline 3787, part 1, environmental meteorology-climate and air pollution maps for cities and regions. Beuth Verlag, Berlin

Yang T (2012) High-rise residential district wind environment spatial distribution adaptability research in hot summer and cold winter zone. The Dissertation for Master Degree of Hunan University, Changsha

Research on Spatial Planning Strategy of Low-Carbon Development of Capital Region in China

Xiao Wu and Yue Zhang

Abstract Development of the capital region in China now faces resources, environment, population, traffic, and various kinds of pressure. Spatial development on the capital region requires effective implementation of "human-oriented, low-carbon development". Based on the view of the framework of low-carbon governance strategies for the capital region, it should works on areas as regional synergy, integration, enhance efficiency, maintain ecological, energy saving and emission reduction to achieve the goal of low-carbon and sustainable development. At the same time, also need to study on the management mechanism of regional coordination to promote regional low carbon development.

Keywords Capital region · Sustainable development · Low-carbon · Planning strategy

1 Introduction

The term "low-carbon economy" was first proposed in the white paper "our energy future—creating a low-carbon economy" in 2003 (State for Trade and Industry, UK, 2003). The core of low-carbon economy lies in the higher economic output with less energy consumption. With regard to the "low-carbon city" concept, a unified cognition has not been formed yet at home and abroad. The "low-carbon city" concept involves the innovation of various aspects including energy technology, production method and life style, and urban planning policies and systems. Low-carbon planning as an important decision-making basis for regional development and regulation is a guideline for future regional development, and also

X. Wu (✉) · Y. Zhang
School of Architecture, Tsinghua University, Beijing 100084, People's Republic of China
e-mail: wuxiao@mail.tsinghua.edu.cn

S. Feng et al. (eds.), *Low-carbon City and New-type Urbanization*,
Environmental Science and Engineering, DOI 10.1007/978-3-662-45969-0_26

the tool for policy formation and implementation plays the strong lock-in role on regional development. Low-carbon planning is a technical, professional, and forward-looking public policy.

2 Studies in Low-Carbon Urban Planning

Low-carbon urban planning regards reduction of energy consumption and greenhouse gas emissions as the planning objective and decision-making premise, on the basis of sticking to the original urban planning theories and index system. It gives priority to rational spatial layout of lands and comprehensive deployment and specific arrangement of various facilities.

Professor Peter Hall believes that during the period of paying attention to climate changes, low-carbon urban planning will be a great revolution of planning. He suggests the planning circle rethink the urban planning theories, methods, and principles to address climate changes (Chaolin et al. 2009). On the basis of the land use planning framework, Gustavo and other scholars put forward the coordination mechanism for building the urban planning technology system and the urban planning evaluation system and making spatial development decisions (Arciniegas and Janssen 2012). In his book Effective Practice in Spatial Planning, Janice Morphet proposed the concept that spatial planning is a key factor for alleviating urban climate changes and achieving low-carbon development. Janice (2011) Crawford thinks it necessary to pay attention to the "low-carbon city" concept and strengthen application of low-carbon technologies during spatial planning (Jenny and Will 2008). Edward proposes to achieve the overall development of regional economic environment through reasonable policy adjustment and land use planning (Glaeser and Kahn 2008). In the metropolitan development plan of Chicago, studies on the low-carbon city focus on energy consumption and carbon emission related to traffic and urban density (Chicago Metropolis 2010).

Baoxing (2010) proposed to change existing planning theories and methods and to explore low-carbon city development models with Chinese characteristics. Qingji et al. (2010) systematically summarized the relevance among connotation, characteristics, and construction principles of low-carbon eco-city. Chaolin et al. (2010) summarized the low-carbon urban planning mainly from five aspects: study of low-carbon urban planning theory framework, study of innovation in overall low-carbon urban planning, study of innovation in low-carbon urban special planning, study of technical methods and indicator systems for low-carbon urban planning, and study of the mechanisms for construction and implementation of low-carbon urban plans. Haixiao et al. (2008), on the basis of the outlook on development of low-carbon city, put forward policy suggestions on urban traffic, land use, density control, and function mix from the perspectives of regional planning, overall planning, and detailed planning. Yaping et al. (2008) assessed the urban compact degree through the urban performance scoring method. Chengri (2005) proposed to promote balanced employment and housing by improving land use density, mixing use, and strengthening land use and traffic integration. Zuda (2010) put forward a set of low-carbon urban cost-effectiveness theories and methods to assess the low-carbon urban planning policies and investment decisions.

3 Status of Development of the Capital Region

3.1 Spatial Range

The academic circle has always been discussing about the spatial range of the capital region. From the perspective of regional planning management, to ensure the plan implementation and effectiveness, most scholars always define the range of the capital region according to administrative boundaries.

The author thinks it necessary to take into account the geographical boundaries and economic attributes of the capital region to determine the spatial range. The capital region should be consistent with the scope of "Capital Economic Circle", namely the Beijing-Tianjin-Hebei Region. The Capital Economic Circle mainly covers the "8 + 2" urban agglomeration, namely two mega-cities Beijing (the capital) and Tianjin, as well as eight cities in Hebei Province (Shijiazhuang, Qinhuangdao, Tangshan, Langfang, Baoding, Cangzhou, Zhangjiakou, and Chengde). The capital region has a total area of 188,000 km^2, and a total population of over 80 million (2010) (Fig. 1).

3.2 Status Quo and Problems

Currently, the capital region with a population of more than 80 million has almost reached the verge of ecological carrying limit in total population and density, becoming a world-class urban agglomeration characterized by high density and rapid growth. The capital region faces severe challenges of population, resources, and environment, including sharp increase of population, continuous deterioration of air quality, frequent occurrence of water crises, continuously widened development gap, and concentrated outbreak of urban development issues in a short term, thus facing unprecedented challenges to achieve regional sustainable development. In addition to the region's unique resource endowment, regional development imbalance and low development efficiency become more prominent (Table 1).

On the whole, the capital region still has the following problems in low-carbon development and construction:

First, the traditional high-carbon industrial economy pattern hinders the low-carbon development. Under the extensive development mode, traditional industries have the problems including the high consumption and high emissions,

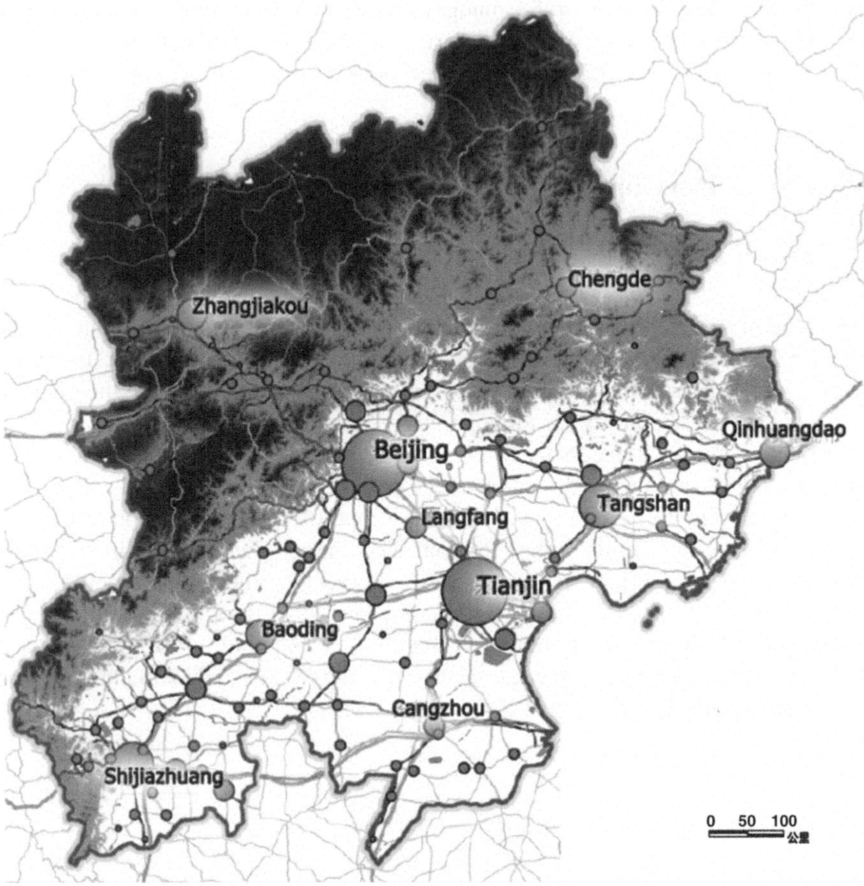

Fig. 1 Spatial range of the capital region

which aggravates the destruction of the regional ecological environment, and the shift from high-carbon industrial economy to low-carbon economy has not been achieved yet.

Second, utilization rate of renewable energy is low. As a traditional industrial region, the capital region has been seriously depending on non-renewable energy sources including coal and oil for a long term, and a large number of industrial enterprises have a single energy structure, thus resulting in the very slow popularization and application of new energy.

Third, population increases rapidly, and environmental resource carrying capacity continuously declines. Currently, the capital region faces increasing demands for expanding the construction scale and increasing pressures for shortage of construction and maintenance funds; problems of resources, environment, and

Table 1 Status quo of cities in the capital region

No.	Name	Total area (km^2)	Resident population (ten thousand)
1	Beijing	16,411	1,961.2
2	Tianjin	11,946	1,293.8
3	Shijiazhuang	20,235	1,016.4
4	Baoding	22,185	1,119.4
5	Tangshan	13,472	757.7
6	Chengde	39,519	347.3
7	Qinhuangdao	7,812	298.8
8	Langfang	6,500	435.9
9	Cangzhou	13,419	713.4
10	Zhangjiakou	36,860	222.5
Total		188,359	8,166.4

Note Excerpted from the data of the sixth population census (2010)

traffic become more and more serious, and regional overall carrying capacity is worrying.

Fourth, regional development coordination is still insufficient. Regional development imbalance remains to be prominent, and it is difficult to establish the regional coordinated development mechanism.

Fifth, citizens' low-carbon awareness is still relatively weak. At present, many citizens in the capital region have somewhat understood the concepts of low-carbon life including conservation of resources, greening, and environmental protection, as well as healthy travel. But it is still necessary to deepen citizens' understanding of low-carbon city, so as to improve their qualities.

4 Framework and Strategy of Low-Carbon Planning for the Capital Region

4.1 Low-Carbon Planning Framework

To achieve the low-carbon development in the region, overall consideration is necessary from the perspective of planning, and targeted planning should be made according to key factors and links affecting regional development. Urban and regional development is complex and integrated, involving many fields such as economy, policy, industry, population, technology, and management. According to resource endowment and development conditions, this chapter analyzes the deep-rooted driving factors and considers possible main planning pressures of all levels of factors, pointing out the direction for planning and design, as well as preparation (Table 2).

Low-carbon planning for the capital region covers various social and economic aspects, but the unified and systematic planning and design frameworks have not

Table 2 Development factors of low-carbon planning for the capital region

Driving force	Real difficulties	Low-carbon planning indicators
Industrial Structure	Extensive growth mode	Proportion of the tertiary industry
Spatial structure	Low land utilization rate	Greening rate, spatial equilibrium
Demographic structure	Excessive concentration of population	GDP per capita
Energy structure	Coal as the main energy	Proportion of renewable energy
Development efficiency	Insufficient promotion of renewable energy	Energy consumption per 10,000 Yuan GDP
Low-carbon life philosophy	Long way to go to advocate the low-carbon concept	Low-carbon living standard index
Low-carbon technologies		
Low-carbon development policy	No unified standard	Low-carbon R&D funding levels
Fund support	Limited special fund sources	Low-carbon development plan

been formed among cities, departments, and subjects. Starting from analysis of the status quo of the current systems and processes for urban and rural planning, this chapter decomposes the objectives of low-carbon development to specific planning types and planning technologies, reflects them in the planning system, and initially builds a design framework for low-carbon urban planning. This chapter first describes the regional coordination and low-carbon objectives, determines the principles and standards of planning, and expounds the low-carbon development concept from the perspectives of region, city, community, and architectural design, according to the characteristics of the capital region (Table 3).

4.2 Low-Carbon Space Development Strategy

1. Control the urban boundary and scale and construct a rational urban system
   The blind expansion mode existing for a long time has resulted in an ineffective control over the development scale and timing sequence of the cities within the capital region. Phenomena of planning failures are rather popular. On the other hand, the blind expansion of urban space has intensified the indeterminacy and repeatability of the population mobility; in most cases, development resources and urban infrastructures cannot be effectively concentrated, which leads to the appearance of "the poverty belt around the capital". Therefore, to promote low-carbon regional development, it is necessary to improve the scientificity and foresight of planning formulation and design, define the urban expansion

Table 3 Design framework of low-carbon spatial planning for the capital region

Regional low-carbon development			
Target system	Livable environment	Intensive use of land	Green traffic
		Improve the land utilization efficiency	Clean energy
Planning principles	Maintenance of ecosystem	Inventory integration TOD development mode	Low-carbon travel
	Design based on natural environment		
Region	Establishment of regional ecological green system	Resources and major implementations	Establish the regional rapid transit system
		Regional co-construction and sharing	
City	Open space design	Enhance mixed land use and job-housing balance	Improve the quantity and level of public transport facilities
	Urban greenways		
Community	Focus on environmental design and communication needs	Keep development intensity and develop functional space	Build a cycle of fast and convenient living
Architectural design	Introduction of green architectural technology and equipment	Composite space, focusing on interface design	Design the space for walk and bikes

boundary, further consolidate the seriousness of the urban planning, and strengthen the urban planning management. Through rational planning and effective control, we should construct a clearly leveled urban system suitable for development of the capital region and form a regional development mode featured by "two cores and multiple centers" (i.e., Beijing and Tianjin as the two cores and the surrounding cities as specialized function centers).

2. Alleviate the capital functions and build a spatial pattern with "one axis and three belts"
   Chinese urban and rural planning system is characterized by distinct political features and administrative properties. Since Beijing is the capital of China, its primacy has been continuously strengthened during the previous urban and regional planning, numerous development resources have been quickly concentrated and the urban functions have been enriched on and on, which results in the excessive aggregation of the population, capital, and transportation as well as the appearance of a very serious "city disease". Along with the formation of the regional development outlook, urban functions of Beijing have been evidently alleviated. In the future, Beijing will still reserve its role as the capital, economic center, and cultural center, and the other non-core functions need to be alleviated to the surrounding cities effectively. On one hand, it can alleviate Beijing's environment resource pressures, and strengthen Beijing's service functions of global significance through integrating the stock land, so as to build a world city image. On the other hand, Beijing can bring development

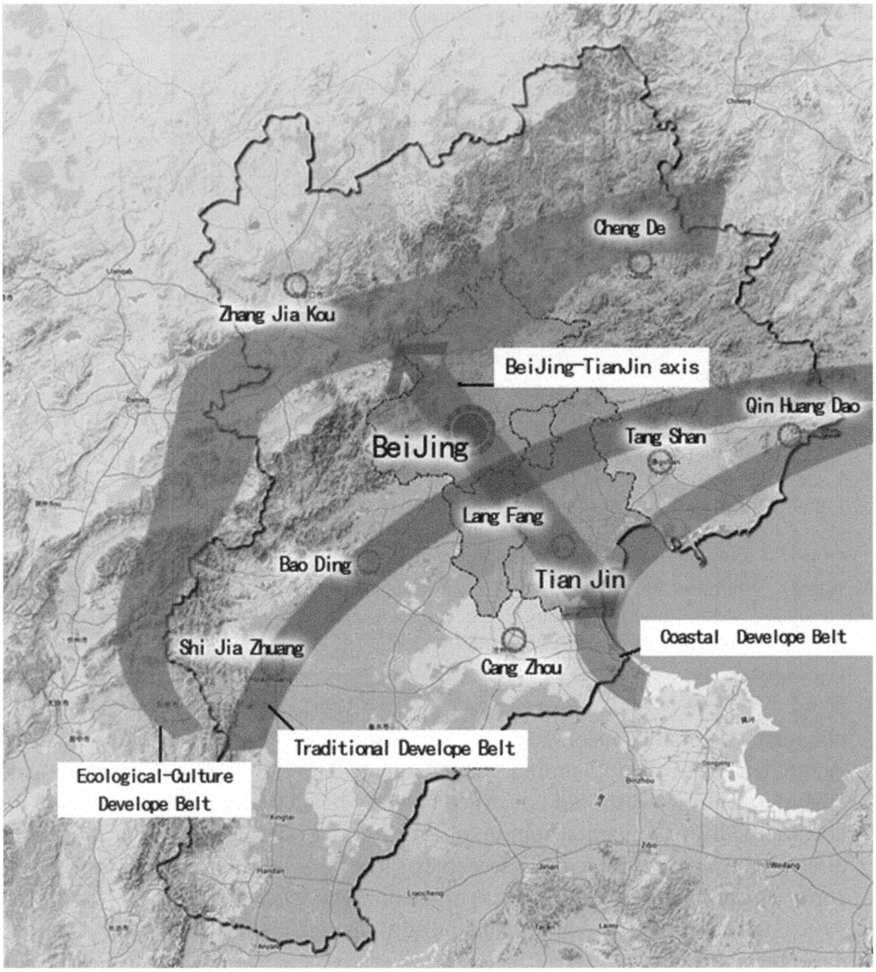

Fig. 2 "One axis and three belts" spatial structure of the capital region

opportunities and possibilities to surrounding cities, reinforce the urban social and economic development and the population aggregation, and promote the regional division and coordination. By doing so, we can gradually build a spatial pattern of "one axis and three belts" in the capital region (Fig. 2).

3. Integrate the four networks and promote the sustainable regional development (Liangyong et al. 2013)
   The so-called four networks integration refers to the integrated development of the urban network, transportation network, ecology network, and culture network. Currently, Beijing has initially owned the population, space, and economic scale necessary for a world-class urban agglomeration, but its

development quality and efficiency still need to be improved significantly. In the context of new urbanization, proper management of the relationships among urban construction, transportation facilities, regional ecology, and cultural inheritance is necessary for regional development. Meanwhile, the integrated development of the four networks can further enhance the integrity and systematicness of the capital region and make it more efficient in energy conservation, emission reduction, the application of green energy, the promotion of energy-saving technologies, and the regional environmental management.

4. Protect the ecology and build a regional green space system
   For the capital region rich in resources, it is of great significance to protect the ecology and intensify the construction and maintenance of the regional green space system. Firstly, we need to accelerate the construction of the second containment areas outside the Beijing center and strictly safeguard Beijing's ecology baseline; second, we need to promote the construction of the green space system in regional cities, effectively control the urban development boundary, and optimize the regional ecological system; third, we need to actively implement the reforesting policy, dredge the river system, improve the water quality, and ensure the ecological safety (Fig. 3).
5. Enhance the carbon sink function and gradually optimize the environment
   Intensify the construction of the urban carbon sink system and reduce the carbon dioxide contents in the air to the largest extent. First, we need to improve the carbon sequestration capability per unit of land area, i.e., reduce the construction of the large-scale grass squares and hard squares, build more ecological green landscape belts which are beneficial to oxygenation, outdoor fitness and relief of the heat island effect, and reserve lake and river waterscapes and natural mountains. Second, we need to improve the carbon sink quality, i.e., cultivate plants with better carbon sink and oxygen release capability in the urban areas of high-carbon emission. For example, we can increase the area of shelter belts in the industrial zones, isolation belts of urban express ways, green belts of the urban roads as well as the green space at the street corners, to better promote carbon sequestration.
6. Promote the construction of the low-carbon demonstration areas and promote low-carbon development
   Promoting the construction of low-carbon demonstration areas and exploring the effective mode for the urban low-carbon development are both effective to promote the low-carbon regional development. The capital region, according to the experience in regional ecology maintenance and construction, builds demonstration areas such as the low-carbon urban communities, low-carbon industrial parks, and low-carbon towns, which can provide experience reference and consultation opinions to the low-carbon development in other cities and regions. Meanwhile, we also need to improve the regional regulations, norms, and standards concerning the low-carbon urban development draw up the low-carbon urban plans, design regional cities, and setup green buildings as samples. During implementation and management of urban and rural planning,

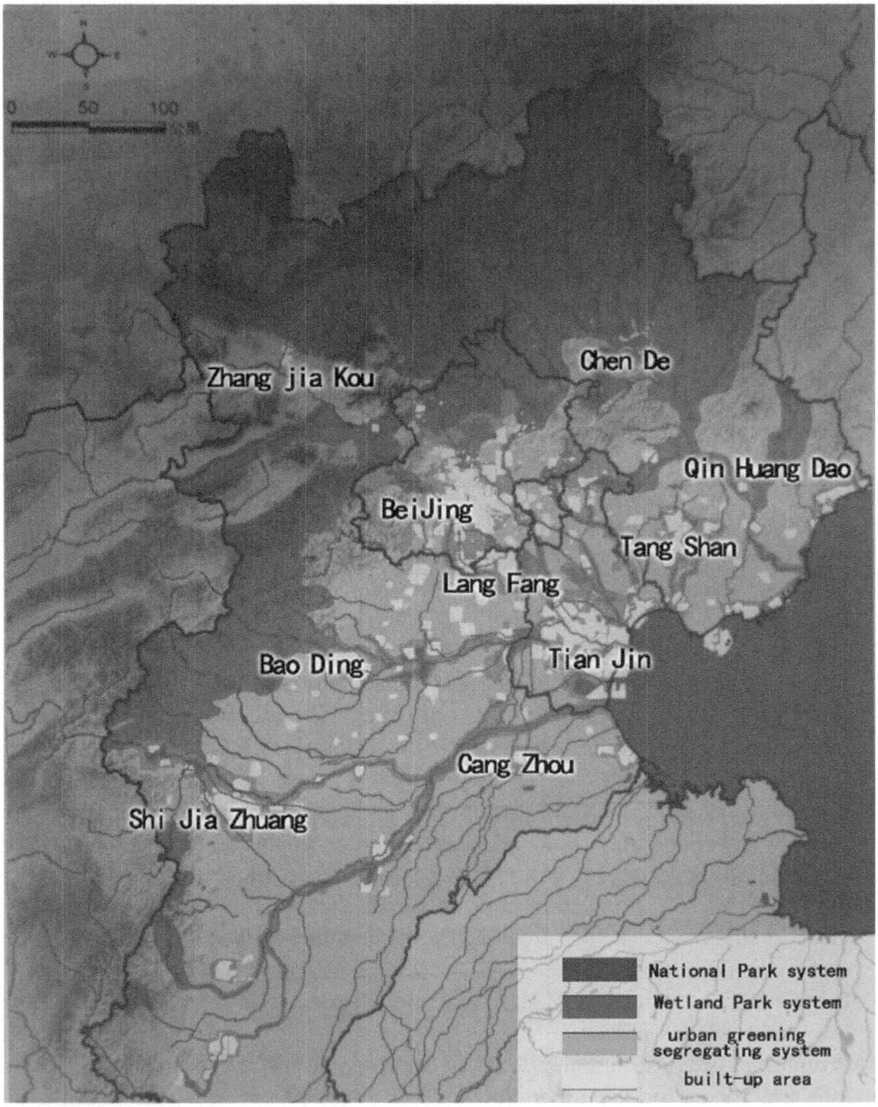

Fig. 3 Green space system of the capital region

we should formulate plans in accordance with the low-carbon regulations, norms, and standards, include carbon emission indicators into the urban space plan, promote energy conservation and carbon emission through incentive policies, and encourage to formulate urban plans, designs, and construction methods which can effectively reduce CO_2 emission, on the basis of the planning samples.

References

Arciniegas G, Janssen R (2012) Spatial decision support for collaborative land use planning workshops. Landscape Urban Plan 107:332–342

Baoxing Q (2010) Challenges and opportunities of China in middle and later periods of urbanization—new trends of urban planning reform. Urban Plan 1:15–23

Chaolin GU, Zongbo T, Wan L (2009) Low-carbon urban planning: seeking for low-carbon development. Constr Sci Technol 15:41

Chaolin G, Zongbo T, Zhilin L et al (2010) Urban planning research framework based on the low-carbon concept. J Urban Reg Plan 2:24–33

Chengri D (2005) Analysis of urban land demands. Foreign Urban Plan 4:19–25

Chicago Metropolis (2010) The metropolis plan choice for the Chicago region. Chicago Metropolics, Chicago 2020

Glaeser EL, Kahn ME (2008) The greenness of cities: carbon dioxide emissions and urban development. Policy briefs of rappaport institute for greater Boston. J Urban Econ 3:1–12

Haixiao P et al (2008) Spatial planning strategy for China's "low-carbon city" construction. Urban Plan Forum 6:57–64

Janice M (2011) Effective practice in spatial planning. Routledge 2 Park Square, Abingdon

Jenny C, Will F (2008) A low-carbon future: spatial planning's role in enhancing technological innovation in the built environment. Energy Policy 12:4575–4579

Liangyong W et al. (2013) Study of urban and rural spatial development planning for Beijing-Tianjin-Hebei Region. Tsinghua University Press, Beijing pp 97–105

Qingji S, Chao A, Changshou L (2010) Discussion about the connotation and characteristics of low-carbon eco-city, and basic principles of planning and construction. Urban Plan Forum 5:48–56

Yaping W, Min Z, Jinbo W (2008) Measurement of compact urban development and land use performance—expansion and application of Clara Alonso model. Urban Plan Forum 3:32–40

Zuda Y (2010) Low-carbon urban planning and construction: cost-benefit analysis. Urban Plan 8:33–38

Experimental Study on Performance of Lightweight Radiant Floor Cooling Combined with Underfloor Ventilation System

Xue-ying Xia and Xu Zhang

Abstract Experimental investigations were carried out to study the performance of lightweight radiant floor cooling combined with underfloor ventilation (LRFCUV) system. Operating parameters of LRFCUV system were recorded. Operating characteristic, heat-transfer capability, and thermal comfort of LRFCUV were analyzed based on experimental data. Experimental results indicate that thermal resistance of lightweight floor structure is considerable. Radiant heat-transfer capability occupies approximately 30 % of the total, and LRFCUV system shows fine thermal stability and thermal comfort.

Keywords Experimental study · Lightweight radiant floor cooling · Thermal comfort · Underfloor ventilation

1 Introduction

Radiant floor cooling combined with underfloor ventilation air-conditioning system has been developed rapidly, and currently, there are quite a number of applications in China. The construction method of lightweight radiant floor system was introduced to China about 10 years ago (Qi and Lu 2002). For lightweight radiant floor cooling system, the pipes are placed in aluminum foil (Zhang et al. 2013). And lightweight radiant floor cooling combined with underfloor ventilation (LRFCUV) system is characterized by less loss of the floor height and less floor load because of its special floor construction framework.

X. Xia (✉)
College of Mechanical Engineering, Tongji University, Tongji 200092
People's Republic of China
e-mail: sdxxynsd@126.com

X. Zhang
College of Power and Engineering, Nanjing Normal University, Nanjing 210042
People's Republic of China

S. Feng et al. (eds.), *Low-carbon City and New-type Urbanization*,
Environmental Science and Engineering, DOI 10.1007/978-3-662-45969-0_27

There have been several studies on radiant floor cooling systems in literature. Energy consumption of radiant floor cooling combined with dehumidifying ventilation system was compared to that of conventional air-conditioning system during cooling period (Leigh et al. 2005; Song et al. 2008), and the former accounted for only 1/3 of the latter. A comparison of the energy consumption estimates shows that savings of 80 % were possible in case thermal comfort was achieved through radiant cooling instead of conventional air-conditioning (Memon et al. 2008). Tian and Love (2008) analyzed thermal comfort of 82 subjects in radiant cooling system and revealed that thermal comfort model of radiant cooling system was consistent with PMV model, and discomfort caused by vertical temperature gradient and blowing was effectively improved in radiant cooling system. Conceição and Lúcio (2011) compared thermal comfort of radiant floor cooling system and radiant ceiling cooling system under the same condition by numerical simulation and found that the former was better. Strand and Baumgartner (2005) summarized a radiant system model within building energy simulation program. A calculation method for floor surface temperature in radiant floor system was proposed (Jin et al. 2010; Zhang et al. 2012), and the effects of thermal resistance of pipe and water velocity on thermal performance in radiant floor cooling system were analyzed by numerical simulation (Jin et al. 2010). Tye-Gingras and Gosselin (2011) investigated on heat-transfer modeling assumptions for radiant panels with serpentine layout.

A new design in combination with prestressed concrete hollow floor units with a new dry floor structure without the filling layer of concrete was presented by Qiu et al. (2005). And heat-transfer theory was proposed, and effect of the factors on radiant floor cooling capability was analyzed by field test (Qiu et al. 2005). Weitzmann and Svendsen (2005) developed a method for calculating thermal properties of lightweight radiant floor panels based on an experimental setup. Hu and Niu (2012) reviewed the application of radiant cooling and heating systems in Mainland China and presented the associated considerations in system design. Up to now, seldom researches have been conducted on thermal performance and thermal comfort of LRFCUV system, which has been analyzed by experimental investigations in this paper.

2 Experimental System and Procedure

2.1 Test Room Layout

A room was equipped with LRFCUV system. The laboratory sketch is shown in Fig. 1. The height of the room is 2.8 m. Outside the test room is a compensate room. The units in the compensate room are used to support the environment that the experiments demand. The thickness of the wall is 240 mm. Heat-transfer coefficient of the wall is 1.97 W/(m^2 K). The distance between pipes next to each other is 100 mm, and the diameter of the pipe is 16 mm.

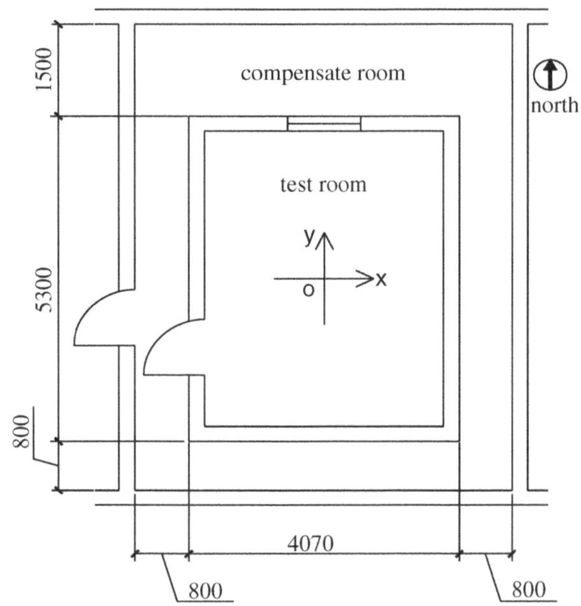

Fig. 1 Laboratory sketch

Chilled water is produced by ground-source heat pump and then distributed to every water loop by manifold to produce space cooling. Water supply temperature is controlled by changing the mixing ratio of primary water supply and backwater. Fresh air and primary return air are mixed and handled by air-handling unit and then transferred by concealed wind pipe to skirting outlet, which is 0.3 m high from floor.

2.2 Measuring Instruments and Test Points' Distribution

In this experiment, temperatures of water supply, return water, outdoor air, indoor air, upper floor surface, interior surface of wall, ceiling, window, and water flow are recorded. Table 1 shows testing parameters, measuring instruments, and the parameters of measuring instruments.

As presented in Fig. 2, temperatures of five points on upper floor surface and interior ceiling surface are measured. Average temperature of the five points represents the temperature of the surface. Three classical test points of indoor air temperature are selected in the middle of the test room as follows: Point 15 is 0.1 m high from floor, which represents human ankle position; point 16 is 1.1 m high from floor, which represents human respiratory position when sitting; and point 17 is 1.7 m high from floor, which represents human respiratory position when standing.

Temperature variation is not obvious in short time due to large thermal inertia of radiation system, so temperatures are measured per half an hour. Velocities of fresh

Table 1 Testing instruments and parameters

Testing parameters	Surface temperature	Air temperature and humidity		Air velocity	Water flow	Water temperature
Testing instruments	Infrared thermometer	Psychrometer		Hot wire anemometer	Ultrasonic flowmeter	Thermometer
Measuring range	−32 to 535 °C	−20 to 60 °C	10–95 %	0–20 m/s	0 to ±32 m/s	0–100 °C
Measuring precision	±1 °C	±0.8 °C	±3 %	<5 %	±1 ‰	–

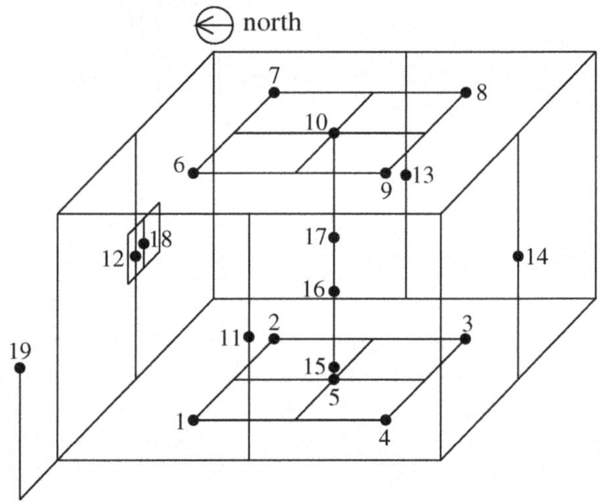

Fig. 2 Test points' distribution sketch

air, primary air, supply air, and water flow are measured as well. LRFCUV system operates from 9:00 to 17:00 in this experimental investigation.

3 Experimental Results and Analysis

3.1 Operating Characteristic Analysis

Temperatures of outdoor air, water supply, return water, floor, wall, and indoor air are recorded during operation. Variation of temperature during operation is shown in Fig. 3, where temperature of wall is the mean temperature value of building envelope. As presented in Fig. 3, indoor air temperature increased in 1.5 h after the system started operating and then became stable, approximately 26 °C. Under stable operation, temperatures of water supply, return water, and floor are approximately 15, 19 and 25.4 °C, respectively, which indicates that thermal resistance of light-weight floor structure is considerable.

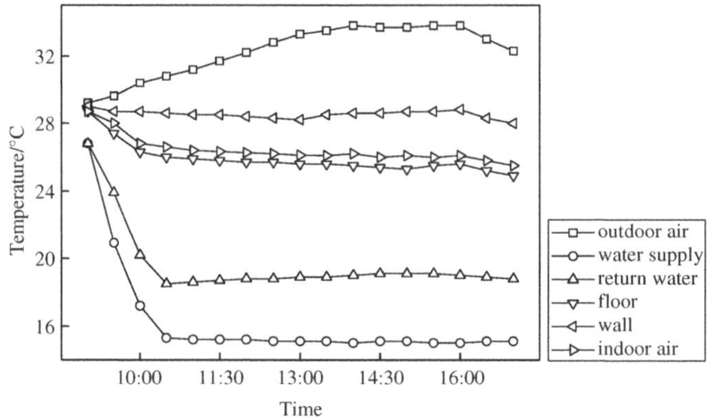

Fig. 3 Variation of temperature

3.2 Heat-Transfer Capability Analysis

Total heat-transfer capability equals refrigerating capacity supplied by both underfloor ventilation system and radiant floor cooling system. Refrigerating capacity supplied by underfloor ventilation system is calculated by Eq. (1).

$$q_c = G_n(h_2 - h_1) \tag{1}$$

where G_n is air supply volume; h_2 is return air enthalpy; and h_1 is supply air enthalpy.

Operating parameters of underfloor ventilation system are presented in Table 2.

Heat-transfer capability of floor by radiation is calculated by Gebhart method Yu (1990). In a closed system, heat-transfer capability by radiation of surface j is calculated by Eq. (2).

$$Q_j = E_j F_j - \sum_{i=1}^{n} B_{i,j} E_i F_i \tag{2}$$

where E_i and E_j are radiant heat-transfer power of surface i and surface j, respectively, $E_i = \sigma_b \varepsilon_i T_i^4$; F_i and F_j are the area of surface i and surface j, respectively; and $B_{i,j}$ is the absorption factor of surface i and surface j, which represents quotient of energy absorbed by surface j from surface i. $B_{i,j}$ is calculated by Eq. (3).

Table 2 Underfloor ventilation operating parameters

Testing parameters	Air velocity (m/s)	Air temperature (°C)	Relative humidity (%)
Air inlet 1	1.2	20.6	72.8
Air inlet 2	1.1	21.2	70.6

Fig. 4 Variation of floor heat-transfer quality

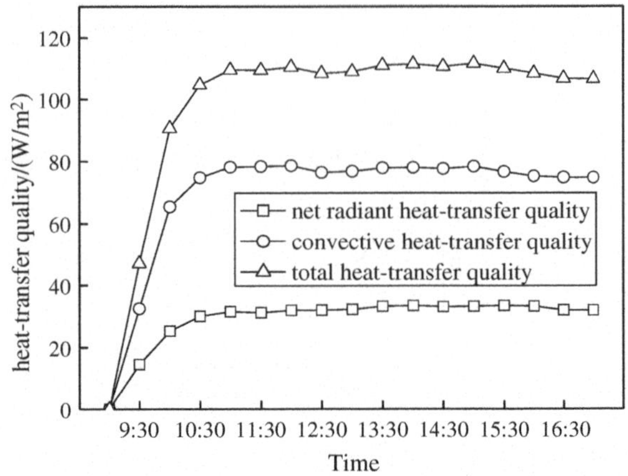

$$\begin{bmatrix} \varphi_{1,1}\rho_1 - 1 & \varphi_{1,2}\rho_2 & \cdots & \varphi_{1,n}\rho_n \\ \varphi_{2,1}\rho_1 & \varphi_{2,2}\rho_2 - 1 & \cdots & \varphi_{2,n}\rho_n \\ \cdots & \cdots & \cdots & \cdots \\ \varphi_{n,1}\rho_1 & \varphi_{n,2}\rho_2 & \cdots & \varphi_{n,n}\rho_n - 1 \end{bmatrix} \begin{bmatrix} B_{1,j} \\ B_{2,j} \\ \vdots \\ B_{n,j} \end{bmatrix} = \begin{bmatrix} -\varepsilon_j\varphi_{1,j} \\ -\varepsilon_j\varphi_{2,j} \\ \vdots \\ -\varepsilon_j\varphi_{n,j} \end{bmatrix} \quad (3)$$

where $\varphi_{i,j}$ is angle factor for surface i and surface j; ρ_i is reflectivity of surface i; ε_i is emissivity of surface i.

Calculated results are shown in Fig. 4. Heat-transfer capability of floor increased in 1.5 h after the system started operating and then became stable. LRFC system shows fine thermal stability because of large thermal inertia due to radiation. When LRFCUV system reached stable operating condition, net radiant heat-transfer quality is approximately 33.2 W/m^2, which occupies 29.7 % of the total, and the remaining sensible cooling load and latent load are handled by underfloor ventilation system.

3.3 Thermal Comfort Analysis

Indoor air temperature is one of the key factors that affect thermal comfort. Indoor air temperatures of test points were recorded under stable operating condition. As presented in Figs. 5 and 6, both horizontal and vertical temperature gradients are less than 0.3 °C/m. Indoor air temperature distribution was uniform. Vertical temperature gradient is positive due to convective heat effect.

In LRFCUV system, thermal comfort is greatly affected by radiant heat transfer between interior room surface and human body. So besides predicted mean vote

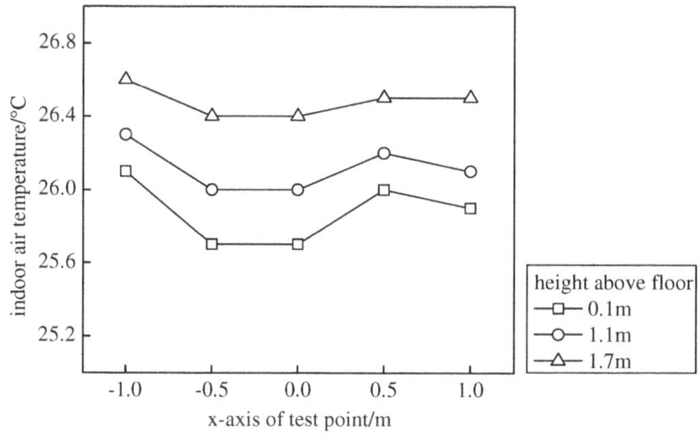

Fig. 5 Indoor air temperature distribution in horizontal direction

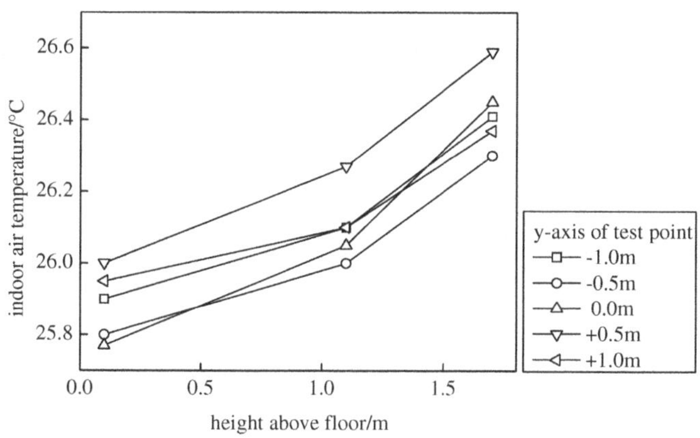

Fig. 6 Indoor air temperature distribution in vertical direction

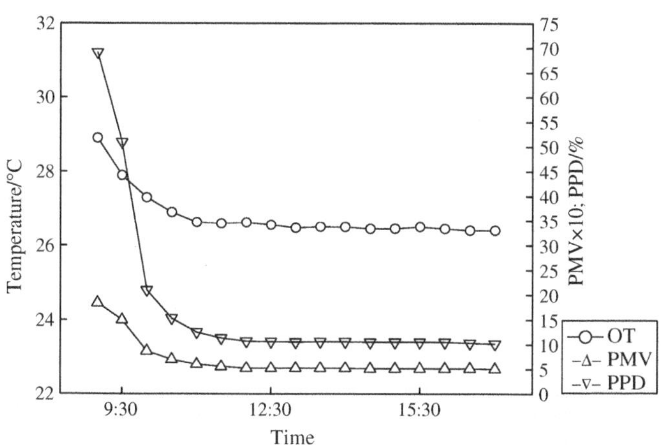

Fig. 7 Thermal comfort indexes' calculation

(PMV) and predicted percent dissatisfied (PPD) indexes, operating temperature (OT) is proposed to evaluate thermal comfort in addition.

As shown in Fig. 7, OT decreased in 2 h after the system started operating and then became stable. Three hours after the system started operating, both PMV and PPD values are in ISO7730-recommended value range.

4 Conclusions

In this paper, operating characteristic, heat-transfer capability, and thermal comfort of LRFCUV were analyzed by experimental investigations.

In summary

1. Thermal resistance of lightweight floor structure is considerable.
2. Heat-transfer capability by radiation occupies approximately 30 % of the total under stable operating condition.
3. Indoor air temperature shows uniform distribution and meets thermal comfort request under stable operating condition.
4. Three hours after the system starts operating, both PMV and PPD values are in ISO7730-recommended value range.

References

Conceição EZE, Lúcio MMJR (2011) Evaluation of thermal comfort conditions in a classroom equipped with radiant cooling systems and subjected to uniform convective environment. Appl Math Model 35:1292–1305

Hu R, Niu JL (2012) A review of the application of radiant cooling and heating systems in Mainland China. Energy Build 52:11–19

Jin X, Zhang XS, Luo YJ (2010a) A calculation method for the floor surface temperature in radiant floor system. Energy Build 42:1753–1758

Jin X, Zhang XS, Luo YJ (2010b) Numerical simulation of radiant floor cooling system: the effects of thermal resistance of pipe and water velocity on the performance. Energy Build 45:2545–2552

Leigh SB, Song DS, Hwang SH et al (2005) A study for evaluating performance of radiant floor cooling integrated with controlled ventilation. ASHRAE Trans 111:71–82

Memon RA, Chirarattananon S, Vangtook P (2008) Thermal comfort assessment and application of radiant cooling: a case study. Build Environ 43:1185–1196

Qiu L, Liu ZS et al (2005) Feasibility study on floor heating by application of prestressed concrete hollow floor and dry floor structure. Build Sci 21:44–47 (in Chinese)

Qi ZX, Lu YJ (2002) Dry modus of water heating system. Build Energy Environ 1:32–34 (in Chinese)

Qiu L, Wang WH, Zheng SH (2005) Experimental research on radiant cooling of a dry structure floor system. Fluid Mach 33:68–71 (in Chinese)

Song D, Kim T, Song S et al (2008) Performance evaluation of a radiant floor cooling system integrated with dehumidified ventilation. Appl Therm Eng 28:1299–1311

Strand RK, Baumgartner KT (2005) Modeling radiant heating and cooling systems: integration with a whole-building simulation program. Energy Build 37:389–397

Tian Z, Love JA (2008) A field study of occupant thermal comfort and thermal environments with radiant slab cooling. Build Environ 43:1658–1670

Tye-Gingras M, Gosselin L (2011) Investigation on heat transfer modeling assumptions for radiant panels with serpentine layout. Energy Build 43:1598–1608

Weitzmann P, Svendsen S (2005) Method for calculating thermal properties of lightweight floor heating panels based on an experimental setup. Int J Low Energy Sustain Build 3:1–15

Yu QZ (1990) Radiant heat-transfer foundation, 1st edn. Higher Education Press, Beijing (in Chinese)

Zhang DL, Cai N, Wang ZJ (2013) Experimental and numerical analysis of lightweight radiant floor heating system. Energy Build 61:260–266

Zhang L, Liu XH, Jiang Y (2012) Simplified calculation for cooling/heating capacity, surface temperature distribution of radiant floor. Energy Build 55:397–404

Analysis on Urban Crisis Conduction Under the Background of Big Data

Yan Li and Yun Lu

Abstract Urban crisis mainly includes four categories. They are natural disasters, accidents, public health crisis, and social security issues. This paper points out the conduction mechanism of urban crisis under the background of big data and through empirical study and concludes that information conduction will intensify the panic among citizens and thus affects urban crisis. The current researches on urban crisis focus on crisis control in one city instead of cities. However, based on the rapid information transmission in the era of big data, panic caused by crisis will soon transmit to related cities, and start urban crisis in those cities too. Therefore, urban crisis conduction studied in this paper has practical significance in advanced warning for urban crisis.

Keywords Big data · Urban crisis · Panic conduction

1 Introduction

According to John Parachini, Director of Intelligence Policy Center in RAND Corporation, "The six-hour data size generated by NSA wiretapping is equivalent to the entire amount of information of all collection of books in Library of Congress, the biggest library in the world." This is only a small part of the entire data generation, storage, and employment. With the development of technology, human beings are already living in a world surrounded by data. Human life and production are becoming digital. What is more directly related to the era of big data is cloud computing. With popularization of cloud computing, smart earth, smart city will be

An erratum to this chapter is available at DOI 10.1007/978-3-662-45969-0_36

Y. Li (✉) · Y. Lu
College of International Business, Shanghai International Studies University,
550 Daling Road, Shanghai 200083, China
e-mail: liyan0417@shisu.edu.cn

S. Feng et al. (eds.), *Low-carbon City and New-type Urbanization*,
Environmental Science and Engineering, DOI 10.1007/978-3-662-45969-0_28

gradually realized and innovative techniques and methods will be created on data application and data processing.

In the era of big data, society will be more networked, digitalized, dispersed, and open. Traditional management on urban crisis will be no longer adapted to the development of society. "As the Internet and ubiquitous computing is much embedded into human life, it is so difficult for governments to prevent data from being disclosed and avoid two-way information transmission. Information technology has profoundly changed the political ecology." With the development of big data technology, not only government work has improved in openness, interactivity, scientific level, and service, but also citizens have received information more promptly. This rapid information transmission will change urban crisis conduction.

2 Proposal of Research on Urban Crisis Conduction Under the Background of Big Data

Urban crisis refers to all kinds of phenomenon and processes produced by interactions between natural and social factors, which have caused the city to stop its normal function of operation and coordination, thus affecting human life, property, and surviving environment. As either a political, economic, or cultural center of a country or a region, city is concentrated with modern factors like commerce or wealth. Therefore, the types, frequency, continuous occurrences, damage, and influence of accidents in the city are far more than places which are not urban. There are four categories of urban crisis according to different generating process, nature, and mechanism: natural disaster crisis, accidental crisis, public health crisis, and social security crisis.

City is an open and complicated macrosystem. Its development almost parallels the process of globalization. Since the city has greatly extended itself, urban crisis will spread accordingly beyond geographical boundaries. It will start and end in the city, and then spread outside the city and even over the national boundaries and extend globally. Urban crisis warning is the initial stage of urban crisis management, and also the first defense line. Building a system of advanced warning for urban crisis is not only a practical request to strengthen urban defense against crisis and secure the consistent development of economic society, but also an inevitable requirement for building safe environment and improving competitiveness of the city. However, due to the wide range of urban crisis, it is difficult to establish a warning system. Consequently, research on crisis conduction for one urban crisis has great significance for those cities which have not had crisis yet.

2.1 Improvement on Intercity Crisis Conduction Research: Demand from the Globalization of Crisis

Due to economic globalization, the world has become a huge net, and every city is one node in the net, which makes the connection between cities even closer.

Therefore, crisis in one city will not only harm that particular city, but also transmit to other cities or even other countries through all kinds of channels, which will leave great impacts on the politics and economies worldwide.

When crisis spread globally, it will spread from social level to economic, political, and cultural levels and lead to considerable consequences. Take SARS for example. It spread from Fo Shan, Canton province, to other parts in China, which affected the international activities of China. Visit of senior management from Intel Corporation was postponed. Many international events, sports games, conferences, and exhibitions were canceled. External economic relations of both Mainland China and Hong Kong were affected at different levels.

Although peaceful development is the main stream in the world, the imbalance in world economic and political development still exists. Hegemonism and terrorism, racial discrimination and extreme poverty, and disasters and diseases still threaten the world's security at any time and also the security of the city. Globalization enables frequent exchanges of staff, resource, and capital. When crisis breaks out in one area, it is easy to transmit to another. Therefore, in the process of globalization, urban management should both consider the benefits and the risks globalization has brought with. In this sense, city does not only need an emergency system, but also need a crisis warning system to control crisis conduction.

2.2 Improvement on Intercity Crisis Conduction Research: Demand from Big Data Technology

Professor Zhai Kun, director of international division, crisis management center of modern international relations research institute, believed the first manifestation of macrourban crisis is mass panic. When domestic situation has undergone great changes or when one country has experienced outbreak of serious social crisis, public fear and unrest will turn into uncontrolled panic. One random accident will start a crisis, which is likely to spread into a more serious one.

As big data technology has developed so fast, rapid transmission of information should be under control, or it will intensify the transmission and spreading of collective panic, thus playing an important role in urban crisis conduction. Scholars studying this field both home and abroad have carried out researches from information perspective, but they have not conducted systematic quantitative research from the perspective of crisis conduction speed. Calvo (1999) believed when investors who enjoyed complete information sold securities in developing markets to satisfy margins, those with incomplete information would mistake it as a signal of dropping returns in developing markets, and then withdrew their investment and gave rise to a crisis. Calvo and Mendoza (2000) claimed globalization would weaken the motives to collect high-cost information and reinforce the motives to imitate random market portfolio, thus encouraging the contamination of financial crisis. He and Xie (2002) regarded adverse selection and moral risk caused by information asymmetry as the underlying reasons for intensification of financial risks. Wei and Qi (2008), from the research on information transmission mechanism in financial market, found out market return and fluctuation, and their influence on correlations of markets was the foundation to test whether financial crisis had been contaminated.

Advanced warning on urban crisis has been a heated research focus for researchers in this area, but most of the researchers have studied this issue from the perspective of urban management, while few have studied from the perspective of information. And there is almost no research carried from the perspective of information transmission based on big data. This paper studies the influence of information control brought by big data on urban crisis conduction and demonstrates the necessity of researching urban crisis conduction under the environment of big data. The theoretical value of this research lies in the cross-disciplinary exploration of the issue from both urban crisis and information technology fields. And in practice, this research points out a new direction for future urban crisis response.

3 Research on Conduction Mechanism of Urban Crisis Under the Background of Big Data

The process of urban crisis conduction can be divided into two steps. The first conduction refers to the expected results made by investors or other financial brokers. This crisis conduction has nothing to do with the observed macroeconomy or other fundamentals. This is what Kaminsky and Reinhart called "real contagion" and Masson "pure contagion." This kind of conduction usually happens in the initial stage of the crisis and is considered as the result of irrationality, such as financial panic, herd behavior, loss of confidence, and increase of risk disgust. The second conduction emphasizes spillover effect caused by interdependence between different markets. This interdependence refers to the fact that the connections between real economies enable local or global impacts transmit internationally. Calvo and Reinhart regard this kind of crisis conduction as conduction based on the fundamentals.

3.1 Analysis on Influence of Big Data Information in the First Conduction of Urban Crisis

The initial application of big data is the transmission of massive information. Analysis on the influence of massive information will be made from Masson's "Pure Contagion Effects," and thus further analyzes the impact of big data.

When a crisis breaks out in a city, citizens' expectant psychological change toward similar cities and confidence crisis cause emotional changes, the result of which is speculative attacks on self-realization of the city. This produces pure contagion effects. A typical example is herd effect.

Decisions of investors are often based on behaviors of other investors because they are influenced by conformity. This is what we call herd effect. When this effect happens, individuals tend to behave in conformity, and one small impact on the overall society will lead to a huge deviation in human behavior. Under particular situations, individuals might abandon their own information and conform to the others', though the information they hold points to a different direction. The process of citizens' psychological panic analyzed by herd effect is demonstrated in the following chart.

The amount of information in Message 1 and Message 2 in Fig. 1 is decisive to crisis conduction. If the amount of information in Message 2 is more than that in Message 1, it will cause great panic among citizens. This conformity will turn into herd effect and thus will spread crisis to a large scale. However, if information transmission technology can be used to control the amount of information in Message 1 and Message 2, then crisis conduction can be interfered.

Some scholars also pointed out the function of information in explaining the herd effect. Calvo (1998) created herd behavioral model on the basis of contagion effect caused by international liquidity shortage. He thought when investors with complete information increased margins, investors with incomplete information would mistake it as a signal to withdraw investment, and thus withdrew their money too, which would lead to a currency crisis in the new market. Froot et al. (1999) considered those investors who made decisions according the noises in the stock exchange as heard on the street. These researchers claimed that institutional investors were highly homogeneous. They usually focused highly on the same market information and adopted similar economic models or information processing techniques. Under these circumstances, institutional investors might react similar to profit forecast or stock analysts' advice. This was shown as herd effect in transactions. Calvo and Mendoza (1997) thought when investors made decisions at the same time, it was also likely to generate herd effect. As world economy was more globalized, it would cost more to find out whether the information was true or not

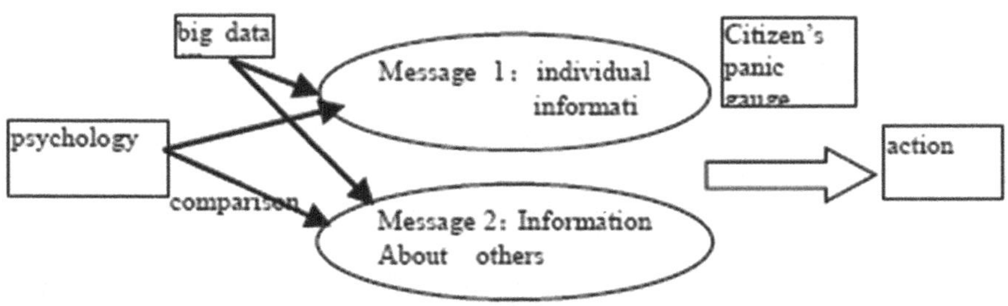

Fig. 1 Herd effect

by collecting information in particular countries or regions. Due to the lack of information, investors would imitate asset portfolios of other investors in the market, hence the herd effect.

3.2 Analysis on Influence of Big Data Information in the Second Conduction of Urban Crisis

The second step in crisis conduction emphasizes the spillover effect caused by interdependence between different markets. This interdependence refers to the fact that the connections between real economies enable local impact to transmit between cities.

Big data information has an impact on crisis conduction between urban organizations. Modern urban organizations are closely connected. For example, interbank payment and clearing system has tied all banks together. Any trivial payment issue in payment and clearing system on multilateral difference will cause an overall liquidity crisis. Electronic payment based on information technology and Web service of all banks has accelerated information transmission between organizations and further influences crisis conduction.

4 Empirical Test of the Impact of Big Data Information on Urban Crisis Conduction

4.1 Time Selection of Crisis Samples

The impact of big data technology on urban crisis conduction is mainly demonstrated in information transmission. At present, the major tool for global transmission is the Internet. The Internet started in the USA and came to influence China in 1995. From 1997, China has had statistics (CNNIC). According to the yearbook, the Internet technology started in 1995 in China, rapidly developed in 1999, and matured from 2005. There are two phases in the development of China's Internet technology: 1995–2004 is the immature phase, while 2005–2014 is the mature phase.

Financial crisis will lead to urban crisis. Between 1990 and 2014, there are three financial crises worldwide: European currency crisis in 1992, Asian financial crisis in 1997, and American subprime mortgage crisis.

Table 1 is the list of three crises and information technology conditions in China.

From Table 1, the time of the three crisis samples used in this empirical test corresponds with the three conditions of China's Internet technology application, so these samples are representative. On selecting the time of crisis, this research can test how information transmission affects the velocity of crisis conduction.

Sample data come from Wind database.

Table 1 Sample time selection

Sample time selection	Name of crisis	China's information technology situations
1991.1.1–1992.12.31	1992 European currency crisis	No application
1997.6.1–1998.12.31	1997 Asian financial crisis	Initial application
2007.9.1–2009.2.31	2007 American subprime mortgage crisis	Wide application

4.2 Variable Selection

Currently, Chinese capital market is not fully open and RMB cannot be exchanged freely. This enables China to have a natural protection against the outside world. In fact, this also helps to avoid risks. On February 19, 2001, China opened its B share market to domestic capital. In December 2002, China launched QFII system, and QDII system was established in 2007. Currently, except for QFII quota, foreign investors can only buy B shares and Chinese stocks listed abroad. They cannot buy A shares and other RMB bonds inside China. Domestic agencies and individuals, except for the quota of QDII, cannot purchase, sell, or issue stocks, bonds, money market instruments, or other derivatives outside China. According to statistics, QFII and QDII both have limited quotas. Consequently, the impact of crisis on Chinese capital market is mainly on psychological aspect. Fluctuation in stock exchange will interact with real estate market, and then further influence the public decision making and urban crisis conduction.

4.3 Empirical Test Methodology

This research uses cointegration test and Granger causality test to empirically analyze the situations of three financial crises conduction into China. The research methodology is as follows:

In 1992, when the first crisis happened, China barely knew about the Internet. In 1997, when the second crisis occurred, the Internet technology just started in China and was hardly used in Chinese stock market. In 2007, when the third crisis started, the Internet technology became mature in China and was widely applied in the stock market. All these three crises did not happen in China. Selecting the data when crises happened and using it to test the impact of US stock exchange on Chinese stock market at the time of the crises can reflect the situations of crisis conduction in China. Through the impact of US stock exchange on Chinese stock market, impact of the Internet technology on the velocity of financial crisis can be tested and further, it can be demonstrated that information has effect on conduction.

4.4 Empirical Test

1. Model selection

This paper employs a bivariant VAR model (Shanghai composite index, S&P500). Shanghai composite index is used as the index to measure fluctuation in Chinese stock market, while S&P500 is used to measure fluctuation in US stock exchange.

2. Unit root test

Index fluctuation in both Chinese and US stock markets in the three crises is shown in Fig. 2.

As shown in Fig. 2, Shanghai composite index and S&P500 are both non-stationary time series, while its first-order difference (return) is stationary series. Therefore, cointegration test and Granger causality test can be used (Table 2).

3. Cointegration test

This paper uses Johansen's maximum likelihood method to test the cointegration between the estimated indexes of both China and the USA. See Table 3 for detailed results. The optimal lag intervals are chosen according to AIC, SC, and HQ. The optimal lag intervals are chosen according to AIC, SC, and HQ standards.

Table 3 demonstrates that during 1992 European currency crisis, Shanghai composite index and S&P500 have one cointegrated vector at a 5 % significant level. During the periods of the other two crises, the two indexes do not have cointegrated relationship. If there are two cointegrated relations, equation needs to be provided to judge the degree of cointegration. However, since there is only one cointegrated relation, no equation needs to be given.

4. Granger causality test

This paper adopts a bivariant VAR model and uses Granger causality to test the returns of both Chinese and US stock indexes to see whether there is any contagious relationship. Test results have been shown in Table 4.

Test results from Table 4 have shown that during 2007 American Subprime Mortgage Crisis, change D (lnsp3) in US stock exchange is the Granger reason for change D (lnsh3) in Shanghai stock market. Therefore, this crisis had influenced Chinese stock market in a short period. The change D (lnsp3) in US stock exchange

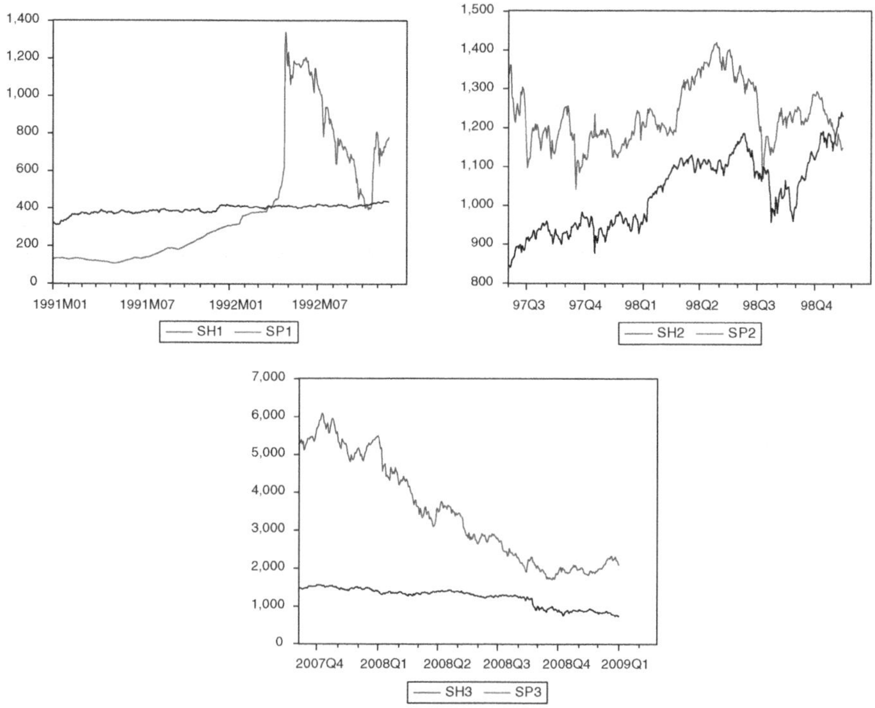

Fig. 2 Index fluctuation chart in Chinese and US stock markets in the three crises. 1991.1.1–1992.12.31 index fluctuation in Sino-US stock markets. 1997.6.1–1998.12.31 index fluctuation in Sino-US stock markets. 2007.9.1–2009.2.28 index fluctuation in Sino-US stock markets

Table 2 Unit root test results

Data	Variant	ADF value	*P* value	Results
1991.1.1–1992.12.31 data	SH1	−2.2706	0.4488	Nonstationary
	SP1	−3.6662	0.0255	Stationary
	rSH1	−18.8849	0.0000	Stationary
	rSP1	−22.3569	0.0000	Stationary
1997.6.1–1998.12.31 data	SH2	−2.6205	0.2714	Nonstationary
	SP2	−2.4591	0.3485	Nonstationary
	rSH2	−17.8735	0.0000	Stationary
	rSP2	−20.1002	0.0000	Stationary
2007.9.1–2009.2.31 data	SH3	−1.7953	0.7052	Nonstationary
	SP3	−2.4937	0.3311	Nonstationary
	rSH3	−19.4816	0.0000	Stationary
	rSP3	−16.2532	0.0000	Stationary

Table 3 Cointegration test results

	Variant	Hypothetical cointegrated vectors	Trace statistics	5 % critical values	Prob.	Results
1991.1.1–1992.12.31 data	rSH1 and rSP1	None	19.5461	15.4947	0.0116	Cointegration exists
		At almost 1	1.0240	3.8414	0.3116	
1997.6.1–1998.12.31 data	rSH2 and rSP2	None	9.8347	15.4947	0.2937	No cointegration
		At almost 1	1.1496	3.8414		
2007.9.1–2009.2.31 data	rSH3 and rSP3	None	5.5318	15.4947	0.7500	No cointegration
		At almost 1	1.5288	3.8414		

Table 4 Granger causality test results

Granger causality	Lag intervals	1991.1.1–1992.12.31		1997.6.1–1998.12.31		2007.9.1–2009.2.31	
		F value	*P* value	*F* value	*P* value	*F* value	*P* value
Reasons why Shanghai composite index is not S&P500	1	1.0298	0.3107	0.1330	0.7155	1.5158	0.2839
	2	0.8245	0.4390	1.9353	0.1458	0.5015	0.6061
	3	0.5975	0.6169	1.9224	0.1255	0.8458	0.4694
	4	0.4703	0.7575	1.4779	0.2082	1.7284	0.1433
Reasons why S&P500 is not Shanghai composite index	1	0.1101	0.7401	2.1789	0.1407	17.5181	0.0001
	2	0.1699	0.8437	1.1172	0.3283	8.9911	0.0002
	3	0.1695	0.9169	0.8501	0.4672	6.3119	0.0004
	4	0.2158	0.9296	0.7160	0.5184	5.1126	0.0005

in other two crises was not the Granger reason for change D (Insh3) in Shanghai stock market; thus, these two crises did not affect Chinese stock market in a short period of time.

5 Conclusion

5.1 Empirical Test Results

Through empirical study, this paper discovers the empirical evidence on how the Internet affects the conduction speed of financial crisis. The result of this empirical study is shown in Table 5.

Table 5 Empirical study results

Number	Name of crisis	The internet application situations in China	Cointegration	Granger causality test
1	Urban crisis caused by 1992 European currency crisis	Not applied	Chinese stock market was highly stable	US stock exchange had no impact on Chinese stock market at the time of crisis
2	Urban crisis caused by 1997 Asian financial crisis	Initial stage of application	Chinese stock market was fairly stable	US stock exchange had almost no impact on Chinese stock market at the time of crisis
3	Urban crisis caused by 2007 American subprime mortgage crisis	Wide application	Chinese stock market was unstable	US stock exchange had an impact on Chinese stock market at the time of crisis

Table 5 shows that in 1992, since the Internet was not used in financial field, fluctuations in US stock exchange did not transmit quickly to China, and Chinese stock market was stable during the crisis. In 1997, the Internet technology just started in China; thus, fluctuations in US stock exchange could barely transmit to China through the Internet, and Chinese stock market was quite stable during the crisis. In 2007, as the Internet technology in China was quite mature, fluctuations in US stock exchange would soon transmit to China through the Internet and caused relative fluctuations in Chinese stock market during the crisis. These results show that the Internet technology can accelerate financial crisis conduction.

5.2 *Value of Studying Urban Crisis Conduction Under the Environment of Big Data*

This paper uses empirical test to prove that information technology has an effect on financial crisis conduction. Under the background of big data, since big data technology, cloud computing, and the Internet have been widely used, it will inevitably bring about influence on urban crisis conduction due to the power and extensive applications of these technologies. Because at present, there is no precise prediction on how urban crisis happens or conducts, advanced preparation should be adopted. Therefore, studying urban crisis conduction under the environment of big data can prepare for future situations after crisis happens, and it is where the theoretical and practical values of this paper lie.

Acknowledgments Financial support: Youth Foundation Program in Shanghai International Studies University, "Financial Crisis Control System in the information age" (Program number: Kx181102).

References

Akhmat G, Zaman K, Shukui T, Sajjad F (2014) Does energy consumption contribute to climate change? Evidence from major regions of the world. Renew Sustain Energy Rev 36:123–134

Calvo GA (1998). Capital flows and capital-market crises: the Simple Economics of Sudden Stops. J Appl Econ 1(1):35–54

Calvo GA (1999). Contagion in Emerging Markets: When Wall Street is a Carrier. J Int Econ (5)

Calvo GA, Enrique G (1996) Mexico's balance-of-payments crisis: a chronicle of a death foretold. J int econ

Calvo GA, Mendoza EG (2000). Capital-Markets Crises and Economic Collapse irn Emerging Markets: An Informational-Frictions Approach. Papers and Proceedings of the One Hundred Twelfth Annual Meeting of the American Economic Association, May, 2000, pp 59–64

Chang SJ (2013) Solving the problem of carbon dioxide emissions. Forest Policy and Econ 35:92–97

David G (1984) Streets. Acid rain control and United States energy policy. Energy 9(11-12): 1049–1063.

Energy statistics division of the national bureau of statistics of the People's Republic of China (2011). China energy statistical yearbook 2010 [M]. China Statistics Press, Beijing

Froot et al (1999) How are stock prices affected by the location of trade? J Financ Econ 189–216

He G, Xie X (2002) Information Asymmetry and currency contagion model. Commercial Research (8)

Khanna N, Fridley D, Hong L (2014) China's pilot low-carbon city initiative: A comparative assessment of national goals and local plans. Sustain Cities and Soc 12:110–121

Steinfield H, Gerber P, Wassenaar T, Castel V, Rosales M (2006). Livestock's Long Shadow: Environmental Issues and Options [R]. Food & Agriculture Organization of the UN (FAO)

Wei Y, Qi S (2008) Research on infection problems emerging Asia financial crisis—Tested by Copula theory methods. Stud int financ (9)

Zhang N, Lior N, Jin H (2011) The energy situation and its sustainable development strategy in China. Energy 36(6):3639–3649

A GIS-Based Approach to Urban Community-Scale Carbon Emissions Modeling

Yang Li, Yao-yu Lin and Yong-xi Gong

Abstract The paper reports on a GIS-based approach to modeling carbon emissions through an urban community. The approach uses bottom-up modeling of urban form components (land use, buildings, and transportation network) to assess the community-wide carbon emission system. Model inputs (land cover, population, building types, vegetation structure) were integration of spatial and non-spatial data derived from diverse sources. Results are quantified and illustrated spatially as a whole system, and component carbon emissions are attributed to (i) buildings, (ii) transportation, (iii) residence, and (iv) vegetation.

Keywords Carbon emission modeling · Urban planning for climate change · GIS

1 Introduction

Approximately urban areas account for 75 % global energy consumption and related CO_2 emissions (Tanaka 2008), which is largely attributable to urban form characteristics (Satterthwaite 2008). The size and morphology of urban blocks impact the density, shape, orientation, and skin area of buildings and therefore energy demand for heating, cooling, and ventilation (Santin et al. 2009; Harvey 2009). Similarly, the urban form variation can influence vehicles' carbon emissions by affecting inhabitant trip demands and trip mode (Steemers 2003; VandeWehge and Kennedy 2007). Likewise, in energy structure, urban form also significantly impacts opportunity for renewable energy strategies and technologies, such as solar thermal, photovoltaic electricity, and passive design (Salat 2009).

Y. Li · Y. Lin (✉) · Y. Gong
Shenzhen Key Laboratory of Urban Planning and Decision Making Simulation,
Shenzhen Graduate School, Harbin Institute of Technology, HIT E422, Xili University Town,
Shenzhen 518055, People's Republic of China
e-mail: linyy@hitsz.edu.cn

S. Feng et al. (eds.), *Low-carbon City and New-type Urbanization*,
Environmental Science and Engineering, DOI 10.1007/978-3-662-45969-0_29

Understandings of the relationships and interactions between urban form, energy demand, and carbon emissions at community and greater scales are crucial to cities for effectively planning for climate change, improving their energy efficiency, and reducing their carbon footprints to meet challenging legislated targets (Calthorpe 2010; Davidson et al. 2011). However, the complexity of the carbon emission activities of cities presents many conceptual and methodological challenges to that understanding. Chief among them is the challenge of assembling and integrating the diverse types and sources of necessary data to undertake energy and emissions modeling at a neighborhood scale. Collected by many different agencies for different purposes, these data are normally diverse in scale and resolution, such as dwelling unit, ownership parcel, postal code, or municipality. Much of these data are non-spatial, like energy demand tied to building type, construction, and age without location. Consequently, these data are difficult to reconcile and correlate with urban form attributes, energy, and GHG emissions, also incapable of informing climate change planning policy and regulation. In order to define appropriate local climate change planning strategies and actions, cities will need to establish emission baselines and calibrate targets for these same community attributes and their potential to mitigate emissions. However, as the existing approach to carbon emissions analysis is lacking a method to deal with the data comprehensiveness and resolution problem, few cities have sufficient expertise or resources to meet that challenge.

The expanded logic operation and data processing function of GIS make this tool an appropriate solution to integrate non-spatial energy and carbon emission data with urban form attributes, assembling a systemic view of community-scale energy and GHG emissions from diverse data sources. With the same extraordinary data visualization function, GIS tools are to be accomplished in quantifying, aggregating, and communicating the contribution of system components in an elaborate way, easier to understand. GIS tools bring new opportunities for carbon emission modeling and visualization at a community scale, and this paper discusses a GIS-based approach to assess the potential community-developing scenarios for climate change mitigation.

2 Method

2.1 Carbon Emission System

IPCC Guidelines for National Greenhouse Gas Inventories recommends five key source categories: energy, industrial processes, product use, agriculture forestry, and other land use. To help cities quantify their GHG emissions, considering both local decision-making needs and the IPCC Guidelines for National GHG Inventories, the Global Protocol for Community-Scale GHG Emissions (GPC) includes six main categories: dividing the energy category in IPCC Guidelines into two groups, stationary units and mobile units. In this paper, in order to integrate

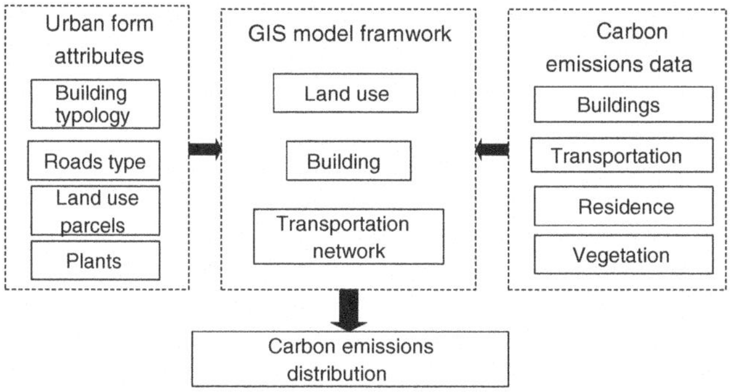

Fig. 1 GIS-based carbon emissions model framework

emission data and urban form attributes, we decompose the community carbon emission system into four sectors: buildings, transportation, residence, and vegetation. Vegetation sector can be regarded as negative emissions.

Carbon emissions in each of four sectors are considered (Fig. 1). Building sector includes stationary combustion sources transforming mostly imported fossil fuel carbon (natural gas, oil, coal) into CO_2 for space heating or cooling, hot water demand and cooking, etc. Transportation sector includes mobile combustion sources, mostly kinds of vehicles that transform imported fossil fuels (gasoline, diesel) into CO_2 within the study area. Residence sector includes emission of CO_2 respiration as a result of the human metabolism and heterotrophic (microbial) decomposition of waste, as well as export of carbon due to waste disposal outside the system's boundary. Vegetation sector includes uptakes or emits CO_2 through photosynthesis and respiration which causes carbon storage in biomass and soils to change. The modeling approach taken in each of these four sectors is elaborated below.

2.2 Carbon Emission Boundary

The complexity of carbon emissions inside and outside the system boundary is dependent on the space scale. In terms of a community-scale system, except the emissions from sources within the community boundary, there are emissions that are a consequence of the activities within the community boundary, but occur at sources outside the community as well (Prairie and Duarte 2006). To confirm the appropriate emission scope for community-scale carbon modeling, we explain the space differences of the four carbon emission sectors (building, transportation, residence, and vegetation) in three situations.

Firstly, local emissions due to local activities (LDL) refer to the release (or uptake) of carbon dioxide within the community (system boundaries) due to combustion, respiration, or photosynthesis.

Table 1 Carbon emission boundary

Sector	LDL	EDL	LDX
Buildings	Power generation emissions associated with power consumption within the community	Natural gas combustion of buildings within the community	None
Transportation	Trips out of the community to work, school, recreation, etc.	Trips within community	Through traffic
Residence	Human respiration outside the community by residents	Human respiration	Human respiration of people traveling through
Vegetation	None	Photosynthesis and respiration of plants within the community	None

Secondly, external emissions due to local activities (EDL) refer to the release of carbon dioxide outside the community (system boundaries) due to combustion or respiration associated with activities within the community.

Thirdly, local emissions due to external activities (LDX) refer to the release of carbon dioxide within the community (system boundaries) due to combustion and respiration from objects or humans traveling through the community, but not associated with activities in the community (e.g., through traffic).

Table 1 lists for each component (buildings, transportation, residence, vegetation) emission processes in each of the three above cases.

2.3 Carbon Emission Data

The building sector carbon emissions are caused by building energy consumption for heating, cooling, lighting, and cooking. Though there is another definition for building energy consumption considering the life cycle process, including material manufactory, onsite construction, building operation, and waste disposal, this paper discusses only building operation energy consumption. The building energy consumption data can be obtained in two ways: from energy company documentation or building energy simulation. The calculation of building sector emission is estimated as follows:

$$E_{\mathrm{b}} = \sum_{k=1}^{N} C(K) \cdot f(K) \tag{1}$$

E_b refers to carbon emissions by buildings; $K = 1, \ldots, N$ refers to the K energy type; and $f(K)$ refers to emission factor for the K energy type.

Transportation sector carbon emission originates from vehicles transforming fossil fuels into CO_2. Transportation-related data can be obtained from three sources: field investigation, resident trip survey, and community traffic modeling. Note that typically the resident trip survey describes the trip characteristics within the whole city, therefore resulting in some deviation when utilized in a specific community. The calculation of transportation sector emission is estimated as follows:

$$E_t = \text{VKT} \cdot TA \cdot Df \tag{2}$$

E_t refers to carbon emission by transportation; VKT refers to trip distance; T refers to travel times; A refers to the proportion of car trip; D refers to fuel efficiency; and f refers to emission factor.

Residence sector includes emission of CO_2 respiration as a result of the human metabolism and heterotrophic (microbial) decomposition of waste, as well as export of carbon due to waste disposal outside the system's boundary. The carbon emission intensity of human respiration is 342.5 kg year^{-1} cap^{-1} (Prairie and Duarte 2006). The calculation of waste disposal carbon emission is estimated as follows:

$$E_w = Mf \tag{3}$$

E_w refers to carbon emission by waste disposal; M refers to waste produced by residents; and f refers to emission factor.

Vegetation sector includes uptakes or emissions of CO_2 through photosynthesis and respiration. The efficiency of photosynthesis and respiration can be obtained by field surveying or reference consulting. Calculation of the carbon fixation for all kinds of vegetation within the study area respectively is not practical. To increase the working efficiency, we recommend firstly categorizing the type of vegetation area, such as recreational vegetation area, productive vegetation area, and road vegetation area, and then converting the subject from vegetation type to vegetation area type, finally calculating the carbon fixation efficiency of vegetation area. The calculation of vegetation sector carbon uptake is estimated as follows:

$$E_p = P \cdot \text{LAI} \cdot S \cdot \lambda \tag{4}$$

E_p refers to carbon fixation by vegetation, negative carbon emission in other words; P refers to efficiency of photosynthesis; LAI refers to leaf area index; S refers to projection area; and λ refers to parameter.

2.4 Carbon Emission Modeling

In order to combine non-spatial carbon emission data with urban form attributes, we need to build not only the appropriate spatial carriers for these data, but also a systemic database to deal with the complicated analysis operation. We utilize Arc GIS software as a platform to build the model, and it is formed by three basic layers: land use, transportation network, and building (Fig. 1).

The land use layer functions as the spatial carrier for vegetation sector carbon fixation data. Besides the original spatial attributes of each polygon element in this layer, another four attributes are added to the property list: vegetation area type, land area, carbon fixation intensity, and vegetation carbon fixation. The land area calculation is one of the basic functions of the software, and the method to obtain vegetation area type and corresponding carbon fixation intensity data has been discussed in 2.3. Multiplying land area by carbon fixation intensity, the carbon fixation of each piece of vegetation area is achieved.

The transportation network layer performs the spatial carrier for transportation sector carbon emission data. Adding the carbon emission data of each road as a new attribute to this layer, the carbon emission distribution of the whole community road network is expressed clearly through model.

The building layer is spatial carrier for two carbon emission sectors: buildings and residence. It is quite simple to add building sector carbon emission to each single unit of building in this layer. And in order to analyze the correlation between building form and energy performance, building typology based on building use and physical form is needed as well (Ratti et al. 2003). The residence sector carbon emission is combined with building layer by distributing nighttime population accurately to buildings; then, the sources of human respiration and waste disposal can be estimated spatially.

In order to superimpose the three layers to get the general emission resolution, we need to perform a uniform grid processing among these layers. The precision of the grid should be set according to the community area.

3 Result

Based on GIS software, we project a community-scale carbon emissions model of OCT community in Shen Zhen, through which the carbon emissions of both components' contribution and general distribution of the community are illustrated in a spatial way; even the carbon emissions of each single building, road, and green land can be informed in the model.

The total carbon emission of OCT community in 2010 is 280,752.8 ton, 67.33 kg/m^2, 5.7 ton per capital, of which building sector accounts for 64 %, ranking the first, residence sector follows with 15 %, transportation sector accounts for 13 %, and the carbon fixation through vegetation is 1,428 ton. Because of the

Table 2 Carbon emission intensity of building

Building type	Subtype	Carbon emission intensity (kg/m^2)
Residential	Multi-story	35.05
	High-rise	24.33
	House	37.62
Non-residential	Office	150.14
	Retail	62.32
	Hotel	286.23
	Exhibition	67.19
	School	21.18
	Hospital	100.03

high openness of community system, the carbon emission within the community just represents 31 % of the total, and the rest 69 % carbon emission occurred all outside the community.

The relativity between urban form and carbon emissions of the OCT community is analyzed as well from three aspects.

Building type: The carbon emission intensity of different types of buildings varies greatly (Table 2). Residential buildings emit the least greenhouse gas, whereas hotels and office buildings have carbon emission intensity five to ten times higher than residential buildings.

Land use: Index such as population intensity, commercial intensity, public transport accessibility, and entertainment facility accessibility displays some correlation between land use pattern and transportation carbon emission. To get more accurate conclusion, more meta-analysis needs to be done.

Community vegetation: The average carbon fixation intensity of OCT community is 4.7 kg/m^2. Carbon fixation through vegetation only covers 5 % of the total community carbon emission. Since the green coverage rate has reached 70 %, and the greening mode adopts ecological stereo structure as well, there is not much room for improvement. Thus, at least in OCT community, the direct contribution of vegetation to carbon emission reduction is not manifest.

4 Conclusion

The existing approach to carbon emission analysis is lacking a method to inform the relativity between urban form and carbon emissions, and therefore incapable of informing climate change planning policy and regulation. This paper analyzes the sources, boundary, and data of community carbon emissions and explains the community carbon emission system dividing into four sectors: buildings, transportation, residence, and vegetation. On that basis, a community carbon emission model is devised through Arc GIS tool, integrating the carbon emission data and urban form attributes via three layers: land use, transportation network, and building. The model established connections between urban form attributes and corresponding carbon emission. It also helps to set emission baselines and calibrate targets for communities and informs appropriate local climate change planning strategies.

References

Calthorpe P (2010) Urbanism in an age of climate change. Island Press, Washington DC

Davidson G, Theobald K, Walker S (2011) Community energy solutions: involving residents in developing low-carbon urban neighbourhoods. Built Nat Environ Res Pap 4:149–286

Harvey L (2009) Reducing energy use in the buildings sector: measures, costs and examples. Energ Effi 2:139–163

Prairie T, Duarte M (2006) Direct and indirect metabolic CO_2 release by humanity. Biogeosciences Discuss 3:1781–1789

Ratti C, Raydan D, Steemers K (2003) Building form and environmental performance: archetypes, analysis and an arid climate. Energy Build 35(1):49–59

Salat S (2009) Energy loads, CO_2 emissions and building stocks: morphologies, typologies, energy systems and behavior. Build Res Inf 37(5–6):598–609

Santin OG, Itard L, Visscher H (2009) The effect of occupancy and building characteristics on energy use for space and water heating in Dutch residential stock. Energy Build 41(11):1223–1232

Satterthwaite D (2008) Cities' contribution to global warming: notes on the allocation of greenhouse gas emissions. Environ Urbanization 20(2):539–549

Steemers K (2003) Energy and the city: density, buildings and transport. Energy Build 35(1):3–14

Tanaka N (2008) World energy outlook 2008. International Energy Agency, Paris

VandeWehge JR, Kennedy C (2007) A spatial analysis of residential greenhouse gas emissions in the Toronto census metropolitan area. J Ind Ecol 11(2):133–144

Decomposition Analysis of Growth in Transportation Carbon Emissions and Low-Carbon Path Exploration in Beijing

Yan-mei Li and Jian-feng Zhao

Abstract With the development of economy, the service industry has become a new field of energy saving and emission reduction needed to explore. As the most major component of service industry, the energy saving and emission in transportation industry are paid more and more attention. The carbon emission in Beijing from 2005 to 2012 is accounted in this paper. With 2005 as the baseline year, the complete decomposition method is used to analyze Beijing's transportation energy consumption and carbon emission. The conclusions are as follows: The overall transportation energy consumption and carbon emission show an upward trend in Beijing, and scale factor and energy consumption intensity factor are playing the catalytic role in transportation energy consumption and carbon emission. Among them, the industrial scale factor plays a major role on the growth of carbon emission in transportation industry. Energy consumption structure factor plays a smaller role on carbon emission in transportation industry, and it almost has no effect on energy consumption. This paper also proposes recommendations for future low-carbon development in Beijing transportation industry.

Keywords Beijing · Energy consumption of transport · CO_2 emissions of transport · Decomposition

1 Introduction

The global climate change has become a major problem which humans have to face with since 90s last century. Research shows that the carbon emission generated by economic activities is one of the main reasons that caused the climate change. The development mode with low energy consumption, low pollution, and low emission has been the inevitable choice for the future economic development of the world.

Y. Li (✉) · J. Zhao
Institute of Recycling Economy, Beijing University of Technology, Beijing 100124, China
e-mail: liyanmei1979@sina.com

S. Feng et al. (eds.), *Low-carbon City and New-type Urbanization*,
Environmental Science and Engineering, DOI 10.1007/978-3-662-45969-0_30

As the megacity, Beijing is limited by resources, energy, environment, and other factors, so it requires developing a low-carbon development path to achieve the goal of low-carbon city. Currently, Beijing has become a service-oriented city and the service industry's gross domestic product accounts for 75 % of gross regional product. Service industry's low-carbon development has become an important area of urban low-carbon development. As the most important part of the service industry, transportation industry's low-carbon development has been paid more and more attention.

In order to reduce the carbon emissions in economic activities, the calculation of the carbon emission and the influence decomposition has become the focus of the academia in recent years. The input–output analysis of Korea and Japan has been made by Rhee and Chung (2006). The energy consumption and carbon emission of America and countries of OECD have been analyzed by Schipper and Howarth using Laspeyres index method (Schipper et al. 1990, 1993; Howarth et al. 1991). The factor decomposition of the carbon emission trend in 13 members of IEA has been done by Schipper using AWD method, and the result shows that the energy intensity and energy consumption can explain the most motive changes of the carbon emission intensity (Schipper et al. 2001).

On the domestic front, Sun Jianwei and Zhao Rongqin used statistic data from 1995 to 2005 of China, based on greenhouse gas inventory method of IPCC, established the framework of carbon emission estimation system of China, estimated the carbon emission in China from 1995 to 2005, and analyzed the carbon emission, carbon emission intensity, and their changing factors by using factor decomposition mode (Sun et al. 2010). Based on the research and analysis about the characteristics and the law of evolution of CO_2 emission in the past 16 years (1995–2010) in China's eight regions, using the method of LMDI decomposition, the effects of China's CO_2 emission were decomposed into four influencing effects, which were population size effect, economic development effects, energy intensity effect, and energy structure effect. Deng Jixiang and Liu Xiao explored China's carbon emission showing regional differences and reason (Deng et al. 2014). Ji and Zhao (2014) calculated the carbon emissions of transport in Tianjin during 2004–2010. Taking 2004 as the base year, using LMDI decomposition method, the change of CO_2 emissions of transport was decomposed. On the basis of the statistical data of the IPCC reports, Xu and Du (2011) calculated the carbon emissions of the transportation industry of China during 1995–2008. Using STIRPAT model, it decomposed the contributors to the growing carbon emission volume into several factors and studied the impact of two of them, population and economic development, on the transportation carbon emission.

In summary, the current researches mainly focus not only on the accounting for carbon emission from energy consumption but also on the influence factors to decompose. With the service industry in Beijing not only plays a significant role but also occupies a crucial position in the rapid economic development now. More and more resources and environment issues are caused by the service industry. As the main department of energy consuming and carbon emission in service industry, transportation industry is in its infancy on energy saving and emission reduction.

However, the potential and space is giant. In view of this, this paper is preparing to thoroughly analyze the transportation energy consumption and the carbon emission, which will produce an important theoretical and practical significance on finding out the potential, establishing the aim, and enacting the specific objective of energy saving and emission reduction (Wang et al. 2013).

2 Methods

2.1 Carbon Emission Calculation

The method in the IPCC guidelines for greenhouse gas emission inventory was used to evaluate the carbon emission of the transportation energy consumption in Beijing, which is expressed as follows:

$$C = \sum_{i} F_i \times \partial_i \tag{1}$$

where C is the total carbon emission of transportation in Beijing (10^4 t); i represents the type of energy consumption; F_i is the terminal consumption of i energy; and ∂_i is the carbon emission coefficient of i energy (calculated as per t standard coal).

2.2 Decomposition Analysis Models

The decomposition analysis of the energy consumption and the influence factors of the carbon emission increase is one of the research focuses in the energy field, and it was studied in many papers (Liang et al. 2007). There are many decomposition forms, and the main difference lies on how to assign the residual term to each influence factor. Sun presented a complete decomposition model, in which the residual term is assigned to the influence factor equally (Sun 1998). The complete decomposition model is not only simple, but also passes the three acceptability tests of the decomposition proposed by which are time reversibility test, cyclic test, and factor reversible test. In addition, the Laspeyres, Paasche, and Marshall-Edgeworth indexes have the same decomposition. So it is gradually adopted by many researchers (Li et al. 2009). The decomposition is made using this method in the paper.

2.2.1 Decomposition of Energy Consumption

The characteristics of transportation energy consumption are decomposed using the complete decomposition model. The key factors that influence the energy change are further discussed, including the industrial scale factor, energy intensity factor, and energy structure factor. Then, the energy consumption amount can be expressed using the following equation:

$$E = \sum_i E_i = \sum_i Q \times \frac{E}{Q} \times \frac{E_i}{E} \tag{2}$$

where Q is the transport industry production (100 million RMB), E is the total energy consumption (ten thousand tons standard coal), and E_i is the i energy consumption in transportation (ten thousand tons standard coal). Q represents the transportation industry scale, $I = E/Q$ represents the transportation energy intensity, and the $F_i = E_i/E$ reflects the transportation energy structure.

Decompose the Eq. (1) completely:

$$\begin{aligned}
\Delta E = E_1 - E_0 &= \sum_i Q^1 \times I^1 \times F_i^1 - \sum_i Q^0 \times I^0 \times F_i^0 \\
&= \sum_i (Q^0 + \Delta Q) \times (I^0 + \Delta I) \times (F_i^0 + \Delta F_i) - \sum_i Q^0 \times I^0 \times F_i^0 \\
&= \sum_i (Q^0 + \Delta Q)(I^0 + \Delta I) \times (F_i^0 + \Delta F_i) - \sum_i Q^0 \times I^0 \times F_i^0 \\
&= \sum_i (Q^0 + \Delta Q)(I^0 \times F_i^0 + I^0 \times \Delta F_i + \Delta I \times F_i^0 + \Delta I \times \Delta F_i) - \sum_i Q^0 \times I^0 \times F_i^0 \\
&= Q^0 \times I^0 \times \Delta F_i + Q^0 \times \Delta I \times F_i^0 + Q^0 \times \Delta I \times \Delta F_i \\
&\quad + \Delta Q \times I^0 \times F_i^0 + \Delta Q \times I^0 \times \Delta F_i + \Delta Q \times \Delta I \times F_i^0 + \Delta Q \times \Delta I \times \Delta F_i
\end{aligned}$$

2.2.2 Decomposition of Carbon Emissions

Similarly, the key factors that influenced the transportation carbon emission are analyzed, including industry scale, energy intensity, energy structure, and carbon emission coefficient. The carbon emission amount can be expressed as follows:

$$C = \sum_i C_i = \sum_i Q \times \frac{E}{Q} \times \frac{E_i}{E} \times \frac{C_i}{E_i} \tag{3}$$

where C is the total transportation carbon emission, C_i is the transportation carbon emission of i energy (10^4 t), $M_i = C_i/E_i$ reflects the carbon emission coefficient, and other variables are same as Eq. (2).

Equation (3) is completely decomposed using the same method as Eq. (2). The contributions of each factor value are (the middle part is omitted as it is too long):

$$Q(\text{effect}) = \sum_i \Delta Q \times I^0 \times F_i^0 \times M_i^0 + \cdots + \frac{1}{4}\Delta Q \times \Delta I \times \Delta F_i \times \Delta M_i$$

$$I_i(\text{effect}) = \sum_i Q^0 \times \Delta I \times F_i^0 \times M_i^0 + \cdots + \frac{1}{4}\Delta Q \times \Delta I \times \Delta F_i \times \Delta M_i$$

$$F_i(\text{effect}) = \sum_i Q^0 \times I \times \Delta F_i \times M_i^0 + \cdots + \frac{1}{4}\Delta Q \times \Delta I \times \Delta F_i \times \Delta M_i$$

$$M_i(\text{effect}) = \sum_i Q^0 \times I^0 \times F_i^0 \times \Delta M_i + \cdots + \frac{1}{4}\Delta Q \times \Delta I \times \Delta F_i \times \Delta M_i$$

where M_i^0 is the transportation carbon emission coefficient in 2005 (baseline), M_i^1 is the transportation carbon emission coefficient in 2012, and other variables same as Eq. (2).

3 Empirical Analysis

3.1 Data Sources

The transportation energy consumption and carbon emission are completely decomposed using method and model above. The data of transportation energy consumption come from the transportation, storage, and postal industry data in "Beijing Statistical Yearbook" (2006–2013). Conversion factors from physical unit to coal equivalent in the calculation process use the data in the "China Energy Statistical Yearbook" (2008).

The carbon emission coefficient comes from (IPCC 2006). The carbon emission coefficients of electricity and heat are calculated according to the calculation method given by IPCC, using carbon emission produced when electricity and heat are produced (Song and Xu 2011).

3.2 Growth Characters

With the scale of transportation increasing, the total energy consumption increases year by year (see Fig. 1). The transportation production added value increased from 40.33 billion in 2005 to 81.63 billion (yuan RMB), which increased 102 %. However, the transportation energy consumption increased form 5.3311 million tons in 2005 to 11.2003 million tons standard coal in 2012, which increased 110 %. This means that the transportation energy consumption increased faster than its development scale.

Transportation carbon emission has increased year by year (see Fig. 2), form 11.8826 million tons in 2005 to 26.9575 million tons in 2012. The transportation

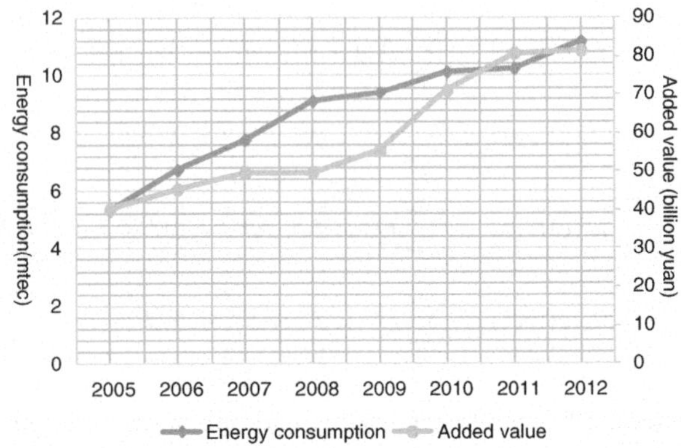

Fig. 1 The total transportation production added value and energy consumption

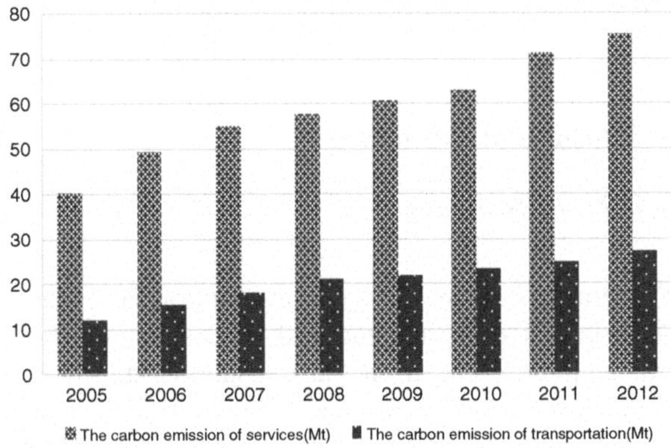

Fig. 2 The carbon emission of services and transportation in Beijing

carbon emission accounts over 30 % in the total emission of service. It means that the transportation has a great impact on the total service carbon emission and has a greater damage to the environment than other parts of service.

3.3 Decomposition Results

As shown in Table 1, the production scale factor has the greatest contribution to the total energy consumption. It is obvious that the industry expending is the main reason that caused the transportation energy consumption to increase. The contributed value to the transportation energy consumption of energy intensity is 10.3719 in 2005–2012, which means that the decrease of energy intensity can effectively relieve transportation energy consumption. During the studied time, the energy consumption structure has nearly zero contribution to the transportation energy consumption, which suggests that the change in the transportation energy consumption structure has hardly impact on the total energy consumption.

Table 1 Contribution value of each factor on transportation energy consumption

Industry	Energy intensity	Energy structure	Total
556.2988	10.3719	−1.21643E−13	566.6707

Table 2 Contributed value of each factor to the transportation carbon emissions

Industry scale	Energy intensity	Energy structure	Carbon emission coefficient	Total
1286.3179	71.1431	139.4809	20.6459	1517.5877

It is shown in Table 2 that the industry scale has the greatest contribution to the transportation carbon emission in Beijing. It is obvious that the industry-scale expending mainly caused the transportation carbon emission to increase. The contributed value of energy consumption structure to the total energy consumption is 139.4809, which ranked second in 2005–2012. It suggests that the improvement of energy structure can efficiently reduce the transportation carbon emission. During the studied time, the contribution of energy intensity to the transportation energy consumption is 71.1431, and it suggests that the transportation energy intensity has an impact on the carbon emission, but the impact is little. The carbon emission coefficient has the little impact on the carbon emission.

4 Conclusions and Suggestions

4.1 Conclusions

By calculating the energy consumption and the carbon emission of Beijing's transportation industry and applying complete decomposition method to the decomposition analysis, we obtain the following conclusions:

1. The overall transport energy consumption and CO_2 emission show an upward trend in Beijing. The growth rate of energy consumption and CO_2 emissions is 102 % and 127 %, respectively.
2. Considering the transportation energy consumption, the industry scale factor and energy intensity factor have a positive impact on the transportation energy consumption, while the energy structure has hardly impact on it.
3. Considering the transportation carbon emission, the industry scale factor, energy structure factor, energy intensity factor, and carbon emission coefficient all have an obvious impact on the carbon emission. The industry scale factor has the main positive impact on the carbon emission, while carbon emission coefficient plays a minimum role in carbon emission.

4.2 Suggestions

The suggestions on the transportation energy consumption and carbon emission are given from three parts as industry scale factor, energy intensity, and energy structure:

1. As the main energy consumption and carbon emission apartment, transportation should pay more attention on the energy saving and emission reduction and establish the macro-mechanism for favorable environment and system to lay the foundation of energy saving and emission reduction.
2. Considering from the industry scale, Beijing has a large population which is rapidly growing, while the scale of transportation cannot be cut short in a short time. However, the low-carbon development can be realized by slowing the speed of transportation scale.
3. Considering from the energy structure, the high-carbon energy should be substituted by the low-energy resources, improving the proportion of gas and low-carbon fuel in the transportation industry. In addition, more investment should be launched in the field of new energy for the development of low-carbon transportation.
4. Considering from the energy intensity, the energy consumption of per unit production added value should be reduced. Besides, energy saving and emission reduction and the technology innovation in the field of new energy should be operated.
5. Considering from urban passenger transportation, the urban public transportation should be operated (Huang et al. 2005). The energy consumption of metropolis passenger transportation is driven by the resident's motorized travel (Li et al. 2008). The carbon emission of private passenger auto is the crucial factor that leads to the growth of carbon emission in the metropolis (Zhang et al. 2012). So, developing the public transportation is the crucial approach to the development of low-carbon transportation.

Acknowledgments This study is supported by the Project of Beijing City Board of Education No sm201410005006 and Ri-Xin Talent Project of Beijing University of Technology.

References

Deng J, Liu X, Wang Z (2014) Characteristics analysis and factor decomposition based on the regional difference changes in China's CO_2 emission. J Nat Res 29(2):189–200

Howarth RB, Schipper L, Duerr PA et al (1991) Manufacturing energy use in eight OECD countries. Energy Econ 13(2):135–142

Huang C, Chen CH, Wang BY (2005) Urban travel modal split and its impact on energy and environment. J Highway Trans Res Develop 22(11):163–166

IPCC (2006) IPCC guidelines for national greenhouse gas inventories: volume [EB/OL]. http://www.ipcc.ch/ipccreports/Methodology-reports.htm.2008-07-20/2013-05-06

Ji W, Zhao T (2014) Decomposition of Tianjin's transport CO_2 emissions based on LMDI method. J Wuhan Univ Technol 36(3):393–396
Li Z, Fu F, Gao D (2008) Decomposition analysis is of motorized passenger transport energy demand changes in selected Chinese cities. J Qinghua Univer (Sci Technol) 48(11):1945–1948
Li Y, Jiang W, Cheng X (2009) Influence of structure and efficiency on energy intensity of inner Mongolia. Areal Res Develop 28(5):27–31
Liang JS, Zheng W, Cai JM (2007) The decomposition of energy consumption growth in China: based on input–output model. J Nat Res 22(6):853–864
Rhee H, Chung H (2006) Change in CO_2 emission and its transmissions between Korea and Japan: using international input-output analysis. Econol Econ 58(4):788–800
Schipper L, Howarth RB, Geller H (1990) United States energy use from 1973 to 1987: the impacts of improved efficiency. Ann Rev Energy 15(11):455–504
Schipper L, Howarth RB, Anderson B (1993) Energy use in Denmark: an international perspective. Nat Res Forum 17(2):83–103
Schipper L, Murtishaw S, Khrushch M (2001) Carbon emissions from manufacturing energy use in 13 IEA countries: long-term trends through 1995. Energy Policy 29(9):667–688
Song D, Xu A (2011) Regional difference and influential factors of China's urban carbon emissions. China Popul Res Environ 21(11):8–14
Sun JW (1998) Changes in energy consumption and energy intensity: a complete decomposition model. Energy Econ 20(1):85–100
Sun JW, Zhao RQ, Huang XJ (2010) Research on carbon emission estimation and factor decomposition of China from 1995 to 2005. J Nat Res 25(8):1284–1295
Wang K, Li J, Tang Y (2013) Carbon emissions accounting and decomposition of energy consumption of service industry in China. China Popul Res Environ 23(5):21–28
Xu Y, Du Z (2011) Measurement and factor analysis of carbon emissions of transportation industry of China. Logistics Technol 30(6):16–18
Zhang Q, Tao X, Yang P (2012) Research on carbon emissions from metropolis urban passenger transport and countermeasures. China Popul Res Environ 22(1):35–42

Analysis on Environmental Costs and Benefits Indexes of Construction of Mountain Town

Yong-hua Yang

Abstract The environmental impact of construction of mountain town may be harmful and negative, such as aggravating environmental pollution, resource, or ecosystems, while may also be beneficial and positive, such as reducing environmental pollution, saving resources, or improving the environment. Firstly, the paper defines the intension and extension of the environmental cost and benefit in the construction of mountain town. Based on this, the paper analyzes the composition, indicators, and accounting methods of the environmental cost and benefit in the construction of mountain town.

Keywords Environmental cost · Environmental benefit · Construction of mountain town

1 Environmental Problems of Construction of Mountain Town

In principle, the environmental impact of construction of mountain town involves all the environmental problems, including air pollution, water pollution, noise pollution, and solid waste pollution to various types of resources and ecosystems. The environmental impact of construction of mountain town may be harmful and negative, such as aggravating environmental pollution, resource, or ecosystems, while may also be beneficial and positive, such as reducing environmental pollution, saving resources, or improving the environment. Different types of impacts will reflect various forms and characteristics. From the perspective of economic analysis, the economic significance of environmental influence is negative or positive environmental impacts on the economy caused by the losses or gains.

Y. Yang (✉)
School of Economics and Management, Yunnan Normal University,
Kunming 650500, China
e-mail: yhyang2003@163.com

S. Feng et al. (eds.), *Low-carbon City and New-type Urbanization*,
Environmental Science and Engineering, DOI 10.1007/978-3-662-45969-0_31

Some environmental impacts appear during the construction such as air pollution, noise pollution, and water pollution. Air pollution is mainly resulted from earthwork construction dust, vehicle dust, and wind generation of fugitive dust. Noise pollution is caused by machine, noise generated by vehicles, and construction workers. Water pollution is attributable to the construction site of sewage and wastewater pollution from construction. The impact on environmental sustainability of the construction of mountain town has three aspects: consuming large amounts of energy and resources, environmental impacts caused by constructions, and lots of problems such as environmental pollution and ecological deterioration.

2 Intension and Extension of Environmental Costs and Benefits of Construction of Mountain Town

Jasch (2006) argued that all the costs associated with environmental destructions and environmental protections constitute the environment cost, including the internal and external environmental costs. Fichter defined environmental cost as the total costs directly or indirectly related to the use of materials and the resulting environmental impacts. Georgilakis (2011) refers those environmental costs due to prevention or correction of your environmental impact and cost. Vaughn Nestor and Pasurka (2005), interpreted the environmental cost from the perspective of economic and environmental significance. The economic environmental cost refers to the value of environmental goods and environmental services during the course of the use in economic activity. And the environmental costs are the actual or potential costs related to the deterioration of environmental resources caused by human activities. Xueyi (2007), China Mining University Professor, pointed out that environmental cost should contain four components: energy consumption cost, environmental expenditure cost, environmental damage cost, and environmental opportunity cost (1997). Luo Guomin, Jinan University professor, argued that environmental cost was the value of cost elements in the entrepreneurial activity as well as for the restoration of ecological environment of various expenditures.

Currently, many scholars made deep research on environmental benefits. Macauley et al. (2003) argued that the essence of environmental benefit was internalization of externalities ecosystem environmental benefits. With the way of internalizing environmental externalities, he divided the environmental benefit into compensatory income for environmental benefit and the revenue of ecological services. Based on the micro-level, Buran defined the environmental benefits as a variety of benefits that because of firms' reducing pollution, investors prefer the enterprise investment, consumers are willing to consume the product, and companies enjoy preferential government policies. Brulla et al. (2011) held that the meaning of eco-efficiency was the beneficial value of environmental assets in the ecosystem. The monetary value is the different value of environmental quality indicators or of environmental condition indicators before and after the measurement of environmental protection. Tang argued that environmental benefits were the monetary effectiveness brought by environmental assets within a certain period, as well as the revenues due to enterprise environmental behavior. Gang (2011) began with research of the enterprise value creation of the environmental assets and then gradually introduced the concept of enterprise environmental benefits. He pointed out that environmental benefits belonged to a branch of enterprise value, and it was the use of ecological benefits of environmental assets.

2.1 Intension and Extension of Environmental Costs of Construction of Mountain Town

The intension of environmental costs in the mountain town construction is the environmental cost or environmental damage caused by the construction project, as well as the cost of measurement for managing the construction project based on the principle of responsibility.

The extension of environmental costs in the mountain town construction contains two aspects. One includes environmental damage and resource depletion costs and mainly reflects the decline of natural resources quantity and of environmental quality during the course of the project constructions. The other is the actual expense on the environmental protection and controls, and the expenditure to improve the environment and restore the quantity of natural resources.

2.2 Intension and Extension of Environmental Benefits of Construction of Mountain Town

The intension of environmental benefits in the mountain town construction is the effectiveness caused by the environmental protection and pollution control during the project constructions.

The extension of environmental benefits in the mountain town construction contains two aspects. One includes investment in environmental protection. With the investment of environmental protection, the ecosystem will be remedied and ultimately provided a steady flow of environmental benefits to the economic system by environmental production. Obviously, the investment of environmental protection is the source of environmental benefit during the construction of mountain town, and the intensity of its investment directly affects the environmental benefits. The other is the effect of environmental protection. It is the direct effect of eco-environmental control, maintenance, and protection. Environmental effects mainly contain the restoration of the destructive resources.

3 Definition and Accounting Methods of Environmental Cost Indexes

3.1 Definition and Accounting Methods of Resource Depletion Cost

3.1.1 Definition of Resource Depletion Cost

Resource depletion cost is a unified environment cost for opportunity cost and compensation cost. Due to the utilization of natural resources during the construction of mountain town, the natural resource reserves gradually decrease. The reducing value of resource is the resource depletion cost. Resources are valuable. Human socioeconomic activity and consumption of natural resources must pay costs or pay the price. The amount of the cost should be greater than or equal to the value of the consumed resource. For the construction of mountain town, environmental resources are not only the potential basement of organized production activities but also the factors of production. For project construction, the premise conditions are the reduction of resource and ecological environment changes in the structure of the system. Project construction's depleting and occupying resources lead to changes in the ecological environment system state so that others lose the best chance to utilize land resources and the environment. We evaluate it as the resource depletion cost.

3.1.2 Accounting Methods of Resource Depletion Cost

Accounting of resource depletion cost can be established on the basis of macroeconomic principle of input–output analysis. The accounting methods are as follows.

$$M_i = P_i X_i (1 - \alpha_i) \tag{1}$$

M_i is the resource depletion cost of source i; X_i is the application amount of source i in the various activities during the construction of mountain town. P_i is the restoration cost of source i; α_i is the restoration ratio of source i.

3.2 Definition and Accounting Methods of Environmental Prevention and Protection Cost

3.2.1 Definition of Environmental Prevention and Protection Cost

The construction of mountain town is bound to lead to resource depletion, environmental pollution, and ecological destruction of environmental resources. It is necessary to invest considerable human, material, and financial resources in order to protect and prevent resources from damage, to avoid environmental pollution and environmental accidents, to improve the efficiency of resource use, and to reduce the extent of environmental pollution and ecological destruction of environmental resources. These constitute the environmental prevention and protection costs. Expenditures for environmental prevention costs are mainly for protection beforehand, such as investment in environmental monitoring, personnel expenditure, and expense on introducing equipment and technology.

3.2.2 Accounting Methods of Environmental Prevention and Protection Cost

Based on the Labor Value theory, enterprise value chain can be regarded as actually consisting of a series of sequential operations by the use of Activity Cost theory. With the analysis and the monetization of environmental influence, we can construct an econometric model of environmental protection cost. Each econometric model of operating unit's environment cost can be expressed using the following equation.

$$S_j = W_j + H_j \tag{2}$$

S_j is the environmental protection and prevention cost of environment j; W_j is the environmental cost for preventing environmental impact in phase j; and H_j is the environmental cost for operating environmental equipment in phase j.

3.3 Definition and Accounting Methods of Environmental Management Cost

3.3.1 Definition of Environmental Management Cost

Environmental management cost is the inherent value of ecological environment, which can be directly used for human utilization and consumption. On the other hand, the ecosystem provides the value of a service to human living environments. It is an intrinsic value of the human living environment in embodiment of natural resources. That is functional value of ecosystem of natural elements, also known as the ecological environment value. With the improvement of social civilization and economic development, this value is expected to grow. In the process of construction, land resources will be occupied, meanwhile waste water, dust, and solid wastes will be emitted. That will bring out destruction of the eco-environmental resources, such as air and water resources, the destruction of land resources, and then inevitably cause secondary environmental disasters. Then, it will cause the loss of ecological environment value. In order to restore and improve damaged eco-environment and bring it to satisfy with human needs of social and economic development, these costs and expenses must be compensated by the environmental costs. They are mainly composed of management expenses and compensation for the damage and losses caused by the pollution. Here, we define the compensation cost caused by the ecological destruction and loss of ecological value during the construction of mountain town as the environmental management cost. It is measured by the industrial waste gas costs, industrial wastewater treatment, and industrial solid waste disposal costs, green and mud-rock flows.

3.3.2 Accounting Methods of Environmental Management Cost

$$G_h = c_g + c_w + c_d \tag{3}$$

G_h is cost "h" of environmental management; c_g is the environmental management cost of waste gas in phase h; c_w is the environmental management cost of wastewater in phase h; and c_d is the environmental management cost of solid waste in phase h.

3.4 Accounting Methods of Total Environmental Cost

This paper above has analyzed various components of environmental costs during the construction of mountain town. Environmental costs are the summary of various components of specific cost index. Therefore, in order to facilitate the measurement of environmental cost and to clearly reflect the ratio of environmental cost elements, the paper constructs the econometric model of environmental protection cost. Equation is shown as follows.

$$C_E = \sum_{i=1}^{n_1} M_i + \sum_{j=1}^{n_2} S_j + \sum_{h=1}^{n_3} G_h \tag{4}$$

C_E is the total environmental cost; M_i is the depletion cost of resource i; S_j is the protection and prevention cost of environment j; G_h is the management cost of environment h; n_1 is the amount of resources consumed and occupied; n_2 is the amount of investment in environmental safety; and n_3 is the amount of investment in environmental management;

4 Definition and Accounting Methods of Environmental Benefits Indexes

4.1 Definition and Accounting Methods of Comprehensive Treatment Rate of Industrial Three Wastes

4.1.1 Definition of Comprehensive Treatment Rate of Industrial Three Wastes

The comprehensive treatment rate of industrial three wastes can reflect the capacity and level of waste management during the construction of mountain town. Effective treatment and prevention of industrial waste is one of the objectives in the process of construction, and it can really reflect the advantages and disadvantages of environmental benefits.

4.1.2 Accounting Methods of Comprehensive Treatment Rate of Industrial Three Wastes

$$K_1 = \sqrt{[3]}W * A * G \tag{5}$$

K_1 is the comprehensive treatment rate of industrial three wastes in constructed mountain town; W is the annual treatment rate of industrial wastewater in constructed mountain town; A is the annual treatment rate of industrial waste gas in constructed mountain town; and G is the annual treatment rate of industrial solid waste in constructed mountain town.

4.2 Definition and Accounting Methods of the Proportion of Environmental Protection Investment in GDP

4.2.1 Definition of the Proportion of Environmental Protection Investment in GDP

The proportion of environmental protection investment in GDP refers to the ratio of investment in environmental pollution control and ecological environment protection and construction investment for GDP. Environmental protection is the key factor to sustainable development of constructed mountain town, and the investment in environmental protection is one of the indicators whether environmental benefits are advantage or not. Economic development and environmental protection in constructed mountain town are interdependent and mutually conditional. Environmental protection should not damage development function, human health, and the environment in constructed mountain town.

4.2.2 Accounting Methods of the Proportion of Environmental Protection Investment in GDP

$$K_2 = P/\text{GDP} \tag{6}$$

K_2 is the proportion of environmental protection investment in GDP in constructed mountain town, P is the environmental investment in constructed mountain town, and GDP is the GDP in constructed mountain town.

Paul Samuelson, prominent economist, has pointed out that under the constraints of limited resources, economic growth could drive GDP growing from phase I to II. But if the overall technological level of the society also remains in the tradition of the heavily polluted technology group, and in the absence of effective environmental protection and environmental investments, GDP growth would lead to further deterioration in environmental quality. This value higher than 1 % will be conducive to sustainable development, and this value lower than 1 % would not be conducive to sustainable development. With the continuous increase of China's investment in environmental protection, the share of investment in environmental protection amounted to GDP rose steadily, from 0.7 % during the Seventh Five-Year Plan to 0.8 % during the Eighth Five-Year Plan. It firstly made a breakthrough 1 % during the Ninth Five-Year Plan. In 2005, the proportion had risen to 1.2 %. International experience shows that if the ratio is within 1–1.5 %, the trend of worsening pollution would be just basically controlled, but if the ratio is within 2–3 %, gradual improvement of the environment would be achieved.

4.3 Definition and Accounting Methods of the Rate of Reducing Energy Consumption Per Unit

4.3.1 Definition of the Rate of Reducing Energy Consumption Per Unit

The rate of reducing energy consumption per unit in constructed mountain town can directly reflect the energy efficiency of a country or region. For the constructed mountain town, increasing economic efficiency and reducing energy consumption is also one of the objectives of ecological construction, and it can reflect the advantages and disadvantages of environmental benefits.

4.3.2 Accounting Methods of the Rate of Reducing Energy Consumption Per Unit

$$K_3 = E_1/\mathrm{GDP}_1 - E_0/\mathrm{GDP}_0 \tag{7}$$

K_3 is the rate of reducing energy consumption per unit; E_1 is the energy consumption in constructed mountain town; GDP_1 is GDP in constructed mountain town; E_0 is the energy consumption in pre-constructed mountain town; and GDP_0 is GDP in pre-constructed mountain town.

According to international statistics, the reducing energy consumption per unit of most major countries such as the USA, China, and Russia declined in the decade from 2001 to 2009. The reduced rate ranged from 0 to 6.72, and the average rate of decrease was 2.035 %. According to domestic statistics, national energy consumption for per GDP is 1.077 tons standard coal every ten thousand Yuan. The reduced rates of all provinces ranged from 0.83 to 9.73, and the average rate of decrease was 3.61 %.

4.4 Definition and Accounting Methods of the Rate of Land Reclamation

4.4.1 Definition of the Rate of Land Reclamation

The reclamation rate of destroyed land is a proportion of the restored land accounting for the destroyed land.

4.4.2 Accounting Methods of the Rate of Land Reclamation

$$K_4 = D_1/D_0 \tag{8}$$

K_4 is the rate of destroyed land reclamation in constructed mountain town; D_1 is the restored land of destroyed land in constructed mountain town; and D_0 is the destroyed land in constructed mountain town.

According to statistics, the average rate of land reclamation in the European developed countries is above 50 %. While the average rate of land reclamation in other countries such as the USA, Australia, Canada, and Germany is higher than 70 %. But the situation of land reclamation in China is not optimistic. According to experts from the Ministry of Land and Resources, there are about more than 130 million mu of destroyed land in China and the reclamation rate is only 25 %. Compared with the developed countries in Europe and America, the gap of reclamation rate is large, and land reclamation work has great exploring potential.

4.5 Definition and Accounting Methods of Forestation Rate

4.5.1 Definition of Forestation Rate

The forestation rate in constructed mountain town is a proportion of the area of forests, grass, and other green vegetation area accounting for the total land area, which is the key indicator to reflect natural ecological conditions and the environmental benefits in constructed mountain town.

4.5.2 Accounting Methods of Forestation Rate

$$K_5 = D_3/D_2 \tag{9}$$

K_5 is the forestation rate in constructed mountain town; D_3 is the forestation area in constructed mountain town; and D_2 is the total land area in constructed mountain town.

Forestation rate varies from countries. For example, the rate is 68 % in Japan, 33 % in the USA, 69 % in Finland, 19 % in India, and 14 % in China. Generally, the forestation rate of 30 % can basically meet the needs of national economic construction and people's lives and protective function. The forestation rate should at least reach 40 % so as to be protective in mountainous and hilly areas.

Acknowledgments This research is sponsored by major science and technology for social development projects in Yunnan Province (2012CA024).

References

Brulla A, van Bohemenb H, Costanzac R, Mitschd WJ (2011) Ecological engineering: from concepts to applications Benefits of ecological engineering practices. Procedia Environ Sci 9:16–20

Gang BAO (2011) Environmental assets—in the enterprise value creation and theory of enterprise environmental accounting ownership. Ecol Econ 1:58–62

Georgilakis PS (2011) Environmental costs of distribution transformer losses. Appl Energy 9:3146–3155

Jasch C (2006) Environmental management accounting (EMA) as the next step in the evolution of management accounting. J Clean Prod 14:1190–1193

Macauley M, Palmer K, Shih J (2003) Dealing with electronic waste: modeling the costs and environmental benefits of computer monitor disposal. Appl Energy 16:89–100

Vaughn Nestor D, Pasurka CA Jr (2005) CGE model of pollution abatement processes for assessing economic effects of environmental policy. Econ Model 1:53–59

Xueyi Z (2007) Theory of coal mining area environment cost accounting. Ecol Econ 2:220–223 (academic)

Retracted: A Fast Time Series Shapelets Data Mining Algorithm

Zheng Zhang

Abstract Time series shapelets are a recent promising concept in time series data mining. Shapelets are time series snippets that can be used to classify unlabeled time series. Although shapelets are a useful concept, the current literature illustrates the fact that shapelet discovery is a time-consuming task. In this paper, we propose a fast shapelets discovery algorithm that outperforms the current algorithm; our experimental results demonstrate that the classification accuracy of the proposed algorithm is not significantly different from the accuracy obtained by the current algorithms, but the running time scalability is better.

Keywords Shapelets · Time series · Data mining

1 Introduction

L. Ye introduced the concept of shapelets and designed an algorithm to improve the brute force algorithm which was introduced in Ye and Keogh (2011). They proposed a technique to calculate a cheap-to-compute upper bound of information gain and use it to admissibly prune some candidates. The current best algorithm is given at Mueen et al. (2011). The algorithm can also find the exact shapelet, but does so more quickly. The speedup comes from using a classic pruning technique to prune some candidates, and from some caching tricks. The worst-case running time of the

---

Retraction of this chapter can be found at DOI 10.1007/978-3-662-45969-0_37

Z. Zhang (✉)
Information Center, GanSu Party School, Jianning Road 199#,
Anning District, Lanzhou 730070, People's Republic of China
e-mail: 630533797@qq.com

Z. Zhang
Complex Data Research Center, BeiHang University, Beijing 100083
People's Republic of China

S. Feng et al. (eds.), *Low-carbon City and New-type Urbanization*,
Environmental Science and Engineering, DOI 10.1007/978-3-662-45969-0_32

current best algorithm is O (n^2m^3), where n is the number of time series in the dataset and m is length of the longest time series in the dataset. In this paper, we will propose an O (nm^2) algorithm for finding shapelets.

2 Definitions and Notation

We begin by introducing all necessary notation and definitions. First, we define a time series:

Definition 1 A time series T is an ordered list of numbers, $T = t_1, t_2, \ldots, t_m$. Each value t_i can be any finite number, and m is the length of the time series T.

A local subsection of a time series is called a time series subsequence.

Definition 2 A time series subsequence S is a contiguous sequence of a time series. Subsequence S of length l of time series T starting at position i can be written by $S = T_i^l = t_i, t_{i+1}, \ldots, t_{i+l-1}$.

For the classification task, many time series are grouped together with their corresponding class labels in a container called a dataset:

Definition 3 A dataset D is a set of pairs of time series, T_i, and its class label, c_i. Formally, $D = \langle T_1, c_1\rangle, \langle T_2, c_2\rangle, \langle T_3, c_3\rangle, \ldots, \langle T_n, c_n\rangle$. n is the number of time series inside the dataset D.

To measure the similarity of subsequences, we define the distance between two subsequences:

Definition 4 The distance between subsequence S and $\hat{S}$ of the same length is the length normalized of Euclidean distance between subsequences S and $\hat{S}$. If both subsequences are Z normalized with mean = 0 and std = 1, the distance is defined as follows:

$$\mathrm{dist}(S, \hat{S}) = \sqrt{\frac{1}{l}\sum_{i=1}^{l}(s_i - \hat{s}_i)^2}$$

Shapelets can be of any length up to m. In order to allow meaningful comparisons, candidate shapelets of different lengths and length normalizations must be used.

We therefore define the distance between a time series and a given subsequence:

Definition 5 The distance between subsequence S of length l and time series T is defined as the minimum distance between subsequence S and any subsequence of T of the same length as subsequence S. Formally,

$$\text{dist}(S,\, T) = \min_i \text{dist}\left(S, T_i^l\right)$$

Suppose that dataset D contains n time series from c different classes. The number of time series in class i is n_i and we define class probability $p_i = n_i/n$. Hence, we defined the entropy of the dataset as follows:

Definition 6 An entropy of the dataset D is defined as $E(D) = \sum_{i=1}^{c} p_i \log(p_i)$.

To divide the dataset into two smaller datasets, we define a split.

Definition 7 A split is a tuple $\langle s, d \rangle$ of a subsequence s and distance threshold d which can separate the dataset into two smaller datasets D_L and D_R. The number of time series in D_L and D_R is n_L and n_R, respectively.

We next define the information gain of the given split:

Definition 8 The information gain of a split sp is

$$I(\text{sp}) = E(D) - \frac{n_L}{n} E(D_L) - \frac{n_R}{n} E(D_R)$$

The distance between two different sides of the given split is a separation gap.

Definition 9 A separation gap of a split sp is

$$\text{gap}(\text{sp}) = \frac{1}{n_L} \sum_{t_L \in D_L} \text{dist}(s, t_L) - \frac{1}{n_R} \sum_{t_R \in D_R} \text{dist}(s, t_R)$$

Definition 10 A shapelet is a split that separates the dataset into two smaller datasets with maximum information gain; ties are broken by maximizing the separation gap.

We visually summarize the concept of shapelets with an example as shown in Fig. 1. A candidate is shown in the box for reference. The distances between the candidate subsequence and all time series are calculated; all corresponding objects are placed on the orderline according to the calculated distance. In Fig. 1, three objects whose distances to the shapelet are small are shown as red rectangles on the left-hand side of the orderline. In contrast, other three objects are shown as green triangles on the right-hand side of the orderline, because their distances to the candidate shapelet are larger.

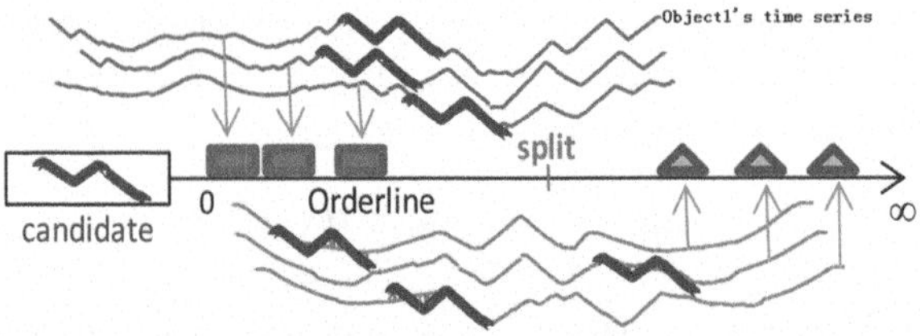

Fig. 1 The orderline shows the distance between the candidate subsequence and all time series as positions on the x-axis. The three objects on the left-hand side of the line correspond to class1, and the other three objects on the right correspond to class2

3 Fast Shapelet Discovery

3.1 Overview of the Algorithm

We propose to solve the shapelet discovery problem with a change of representation. In particular, we will transform the raw real-valued and high-dimensional data into a discrete and low-dimensional representation. Searching over a smaller representation is obviously more efficient; however, more importantly, having a discrete representation will allow us to hash our data, and use the collision history to inform our search.

3.1.1 Generating SAX Words

For each object in the dataset from example in Fig. 1, we transform the time series into a symbolic representation using Symbolic Aggregate approXimation (SAX) (Lin et al. 2007). The top part of the Fig. 2 shows an example of a SAX word **adbacc**, created by the first subsequence of Object1's time series. Multiple SAX words will be generated for a given time series using the sliding window technique (Lin et al. 2007).

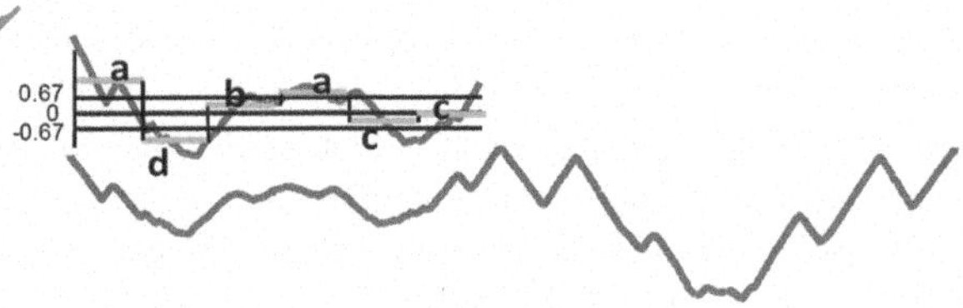

Fig. 2 The SAX word adbacc created from a subsequence of Object1's time series using the sliding window technique

3.1.2 Random Masking

A single time series can create multiple SAX words, and corresponding to the multiple subsequences, we can obtain as we slide the shorter subsequence length across the longer time series as shown in Fig. 3, where each object time series creates two or three SAX words of length five.

Having created the SAX representation of raw data, we have an apparent solution to the shapelet discovery problem. We could conduct a brute force search for the shapelets in the SAX space. Because of the reduced dimensionality of SAX, it would be faster than working with the raw data.

3.1.3 Counting Similar Objects

To avoid all-to-all distance computations, we apply hashing on all our data objects. The idea is to project all SAX words of high dimensionality to smaller dimensionality. All SAX words of dimensionality five have been randomly masked at two positions, reducing the dimensionally to three. Figure 4 shows that after hashing, the SAX words **adbac** and **acbac** share the same signature, ****bac**. All SAX words that have signature **bac** have their counters incremented in the relevant Table shown in Fig. 4. Similarly, in the second iteration, the words **adbac** and **acbac** once again randomly hash to the same word, this time **a**ac**.

3.1.4 Finding the Best Candidates

Continuing with our example, let us assume that we have done random projections for five iterations and the collision table is shown as in Fig. 5a. As shown in Fig. 5b, we can condense the collision table by summing all the object-based counts to the class-based counts, and creating the complementary data structure in Fig. 5c. From

Fig. 3 SAX words of each object

	SAX Words				
Obj 1	a	d	b	a	c
	a	c	a	a	c
Obj 2	a	c	b	a	c
	b	c	c	c	d
	b	d	c	d	d
Obj 3	b	b	a	c	d
	d	c	a	a	c

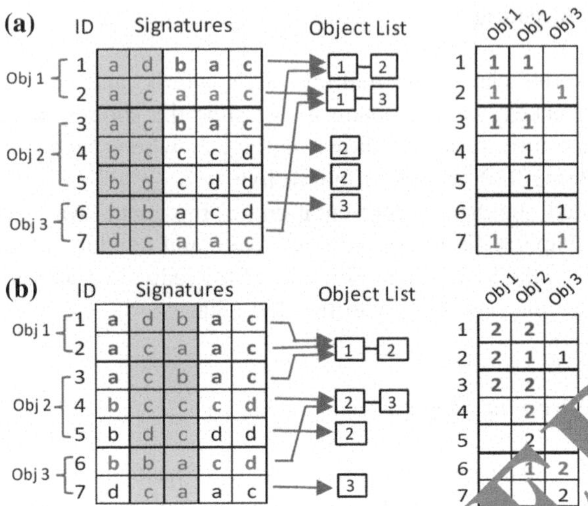

Fig. 4 The first (a) and second (b) iterations of the hashing and counting process. Thus, after r iterations of random projection, we expect the collision table shown in Fig. 4 right to remain mostly sparse, but to contain some locations that have values that are a significant fraction of r, this information can guide shapelet search

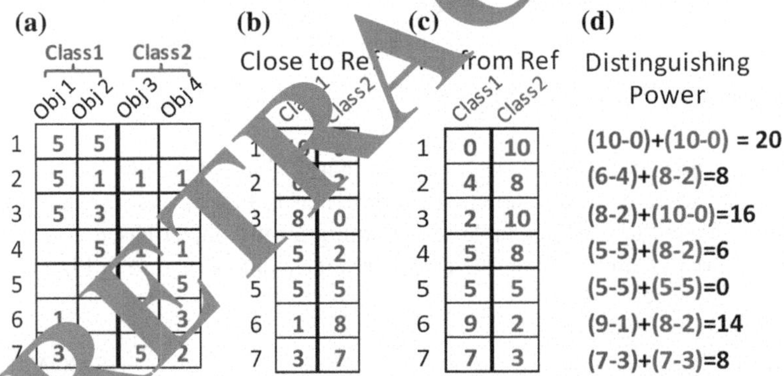

Fig. 5 a The collision table of all words after five iterations. Note that counts show the number of occurrences that an object shares a same signature with the reference word. b Grouping counting scores from objects in the same class. c Complement of (b) to show that how many times objects in each class do not share the same signature with the reference word. d The distinguishing power of each SAX word

these two tables, we can calculate the distinguishing power of each SAX word using the simple equation shown in Fig. 5d. Note that distinguishing power is high if the reference words appear frequently in one class but rarely in another class.

In this example, the highest score is from SAX word1 because it is close to objects in class1 10 times (obj1 5 times and obj2 5 times) and far from objects in class2 10 times; hence, its distinguishing power is 10 + 10 = 20. In contrast, SAX word5 receives power score of zero because this reference word is similar to objects from class1 5 times but also far from objects in class1 5 times, and has the same distribution in class2. Hence, the score is (5 − 5) + (5 − 5) = 0.

This list of SAX words with high distinguishing power is almost a solution for finding shapelet, as it very highly correlated with the quality (i.e., their *information gain*) of the corresponding shapelets in the original raw data space. Empirically, we can be certain that the best shapelet is near the top of this list.

3.2 Fast Shapelet Algorithm

So our shapelet discovery algorithm is shown as below.

Algorithm: FastShapelet
Input: D*: Dataset contain time series and class labels*
r *: number of random iterations*
k *: number of SAX to be candidates*
Output: shapelet*: the final shapelet*

```
[TS,Label] = ReadData(D)
for len = 1 to m
SAXList = CreateSAXList(TS,len)
Score = {}
   for i = 1 to r
   Count = RandProjection(SAXList,TS)
   Score = UpdateScore(Score,Count)
   end for
SAXCand = FindTopKSAX(SList,Score,k,r)
TSCand = Remap(SAXCand,TS)
max_gain=inf, min_gap=0
   for i = 1 to |TSCand|
   cand = TSCand[i]
   DList = NearestNeighbor(TS,cand)
    [gain,gap] = CalInfoGain(DList)
      if (gain>max_gain) ||
      ((gain==max_gain)&&(gap>min_gap))
      max_gain = gain
      min_gap = gap
      shapelet = cand
      end if
   end for
end for
```

The process is split into two phases. In the first phase (lines 3–10), we will select potential subsequences after a search in the SAX space. In the second phase (lines 11–23), we will measure the quality of those potential candidates in the raw data space and return the best candidate as the final shapelet.

To select the candidates, all subsequences of length `len` from all time series are created using sliding window technique, and we create their corresponding SAX word and keep them in `SAXList` (line 3). After the list of SAX words has been created, we will use these discrete representations to do hashing with `RandProjection()`, by creating a hash signature of each SAX word and give one count for each SAX words based on its signature. Then, we update the total score from multiple iterations (line 7).

Next, each SAX word is given a score to show that how many times each word occurs in each object. We then calculate a distinguishing power for each SAX word, and pick top k subsequences that have highest score (line 9). We mapped these SAX words back to their original raw data subsequences (line 10).

The second phase (lines 12–23): We will calculate the information gain for each candidate in the top k list and pick the best one as the shapelet. In detail, each candidate is considered one at a time (line 13). The body of the loop calculates the distance between the candidate subsequence and each time series using the equation in Definition 5. After these calculations, (lines 17–21) we will pick the subsequence which has the highest information gain, breaking ties by maximum gap as the final shapelet.

3.3 Experimental Results

We compared our algorithm with the current best algorithm (Mueen et al. 2011). For fairness, we used the code provided by original authors and set the parameters as they had recommended on their supporting Web page.

We begin by considering the accuracy on 32 datasets from UCR Time Series archives (Keogh et al. 2012) in Fig. 6. The results prove that our algorithm accuracy is no worse than the current best algorithm. The real difference between our algorithm and the current best algorithm is the running time scalability, as the running time comparison shown in Fig. 7.

The running time of the current best algorithm in Fig. 7 increases from 16 s to 8.7 h from $n = 50$ to $n = 800$ (m is fixed at 100). However, our algorithm is significantly faster; the running time is 0.76 s at the beginning and less than 16 s when $n = 800$. Thus, the speedup factor when $n = 800$ is 1,970X. Figure 7 shows that our algorithm achieves similar speedups when m is increased.

This empirical results are not surprising given the time complexity analyses of the algorithms. As described previously that the worst-case running time of the current best algorithm is $O(n^2m^3)$, and in the best case, if the triangle inequality can prune all candidates, the running time can be as low as $O(n^2m^2)$ (Mueen et al. 2011). However, our algorithm is just $O(nm^2)$.

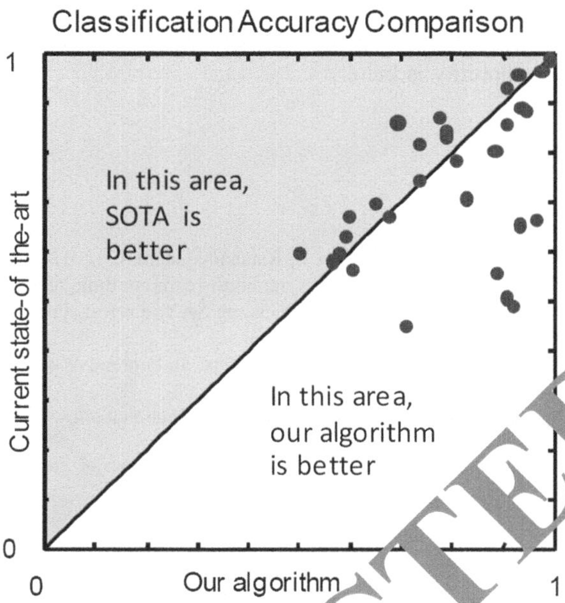

Fig. 6 Classification accuracy of our algorithm and the current best algorithm on 32 datasets from the UCR time series archives

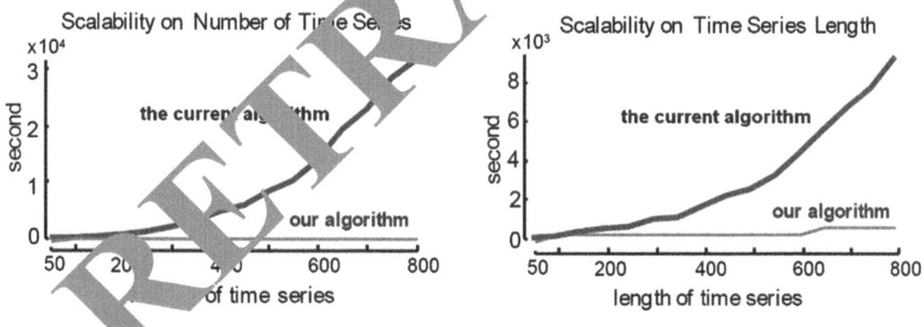

Fig. 7 Running time scalability comparison between our algorithm and the current best algorithm on 32 datasets from UCR time series archives

4 Conclusion

According to time series shapelets data mining problem, we propose a new algorithm for finding shapelets. We exploit a random projection technique on the SAX representation to find potential shapelet candidates. Our experimental results demonstrate that the classification accuracy of the proposed algorithm is not significantly different from the accuracy obtained by the current best algorithms and the running time scalability is better.

References

Keogh E, Zhu Q, Hu B, Hao Y, Xi X, Wei L, Ratanamahatana C (2012) The UCR time series classification/clustering homepage. www.cs.ucr.edu/~eamonn/time_series_data

Lin J, Keogh EJ, Wei L, Lonardi S (2007) Experiencing SAX: a novel symbolic representation of time series. DMKD 15(2):107–144

Mueen A, Keogh EJ, Young N (2011) Logical-shapelets: an expressive primitive for time series classification. In: KDD, pp 1154–1162

Ye L, Keogh EJ (2011) Time series shapelets: a novel technique that allows accurate, interpretable and fast classification. DMKD 22(1–2):149–182

The Design and Implementation of Management Information System of Electric Vehicle Charging Station Based on Cyber-Physical System

Zhi-cong Zhu, Xiao-fei Yang, Cheng-gang Cui, Shi-feng Wang, Ting-yu Gao, Dan Liu and Wei-jun Liu

Abstract Electric vehicles (EVs) charging stations in low-carbon city are widely distributed, and some of which use clean energy power supplies, such as solar cell and wind power; however, most of them are unattended. Thus, how to manage the equipment and operation process efficiently has become a problem that we must resolve as fast as possible. According to the needs for equipment and operation process monitoring in charging station, this chapter introduced the cyber-physical system into the charging station management. On the basis of cyber-physical-related technologies and fully considering equipment features and operation regulation, we designed and implemented a management system for EVs charging station, which has the functions of equipment monitoring, business process management, environmental perception, message notification, etc. The running results of this system show that this system is effective and feasible, which can provide reliable method of EVs station monitoring and management, while improves its operation efficiency as well as the supervision level.

Keywords Business process management · Cyber-physical system · Electrical vehicle charging station · Equipment monitoring · Low-carbon city · Management information system

1 Introduction

In recent years, electric vehicles have developed fast with the improvement of people's consciousness of environment protection and the encouragement of the related policy. More and more people prefer using electric vehicle to traditional vehicles. As an important supporting infrastructure, charging station,which provides

Z. Zhu (✉) · X. Yang · C. Cui · S. Wang · T. Gao · D. Liu · W. Liu
Cultural Heritage Science Conservation Laboratory, Shanghai Advanced Research Institute, Chinese Academy of Sciences, Shanghai 201210, People's Republic of China
e-mail: zhuzc@sari.ac.cn

S. Feng et al. (eds.), *Low-carbon City and New-type Urbanization*,
Environmental Science and Engineering, DOI 10.1007/978-3-662-45969-0_33

convenient and reliable charging service, is the key factor of popularizing electric vehicle better. Electric vehicles (EVs) charging stations are widely distributed, some of which use clean energy power supplies, such as solar cell and wind power; it puts forward higher demands for automation level and maintainability of EVs charging station. Most of the charging stations have related real-time monitoring systems in China; however, these systems mostly focus on the research of the control strategy of power convert part, energy optimization, grid connection control strategy, and vehicle to grid (Yu 2011), and there is lack of efficient management methods on the operation of charging stations. Thus, how to manage the equipment and operation process efficiently has become a problem that we must resolve as fast as possible.

Cyber-physical systems (CPS) provide a novel approach and an effective path to realize distributed electric vehicle charging stations. CPS has been applied to the fields of power grid, transportation, medical care (Kleissl and Agarwal 2010), and so on. Generally, CPS consists of information system and physical system. It is a new complicated system that can combine the computing resources and the physical resources closely (National Science Foundation of the United States 2010). He Jifeng summarized CPS for "3C," that is computing, communication, and controlling (Jifeng 2010). Through the interaction between information system and physical system, information system can perceive, monitor, and control the physical system in a safe and effective way, thus achieving highly fusion of information world and physical world.

EVs charging station is a typical CPS system which consists of information system and physical system. Physical system in charging station contains the devices of fans, solar cells, charging piles, and so on, and it also contains some continuous variable processes produced by charging station operation, such as electric current, voltage, rotate speed of fans, and power. Information system mainly refer to the model, algorithm, and related computing process which are used to monitor device, analyze operation data, and generate command to control the physical system. By integrating the information system and physical system, charging station management information system could promote the mutual cooperation among two systems to make sure the safe and reliable operation of charging station.

According to the needs for equipment and operation process monitoring in charging station, this chapter firstly introduced a cyber-physical system framework based on Service-oriented Architecture (SOA), and then applied it to the design and implement of a management system for EVs charging station; finally, the practical application of this system was discussed.

2 Analysis of the CPS Framework

A conventional CPS is mainly a two-layer structure, which includes physical system and computing system (Meng et al. 2014). Physical system consists of multiple units just as sensor unit, acquisition unit, and execution unit, which is responsible for environment perception, data acquisition, and instruction execution, while computing system is responsible for data analysis, decision making, and command generation. SOA is a distributed module model which provides distributed deployment and composition of service components. Physical system and computing system in CSP are packed in various services, which connect different resources in CPS through well-defined interfaces and contracts that can fulfill various tasks via composition and mutual cooperation between these services. In order to solve the problem of charging stations management, a novel CPS framework is proposed which integrates the SOA approach with traditional CPS structure.

2.1 An Overview of the CPS Framework

The research on the CPS framework is still in a preliminary stage; there is no standard CPS framework which is suitable for all application scenarios. In the early research on CPS framework, event-driven model is usually used to describe the interaction between physical system and information system. Ying Tan proposed a prototype architecture for CPS driven by events (Tan and Goddard 2008), based on which he also proposed a spatiotemporal event model (Tan et al. 2009). In this model, CPS is divided into different components including physical event, physical observation, sensor event, cyber-physical event, and cyber-event. This framework based on event-driven model focuses on the interaction between physical and information systems; however, it has not discussed the credibility evaluation, events combination, and so on. Zhu Quanyan proposed a hierarchical security architecture, in which the security key points of each layer and the cross-layer security strategy were discussed (Zhu et al. 2011). Zhang Kan discussed a framework of trusted CPS and introduced a trust mechanism to ensure high reliability of CPS (Kan et al. 2011). Liu Hanyu proposed six-layer control framework, including application layer, network layer, link layer, coordination layer, adjustment layer, and physical layer, and the functions of each layer are described.

Based on the above studies, this chapter integrated the SOA approach in CPS and proposed a novel CPS framework which is designed for the charging station management information system.

2.2 Design of a CPS Framework Based on SOA

SOA is a distributed architecture model which can deploy, combine, and use loosely coupled coarse-grained application components in distributed mode through Internet as required. Service is the key of SOA (Li and Wu 2009). Figure 1 below shows a typical SOA infrastructure.

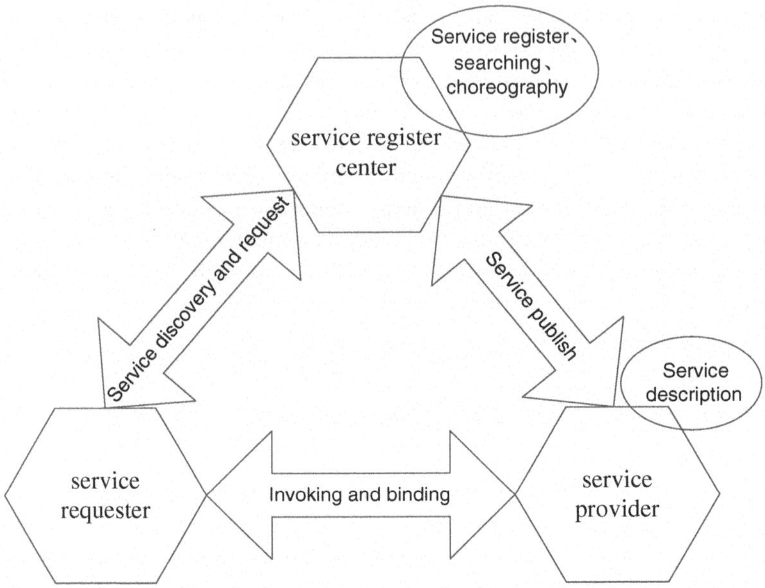

Fig. 1 A typical SOA infrastructure

The infrastructure of SOA consists of three parts which are service provider, service requester, and service register center. The service provider registers services into the service register center, finishes a group of tasks, and delivers the result to the service requester. The service register center provides services which are service registration, service discovery, and service matching. The service requester searches required services through service register center, on which the service requester can get the information of service description and interface definition. By using the well-defined interfaces among the various services of the applications on the heterogeneous platform, the distributed systems integrate the services together in a loose coupling mode, that is, integrate or composite multiple existing services to form new applications through network. At present, there have been some preliminary researches on how to integrate SOA approach into CPS. La HJ applied SOA and cloud computing concepts to CPS (La and Kim 2010), Wang Xiaole proposed a service-based framework of CPS, which includes node layer, network layer, resource layer, and service layer, and introduced the technologies and function module of each layer (Xiaole et al. 2010).

On the basis of above research, this chapter designed a novel four-layer SOA framework of CPS for the charging station management. Figure 2 shows the framework. This framework consists of four layers which are physical/sensor layer, network layer, service component layer, and application layer.

Physical/sensor layer, which consists of sensor unit, control/drive unit, computing unit, communication unit, and interaction unit, is the intelligent terminal of CPS. In this layer, data produced by operation process of the charging station,

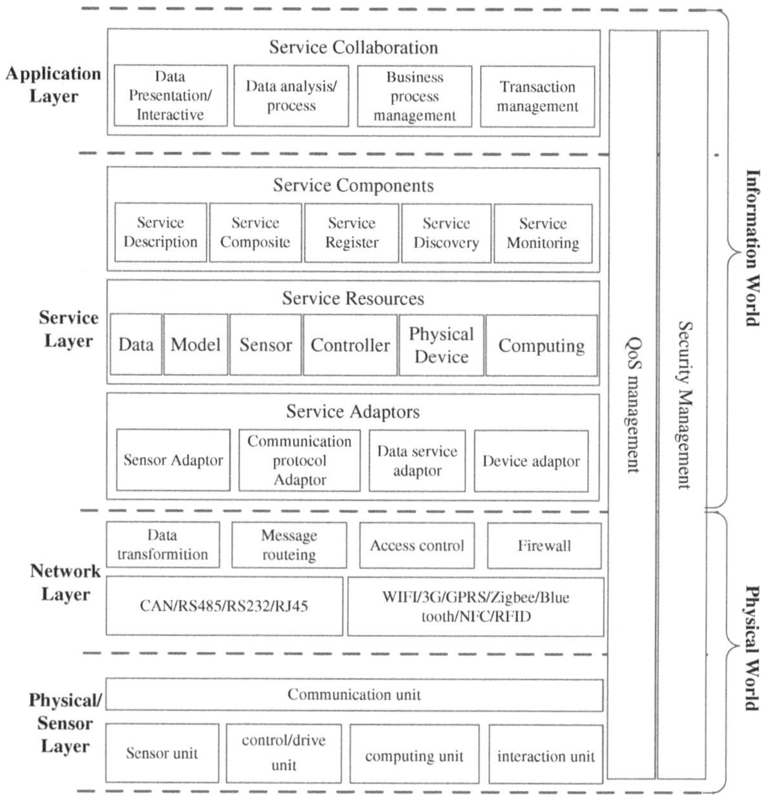

Fig. 2 A CPS framework based on SOA

including current, voltage, power, and environment data, are collected, filtered, and uploaded to the information world. In particular, user behavior information also can be collected by the interaction unit, which converts the user behavior into the standard data for information world to analyze and process. In order to manage the widely distributed physical elements such as sensors, equipment, and the produced data in a unified and standard way, these physical elements should be wrapped into precisely defined and well-encapsulated Web services by adapters and then be registered into the service register center, in which these services can be well managed.

Network layer provides the information transmitting channel in CPS for components communicating and resource sharing. In this layer, various network protocols and standards including wire and wireless communication modes should be supported to connect various sensors and equipment. In order to reduce the communication complexity, these network protocols should be encapsulated into well-defined communication interfaces by a communication adaptor, in which the communication functions of routes, protocol translation, and so on are implemented, while sensors, equipment, and components of Web services with different communication protocols can communicate each other in a unified and normative way through these interfaces without concerning implement details.

Service layer consists of three modules that are service adaptor, service resource, and service supporting component. Service adaptor connects the physical world and the information world. It encapsulates sensors, equipment, and relative data of the physical world into standard services, which can be registered, discovered, and invoked by service requester as service resources. On the other hand, service adaptor also transmits the commands produced by the service components of the information world to control the physical system. Service resource is an abstraction of entities of CPS, includes wrapped sensors and devices in the physical world, as well as the data, models, algorithms in the information world. Service supporting component is a group of supporting services which manages the services in CPS and provides various functions of service discovery, service composition, service monitoring, and so on.

Application layer is a group of applications assembled by services. In this layer, business process is established through mutual cooperation of services to finish a specific task as required. Transaction management module is used to ensure atomicity and consistency of transaction between distributed services. Interaction module receives and processes requests from users and then presents the results in a visualized way. It can also integrate different services or applications into one page using Portal technology.

Quality of service (QoS) management and security management run through the CPS framework based on SOA and play an important role in each layer. They guarantee the CPS system providing high-quality service continuously. The QoS resources and network resources will directly affects the stability, availability, security, and performance of CPS system. Huang Yunfeng utilizes the aspect-oriented technology to model the QoS aspect of real-time systems in cyber-physical network environment (Huang 2011). In the aspect of security management, Clifford (2009).

3 Design and Implement of EVs Charging Station Management Information System Based on CPS

In this chapter, an EVs charging station management information system based on the CPS will be designed and implemented. Firstly, the business process of charging station will be analyzed, and then a charging station management information system will be designed and implemented; finally, the practical application of this system was discussed.

3.1 Analysis of EVs Charging Station Business Process

There are two core businesses of EVs charging station (Fengquan et al. 2010). One is the EVs charging and the other is the traction battery replacing.

1. Work Flow of EVs charging
   Figure 3 gives the detailed work flow of EVs charging.
   (a) Firstly, user puts the RFID charging card onto the RFID reader embedded in charging pile. RFID reader reads the card number and sends it to management system, which queries the related information of this card such as the status and balance, verifies whether the card is available or not, and sends the result back.
   (b) Charging pile will display the charging mode options if the card is available, otherwise some error messages will be shown. The door of charging pile will open when user selects and confirms a charging mode, and then charging pile will start to charge after user inserts the charger.
   (c) The input capacity will be calculated and be billed in real time when the charging is beginning. Charging pile will stop charging when the car is filled up or there is no charging card balance; finally, a mobile short message with charging information will be sent to the user as well.

2. Work flow of the traction battery replacing and maintenance
   Work flow of the traction battery replacing and maintenance is shown in Fig. 4. It includes 5 steps that are given below.
   (a) Firstly, reception staff reads the car number using RFID reader, and then record the information of user such as name and phone number to create a new work form via PDA. According to the car number, management system will query and display related inventory information of traction battery for reception staff to choose. Reception staff selects one of the traction batteries and submits the work form, while an outputting request will be created and sent to storekeeper by management system.
   (b) Maintenance staff reads the battery number of electric vehicle using RFID reader and verifies the battery when he received a work form of traction battery replacing, and then he gets the traction battery from warehouse.
   (c) Storekeeper checks inventory of the traction battery on management system according to the received request, gets the location of the battery via RFID reader, and changes the battery's status in system.
   (d) Maintenance staff replaces the traction battery, records the replacing process, and checks items via PDA as well. The records will be synchronized between the PDA and the management system.
   (e) Finally, user will check the work form and sign to confirm it when the replacing and maintenance is finished.

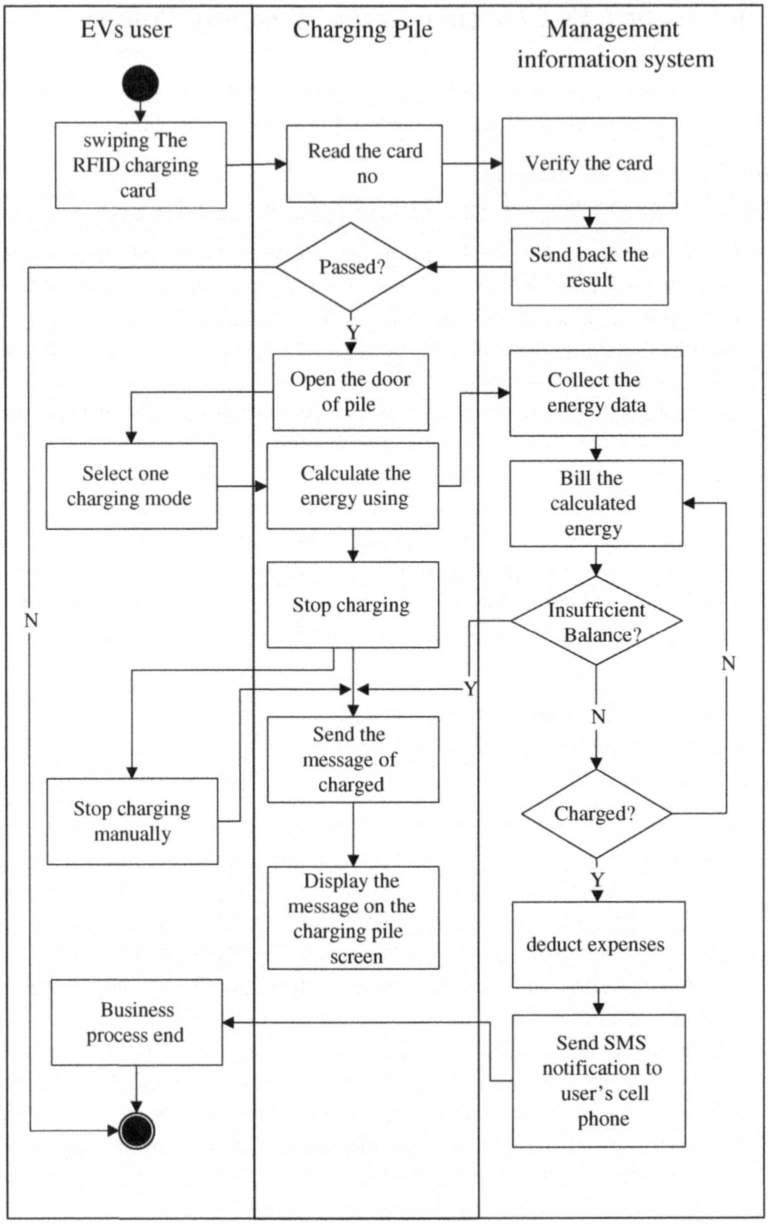

Fig. 3 Work flow of EVs charging

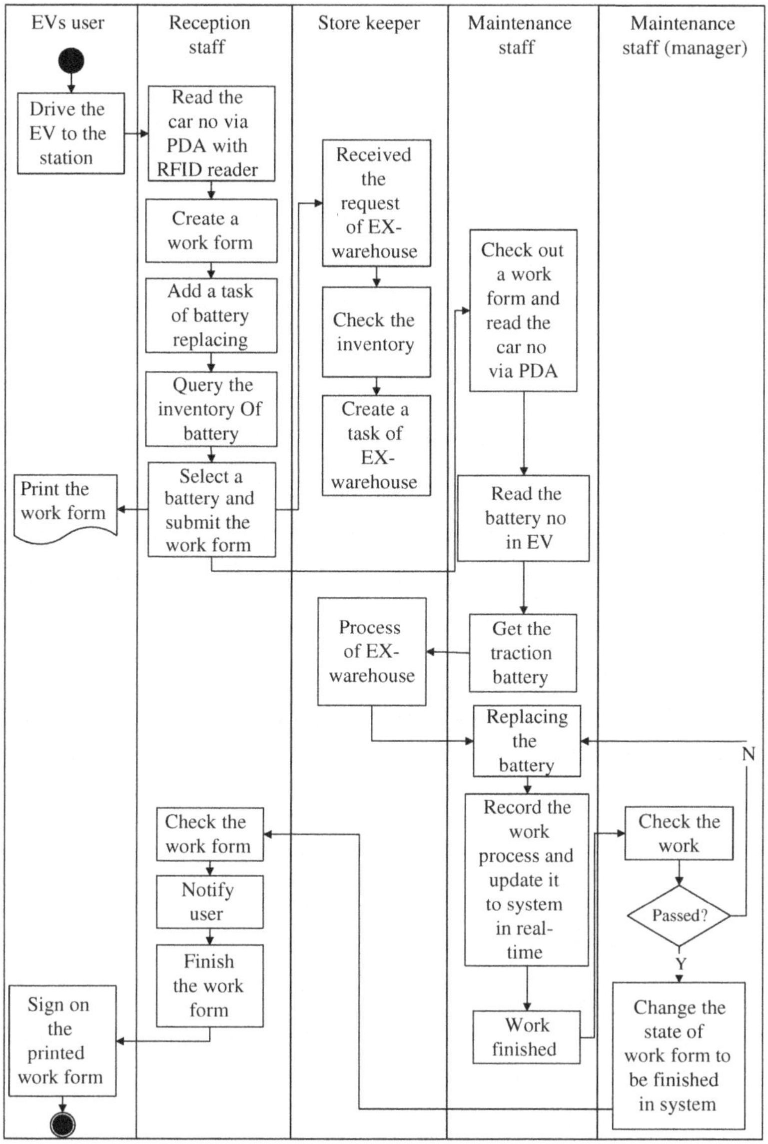

Fig. 4 Work flow of traction battery replacing

3.2 Design and Implement of the Management Information System

From the business analysis above, the tasks of work flow are usually driven by events and messages, and there are frequent interactions between the physical devices (charging pile, RFID reader, PDA, etc) and the management system. Thus, it is well suited for using CPS to resolve problems in charging station management. As shown in Fig. 5, on the basis of the CPS framework based on SOA, a charging station information management system consists of 5 layers—physical/sensor layer, communication layer, infrastructure layer, service bus layer, and business application layer.

1. Physical/sensor layer
   Physical/sensor layer contains data acquisition program, RFID middleware, RFID adaptor, data adaptor, device adaptor, and the physical devices (such as RFID tag, RFID reader, solar cells, and electric vehicle). Data acquisition program acquires the dynamic data of the operating devices via its communication interface (serial ports, Modbus, OLE for Process Control, etc.). Data adaptor and device adaptor encapsulate these data and devices to standard Web services which will be registered to the service register center as kinds of resources. These registered Web services have been well descripted, and the description includes the acquisition time stamp, value, name of the data, and the location information and configuration information. RFID middleware, which will be encapsulated to Web services by RFID adaptor, provides the communication interfaces include reading and writing operations between RFID readers and management system. Adaptors, includes data adaptor, device adaptor, and RFID adaptor, not only wrap the data, devices, and operations into standard Web services to realize the data exchange and communication between different application modules and physical equipment, but also translate the standard execute request to the specific invoking command to control devices.
2. Communication layer
   Communication layer includes several wired and wireless communication protocols, such as RS485, Modbus, WiFi, RJ45, and 3G/4G. Communication protocol adaptor hides the details of communication implements and provides precisely defined and well-encapsulated communication interfaces through which the distributed devices or application modules with different communication protocols can communicate each other in a standard and unified way.
3. Infrastructure layer
   Infrastructure layer, which is the computing foundation for constructing management system, consists of several kinds of middleware, including database, message server, cache server, and application server. MongoDB is used to store non-structural real-time data collected by data acquisition program. Redis, as a cache database, is used to cache the Web pages generated on-demand, or the static data in database, to achieve high scalability and performance. RabbmitMQ

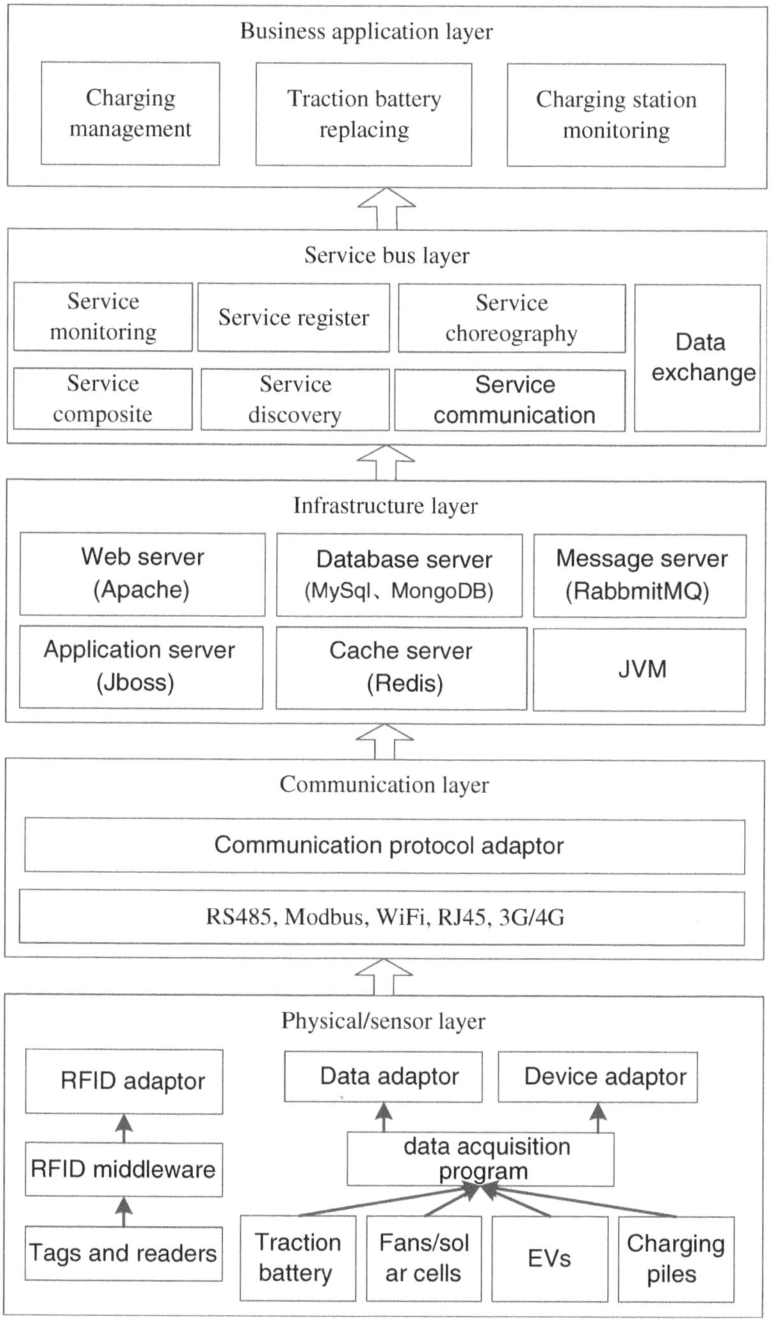

Fig. 5 Framework of the EVs charging station management information system based on CPS

is a message server which creates communication channels for distributed application modules in publish–subscribe message mode. As an application server, JBoss provides a standard J2EE environment for applications of charging management information system.

4. Service bus layer
Service bus layer provides service management functions such as service monitoring, service register, service composite, service discovery, and service choreography. Service monitoring maintains a service catalog and sends the heartbeat detection message at regular intervals. If one service is detected to be unavailable or service failure, it will be replaced by the backup service in the service cluster to ensure the high availability of management system. A service register function provides a group of service interfaces and user interfaces for service provider to register Web services on to service catalog. A service discovery function provides flexible service searching and service binding in the distributed management systems for the service requester. Service choreography and composite is supposed to process available Web services and extract their behavioral information, accept requests for new services, and build a valid composite service for the request, if possible. Hashemian and Mavaddat (2006) discussed how to find useful Web services for a request and how to compose them to obtain the expected behavior using the graph and the process algebra.

5. Business application layer
Business application layer contains applications which are aggregated with fine-grained Web services to finish specific business processes, mainly including functions of charging management, traction battery replacing, and charging station monitoring.

    (a) Charging management module
    In charging management module, charging piles, which are the physical elements in CPS, collect the dynamic charging data and collaborate closely with backend Web services to finish the business processes of user authorization and certification, charging pile control, input capacity calculating, and billing. The communication process and message are encrypted and they cannot be read or tampered with. The communication is encrypted using the Secure HyperText Transport Protocol (HTTPS) which is the most widely used secure Web encryption. Meanwhile, the message is encrypted using RSA cryptosystem, which is the public-key cryptosystem most widely used in practice, and 3DES cryptosystem, which is the block-based cipher system based on the data encryption standard (DES). Charging pile and management system both keeps the 1024 bit-secret-key and public-key of RSA cryptosystem. Charging pile will send a request to management system for a dynamic secret-key of 3DES when the user swipes RFID charging card for authentication. After the request received, management system will generate a 3DES secret-key randomly, encrypt the secret-key using RSA public-key, and then send the encrypted 3DES secret-key to charging pile. By using RSA 1024 bit-secret-key, charging

pile decrypts the encrypted 3DES secret-key, which will be used to encrypt all messages of communication between charging pile and management system.

(b) Traction battery replacing module
In the business process of traction battery replacing, PDA with RFID reader plays an important role, which collects the TAG data of electric vehicle and traction battery, interacts with maintenance staff, and triggers the business event. Management system changes the state of business event, generates commands, and sends it back to the PDA to drive the business process. If it requires human interaction to drive the business process, messages will be sent to related people for notification.

(c) Charging station monitoring module
Devices of charging station to monitor include fans, solar cells, and charging piles. These devices and related data are encapsulated into Web services, which report real-time data of devices to management system through data adaptor. The report data mainly include output voltage, output current, charging time, charging percentage, and fan speed. In order to communicate these devices with different communication protocols in a standard and unified way, a communication protocol adaptor is designed to hide the implementation of different communication protocol, mainly include RS485, Modbus, and Controller Area Network (CAN), and this adaptor provides unified and well-defined interfaces to communicate with other applications. If an exception occurred, the monitoring module immediately informs related people by sending message and mail. To ensure the related people can find out exceptions in first place, WebSocket and HTML5 technology is used to push real-time data to monitoring pages to display data in a visual way.

3.3 *Practical Application of the System*

The EVs charging station management information system based on CPS has been successfully deployed and has been in trial operation phase. Figure 6 shows the deployment structure of this system.

Staff of the charging station can monitor the equipment in real time through the page shown in Fig. 7, including video monitoring information, charging pile monitoring information, environment monitoring information, and carbon reduction information and so on.

Figure 8 shows the work form of traction battery replacing. In this page, reception staff edits the task items and then submits it to the work queue. Maintenance staff will take a work form from the work queue and do the task according to the work form.

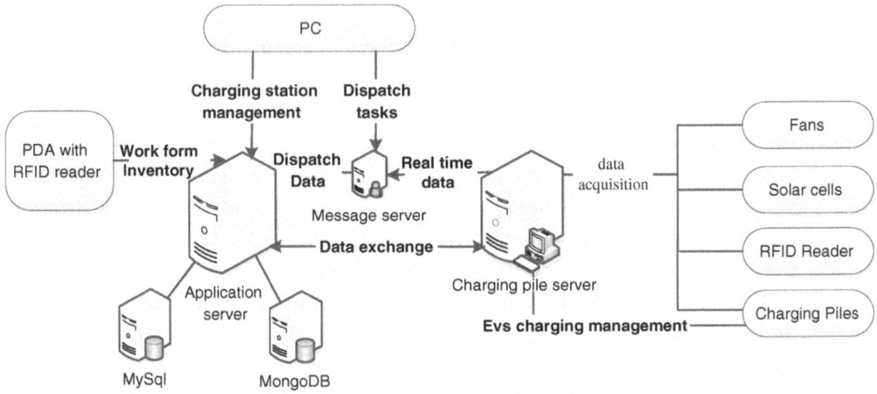

Fig. 6 Development view of the EVs charging station management information system

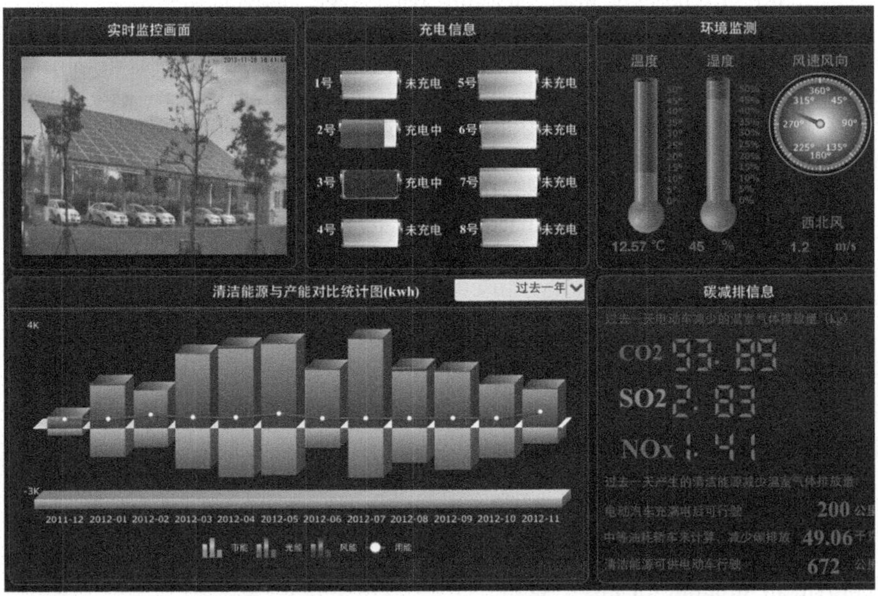

Fig. 7 Page of EVs charging station monitoring

Figure 9 shows the pages of the PDA with RFID reader, through which maintenance staff can read the TAG no, record the work process, and change the work state.

The system has been made a large number of tests and acquired a mass of measurement data. The system operation result shows that introducing CPS into the management information system acquired a great of achievements; compared to

Fig. 8 Page of work form editing

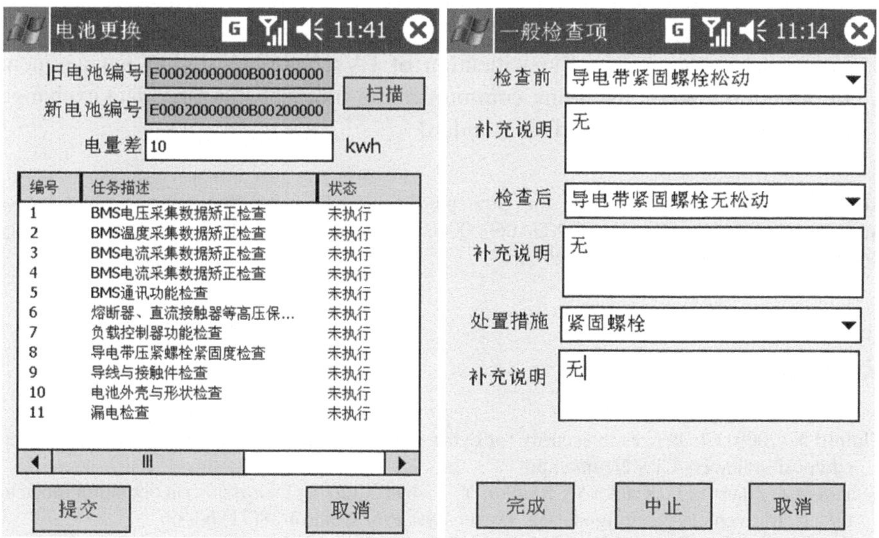

Fig. 9 Pages of traction battery replacing on PDA

traditional management information system, the efficiency of traction battery replacing process has been improved 30 % above; meanwhile, the record accuracy of vehicle number and battery number has reached 99 % above.

4 Conclusions and Further Work

According to the needs for equipment and operation process monitoring in charging station, this chapter introduced the cyber-physical system into the charging station management. On the basis of cyber-physical-related technologies and fully considering equipment features and operation regulation, we designed and implemented a management system for EVs charging station, which has the functions of equipment monitoring, business process management, environmental perception, message notification, etc. The running results of this system show that this system is effective and feasible, which can provide reliable method of EVs charging station monitoring and management, while improves its operation efficiency as well as the supervision level.

With fast development of EVs charging station, we should do more study on the EVs charging station management information system based on CPS.

1. Research on the QoS strategy of Web services for management system to provide high availability, high performance, and continuous service.
2. Research on the energy management system of charging station, which predicting and optimizing the energy consumption to achieve satisfactory emission reduction effects.
3. Research on the standard specification of EVs charging station management information system, including communication protocol standard, data exchange standard, and adaptor interface standard.

Acknowledgments Supported by the Strategic Priority Research Program of the Chinese Academy of Sciences, Grant No. XDA06000000. Supported by the National Key Technology R&D Program, Grant No. 2012BAH43F03.

References

Clifford N (2009) Challenges in security for cyber-physical systems [EB/OL]. http://cimic.rutgers.edu/positionPapers/CPS-Neuman.pdf

Fengquan Z, Zhanwei L, Xiaolei W, Xiaohui Y, Yishan X (2010) Discussion on operation mode to the electric vehicle charging station. Power Syst Prot Control 38(21):63–66

Hashemian, SV, Mavaddat, F (2006) A graph-based framework for composition of stateless web services. In: 4th European conference on web services, ECOWS '06, School of Computer Science, Waterloo University, Ontario, pp 75–86

Huang Y (2011) Aspect-oriented modeling for quality of service of real-time system in cyber physical systems. Guangdong University of Technology, Guangzhou (in Chinese)

Jifeng H (2010) Cyber-physical Systems. Commun CCF (China Computer Federation) 6(1):25–29 (in Chinese)

Zhang K, Zhang G, Zhang M (2011) A preliminary framework for designing the trusted cyber-physical systems. J Comput Res Dev 48(2):242–246 (in Chinese)

Kleissl J, Agarwal Y (2010) Cyber-physical energy systems: focus on smart buildings. In: Proceedings of the 47th design automation conference, ACM, New York City

La HJ, Kim SD (2010) A service-based approach to designing cyber physical systems. In: 2010 IEEE/ACIS 9th international conference on computer and information science, Yamagata, 2010, pp 895–900

Li H, Wu Z (2009) Research on distributed architecture based on SOA. In: 2009 international conference on communication software and networks, Macau, pp 670–674

Meng C, Dong L, Kairui W, Jiang Y (2014) Cyber-physical modeling of wind turbine and photovoltaic cell based on hybrid system. East China Electr Power 42(4):0656–0660 (in Chinese)

National Science Foundation of the United States (2010) Cyber physical system (CPS) program solicitation[EB/OL]. http://www.nsf.gov/pubs/2010/nsf10515.htm. 01 July 2010

Tan Y, Goddard S (2008) A prototype architecture for cyber-physical systems. Sigbed Rev 5(1):51–52

Tan Y, Vuran MC, Goddard S (2009) Spatio-temporal event model for cyber-physical systems. In: Proceedings of the 29th IEEE international conference on distributed computing systems workshops, IEEE, Piscataway, pp 44–50

Xiaole W, Lina C, Hongbin H, Su D (2010) A service-oriented architecture framework of cyber-physical systems. J Comput Res Dev 47(S2):299–303 (in Chinese)

Yu R (2011) Study on key technologies of photovoltaic power generation system and reliability of electric vehicle charging station. Beijing Jiaotong University, Xizhimen (in Chinese)

Zhu Q, Rieger C, Basar T (2011) A hierarchical security architecture for cyber-physical systems. In: 4th international symposium on topic(s): communication networking and broadcasting, pp 13–14

Social Responsibility, Green Finance, and the New Urbanization Construction

Zhi Dong and Ying Li

Abstract The rate of China's urbanization has risen from 18 % in 1978 to 53 % in 2012, especially with a significant speed in the past 10 years, but mainly the land's urbanization, not the person's urbanization. The population of first-tier cities soared, with a "big city disease," such as strained resources and environmental damage. The second- and third-tier cities had a phenomenon of "empty city." Therefore, we must develop low-carbon economy and implement economic transformation and sustainable development. In this paper, based on the research achievements of green finance in recent years, we define the basic connotation of green finance, and elaborate the mechanism of green finance, and put forward the recommendations of green credit, green insurance, green securities, and green risk investment.

Keywords Social responsibility · Green finance · Low-carbon economy · New urbanization

1 Introduction

In recent years, with the continuous development of society and economy, the process of China's urbanization speeds up significantly. However, as in the process of urbanization in developed countries, the energy and environment problem breaks out in China. Energy deficit, environmental pollution, and ecological deterioration become a "bottleneck effect," with the prominent factors affecting social harmony, restricting the construction of the new path of urbanization.

Z. Dong (✉)
Hebei University, No. 180, May 4th East Road, Baoding, Hebei, People's Republic of China
e-mail: dongzhi@hbu.edu.cn

Y. Li
Agricultural University of Hebei, Baoding 071002, People's Republic of China

S. Feng et al. (eds.), *Low-carbon City and New-type Urbanization*,
Environmental Science and Engineering, DOI 10.1007/978-3-662-45969-0_34

With the concept of sustainable development, by means of technical innovation, institution innovation, industrial transformation, new energy development, and other means, develop the low-carbon economy, so as to achieve the win–win stage of urbanization construction and ecological environment protection.

Green finance, as a financial measure of environmental protection, is an important content of environmental protection system; for quite a long time in the present and in the future, economic and social development is facing a major issue (Wang 2013).

Green finance refers to the financial measure of the environmental protection. Capital flows in the process of investment and financing decisions should consider the potential impact on the environment; the potential returns associated with environmental standards, conditions, risks, and costs should be merged into the daily business of the bank. And following the laws of market economy, guided by the construction of ecological civilization, we should use the method of credit, insurance, securities, industrial funds, and other financial derivatives, so as to promote the coordinated development of the energy conservation and emissions reduction, and coordinate with energy, resources, and environment.

2 Low-carbon Economy and New Urbanization

Urbanization has been an important content of modernization construction in China and also has the most potential for stimulating domestic demand. The Central Economic Working Conference (2012) clearly puts forward that the concept and principles of ecological civilization must be integrated into the whole process of urbanization, and ultimately implements the new path of urbanization with intensiveness, intelligence, green, and low carbon. To achieve this goal, we must change the path of traditional industrialization and extensive form of economic growth, and commit to developing low-carbon economy, achieving sustainable development.

In addition, in the process of development, its resources and environment carrying capacity are limited. According to international experience, this stage tends to be concentrated outbreak period of "big city disease." For example, caused by the excessive concentration of industry and population density, constrained resources, traffic congestion, and environmental pollution begin to come out.

Long time of extensive economic growth in China and "GDP incentive mechanism" make the "big city disease" more outstanding. If the process is not good, it may eventually lead to a "morbid urbanization" and "poisonous" GDP (Xia 2013). The new path of urbanization and people-oriented concept have exposed the problems in the process of urbanization in China; we must accelerate transformation of the pattern of urban development and explore the environmental protection and low-carbon economy, as the main characteristics of the new urbanization.

3 The Mechanism of Green Finance

The core of green finance is calculating the stock of natural resources, or the loss of natural resources and environmental damage, caused by human economic activity. It can be calculated by economic value or environmental value measurement, and use in the field of economic resources allocation and financial activity evaluation.

Much attention has been paid to the "green finance," which is derived from the role of incentive and constraint, in strengthening the company's social responsibility, promoting energy conservation and emissions reduction. The role and influence is complicated, direct and indirect, both realistic and potential (Fig. 1).

3.1 Standardize Enterprise Behavior

Environmental pollution has different causes and different forms, and companies tend to try various means to avoid regulation and escape punishment. Government regulation is often not complete, because of the large workload and wide coverage (Qiao 1999).

Therefore, various financial instruments can directly impact on polluting enterprises financing, if the government, social and financial institutions can coalesce together will achieve the goal of pollution abatement.

In addition, the issue of environmental pollution prevention and control of fatigue, in the final analysis, is due to private companies or government departments lack of necessary positive incentive and punishment enough pressure. Once in a

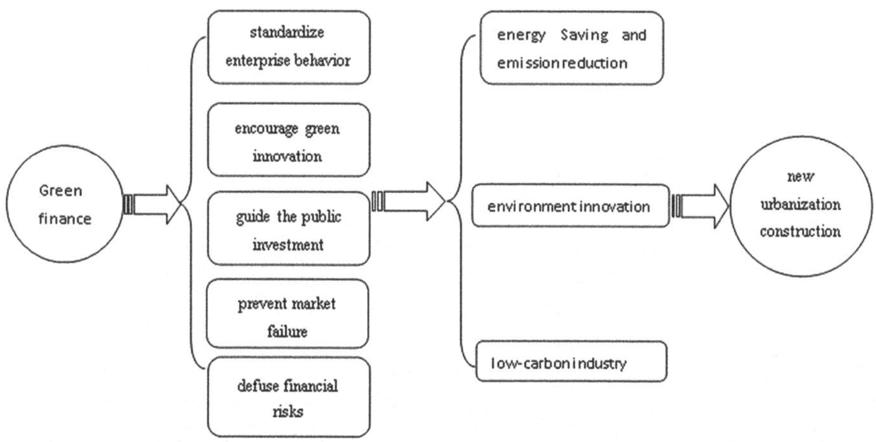

Fig. 1 Connotation and logic of green finance, low-carbon industry, and new urbanization construction

while, in the financial institutions in the prevention and control of environmental pollution problems, it is possible to form a strong incentive mechanism and punishment mechanism (Yang 2011).

3.2 Encourage Green Innovation

Green finance will provide strong financial support for the green industry. Accelerate accumulation of financial institutions, attraction of green finance research and development institutions, financial and carbon assessment certification, and other intermediary organizations in energy conservation and emissions reduction. Establish the guiding fund to attract investment in the field of new energy, energy conservation, environmental protection, and carbon trading funds, and private industry to build support for the development of green industry multi-level financial service system form adapted to the industrial development of green financial culture.

3.3 Guide the Public Investment

Financial institutions should provide risk reputation for loans and financing company caused by the destruction of environment. In addition, the public demand for healthful green environmental protection product, as well as the living environment for people's health, will not only affect the enterprise product sales, but also indirectly affect the interests of the bank, and the public itself is also a bank customer; along with the unceasing enhancement of environmental consciousness, the public will form a "preference" of environmental protection. At the same time, commercial banks provide green credit that is beneficial to attract more customers and gain customer loyalty, enhance brand image, gain more business license, and consolidate relationship with external stakeholders (Brammer and Pavelin 2008).

3.4 Prevent Market Failure

The problem of environmental pollution is mainly caused by "market failure," due to its strong negative externality. In order to solve the problem of "market failure," you need the government's intervention. The government is mainly punishing after the pollution arising, with bureaucracy and low-efficiency, result in "government failure" finally. The emergence of "green finance" is the combination of environmental risk and the financial risk, making full use of financial risk management technology, through the market mechanism, government regulation and social supervision, the news media, non-governmental organizations, and other power, and making punishment for prevention, both to solve the "market failure" and to avoid the "government failure."

3.5 Defuse Financial Risks

As the example of banks, when the bank provides loans for an enterprise, it must carry on the strict censorship with the environmental protection measures. After making loan, if once found that the enterprise has the action of environment pollution, the bank will bear corresponding responsibility. If the loan approval is lax, it means that there is something wrong with the bank's own creditability, to not only alienate bank depositors, but also meet the shareholders of the bank, "vote with their feet."

As the example of insurance, when the applicant buys the insurance, in order to reduce the costs of settlement of claims, the insurance company must be particularly concerned about the environmental and climate change and attention to the environmental protection measures of the enterprise, supervise the emission behavior of the enterprise, to minimize pollution emission activities of enterprises because of the climate and environment change or losses to policyholder, and indirectly, on behalf of the interests of the insurance companies themselves.

As the example of securities, when investors bought an enterprise's stocks or bonds, once the enterprise environmental pollution event happened, it can occur from the victim's legal claims or from the administrative punishment of the competent department of the government and thus affect the enterprise's economic benefit, leading to the enterprise issued by falling stock prices and lower bond yields, which affect the benefit of investors. Therefore, when investors have the securities investment, they will be particularly interested in the environmental protection measures of enterprises or projects, and the environmental protection consciousness of the enterprise staff.

Here, we further build a theoretical model to illustrate the necessity of the development about green finance:

1. Assume that the production function of a single enterprise for the whole economic activity

$$Y_i = \alpha_i f_i(E_i, K_i, L_i) \tag{1}$$

Among them, Y_i denotes the output level of individual enterprise, E_i denotes enterprise production environment of capital investment is needed (in the traditional sense of the environmental resources have been or are on the rise for the reality of economic and social development capital elements.

K_i denotes such as machine and equipment material capital, L_i said in labor, and α_i said a single technical level of the enterprise.

$$\text{And} \quad \partial f_i/\partial(E_i) > 0, \quad \partial^2 f_i/\partial(E_i)^2 < 0 \tag{2}$$

It suggests that environmental capital, material capital, and labor are an indispensable element of enterprise production inputs, and each factor input has marginal productivity which is greater than zero, but the characteristics of diminishing marginal productivity.

The whole economic activities have n similar enterprises; the total output is

$$Y(E,K,L) = \sum_{i=1}^{n} \alpha_i f_i(E_i, K_i, L_i) \tag{3}$$

The Y denotes the whole output of all enterprise

$$E = \sum_{i=1}^{n} E_i, \quad K = \sum_{i=1}^{n} K_i, \quad L = \sum_{i=1}^{n} L_i \tag{4}$$

Respectively shows single enterprise environmental capital, the sum total of material capital and labor input.

2. The condition of maximal output

$$\max Y(E,K,L) = \sum_{i=1}^{n} \alpha_i f_i(E_i, K_i, L_i) \tag{5}$$

$$\text{s.t. } E = \sum_{i=1}^{n} E_i \tag{6}$$

Construct the Lagrange function

$$\text{Lagrange}(E_i, K_i, L_i) = \sum_{i=1}^{n} \alpha_i f_i(E_i, K_i, L_i) + \lambda(E - E_i) \tag{7}$$

To maximize, it strives for the first derivative

$$\alpha_i = \frac{\partial f_i}{\partial E_i} = \alpha_j \frac{\partial f_i}{\partial E_i} = \lambda \tag{8}$$

If λ denotes environmental shadow price of capital, if $\alpha_i > \alpha_j$, and if the social production is the largest, $\frac{\partial f_i}{\partial E_i} < \frac{\partial f_j}{\partial E_j}$.

Because $\partial f_i/\partial(E_i) > 0$, $\partial^2 f_i/\partial(E_i)^2 < 0$, and so $E_i > E_j$.

If the enterprise technical level is higher, the more is the environment of financial capital (in this case, the financial capital and financial resources are equivalent). This is from the society as a whole, the optimal allocation of environmental resources in the condition.

4 The Countermeasures and Suggestions of Chinese Green Finance

Looked from the development of the practice in China and abroad, as part of the environmental economic policy of green finance, its mainstream model mainly includes the following:

4.1 The Green Credit

According to the economic policy and industrial policy, financial institutions should provide support to research and development, of anti-pollution facilities, engage in the ecological protection and construction, the development and utilization of new energy, engaged in the production, green manufacturing and ecological agriculture circular economy of enterprises or institutions; New project for pollution production and pollution enterprise investment loans and working capital loans, and implementation of policy of punitive interest rates (Ye 2012).

4.2 The Green Insurance, Named as the "Environmental Pollution Liability Insurance"

It is widely used in the international affairs. It is based on the compensation liability to the damage caused by enterprise pollution accidents, in accordance with law.

4.3 The Green Securities

The green securities refers to the listed company environmental protection verification system and environmental information disclosure system as the core, through the regulation of society to raise funds, to curb the "play" (high energy consumption, high pollution) enterprise overexpansion, prevent capital risks, and promote continuous improvement in environmental performance of listed companies.

4.4 The Green Risk Investment

It puts the money into dangerous resources-saving enterprises; there is a large market and environmentally friendly orientation of high and new technology development area, in order to obtain high capital gains after a successful business investment behavior. In addition, there are green finance financial bonds and the green repurchase agreements, such as green industry fund form (Liu 2009).

Acknowledgments This paper is supported by the soft science project of Hebei province (134576291), the research of people's livelihood in Hebei province (201201080), and social science fund project of Hebei province (HB11YJ069).

References

Brammer S, Pavelin S (2008) Factors influencing the quality of corporate environmental disclosure. Bus Strategy Environ 17(2):120–136

Liu M (2009) Green risk investment power green GDP. Productivity research

Qiao H (1999) Set up the concept of financial ecology. Ecological economy

Wang L (2013) Green finance is important for the construction of ecological civilization, the financial times

Xia J (2013) Promote new urbanization should attach great importance to developing the green service industry, China's development

Yang Y (2011) Green finance policy, corporate governance and corporate environmental information disclosure, finance and trade research

Ye Y (2012) The implementation of green credit and management. China's finance

Scenario Analysis for Low-Carbon Urban Planning: A Case of Kaiyuan in Yunnan Province of China

Qi He, Hong Fang, Han Ji and Meng Du

Abstract With the world's accelerating economic growth as well as the continuous expansion of the global population, the demand for energy and other resources from human society has correspondingly increased year by year. For the purpose of energy saving and environmental protection, a "low-carbon tide" has swept the globe. Despite of the bargaining in right and obligation associated with carbon emissions abatement, most countries per se have taken the low-carbon path to prepare for strategic competitiveness in the long run. Beyond all question, low-carbon economy would be a significant issue in China's development, which has already orientated by policies such as National twelfth Five-year plan. As a typical resource-based city, Kaiyuan has consolidated and developed several pillar industries including coal, electricity, chemical, building materials, and food industries around its coal energy industry. The well-developed industrial system and especially the heavy industries have contributed a lot to local economy, however, also brought about severe challenges to the sustainable development of Kaiyuan. As a city whose 90 % energy depends on coal, Kaiyuan is suffering serious pressure in reduction of CO_2, SO_2, and oxynitride. Therefore, supportive works for Kaiyuan's low-carbon planning are in sore need and could make profound meaning. This paper firstly establishes an accounting model for measurement of city's carbon emissions in which every category, namely husbandry and energy consumption, which should be taken into account has been recognized. After the definition of accounting scope, the calculation method and standard for the emission of each category has been formulated. Some quantitative researches for the total carbon emissions and their sub-items have accordingly carried out to reveal the contradictions and problems during the development of Kaiyuan. Based on the current status as to carbon emissions, three sceneries, referred as Basic analysis unit (BAU),

Q. He · H. Fang · M. Du
School of Economics and Management, Beihang University, Beijing 100191
People's Republic of China

H. Ji (✉)
School of Agricultural Economics and Rural Development,
Renmin University of China, Beijing 100872, People's Republic of China
e-mail: hans0929@sina.com

S. Feng et al. (eds.), *Low-carbon City and New-type Urbanization*,
Environmental Science and Engineering, DOI 10.1007/978-3-662-45969-0_35

Basic policy (BP) and Low-carbon (LC) scenery, have been set up to check out the development goals for LC urban development of Kaiyuan. According to these analyses, some policy advices have been proposed for reference.

Keywords Carbon emission · Energy consumption · Low-carbon · Economy · Low-carbon urban planning · Scenario analysis

1 Introduction

Nowadays as a result of rapid population growth and economic priorities, there are more and more prominent conflicts between environment and economic development. On the one hand, immoderate energy consumption has brought serious environmental problems (Xia and Xu 2012), including air pollution (Hu et al. 2014), acid rain (Streets 1984), and global warming (Akhmat et al. 2014), heavily challenging worldwide human survival; On the other hand, the traditional way of extensive energy consumption has caused huge waste of resources, resulting in energy intensity especially for populous China (Zhang et al. 2011). In the macro-background of global environment crisis and China's energy intensity, energy conservation and emissions reduction associated with the construction of "low-carbon city" in the country will produce amplifying abatement effect. It is known that a so-called low-carbon city is that keeps energy consumption and CO_2 emissions at low level under the premise of rapid economic development (Khanna et al. 2014). In this context, targeting the carbon emission as a clear indicator of urban development is a precondition in the planning and construction of a low-carbon city.

Kaiyuan is the traffic fortress and the center city locating in the southeast of Yunnan province. It is extremely rich in coal resources, which has the largest open-pit coal mine in Yunnan called "Small longtan lignite," along with abundant grain and sugarcane providing raw materials for the food industry. Given the coal power, cement, chemical fertilizer, sugar, paper, wine, food, and other industries taking shape, how to balance the coal-based economic benefits and ecological benefits becomes a top priority in future development of Kaiyuan. For these reasons, this paper checks computation of previous CO_2 emissions in Kaiyuan based on the carbon accounting model and carries out scenario analysis for future emission levels to capture the dynamics of CO_2 emissions in Kaiyuan and facilitate its carbon reduction.

The remainder of the paper is organized as follows: Section 2 specifically introduces the methodology, namely the carbon accounting model and scenario analysis. In Sect. 3, we discuss the current situation of Kaiyuan in the development of low-carbon city, including the trend of CO_2 emissions, CO_2 emissions per capita, and CO_2 emissions per GDP associated with economic development of Kaiyuan. The results of scenario analysis are reported and discussed in Sects. 4 and 5 correspondingly provides some policy proposals.

2 Methodology

2.1 Carbon Accounting Model of the City

2.1.1 Carbon Accounting Scope in the City

In this paper, the carbon accounting scope of the city mainly involves two parts:

1. Fossil fuel consumption, including: raw coal, gasoline, diesel oil, liquefied petroleum gas, and a series of carbon emissions from fossil energy consumption;
2. Farming and animal husbandry, including: methane emissions in rice planting, greenhouse gas emissions from livestock, and poultry farming.

The data of citywide overall energy consumption in every calendar year are provided by statistical bureau of the city, which comprehensively covers fossil energy and electricity consumption. Moreover, unlike most other cities, Kaiyuan is one of the most important production bases of electricity so that its local electricity consumption is mostly supplied by native power plants.

It is known that electricity can be generated by coal-fired and some other clean energies such as hydropower and wind power. However, the former, thermal power generation, is totally a process of energy conversion so that the large amount of raw coal used to generate the electricity has already been included in the citywide overall energy consumption of Kaiyuan. The consumption of coal-fired electricity per se does not bring CO_2 emissions, so does the hydropower and wind power electricity. Therefore, given that the fossil energy consumption cannot be available by category in details, the electricity consumption will be eliminated from citywide overall energy consumption so as to avoid double counting of carbon emissions due to secondary energy.

2.1.2 Carbon Accounting for Farming and Animal Husbandry

In addition to the fossil energy, agricultural carbon emissions are also an important source of carbon emissions, including emissions from farmland and livestock. In general, the methane gas emitted from the paddy fields can be regarded as the main source of carbon emissions in farmland. Meanwhile, other crops per se do not produce methane during their growth and the CO_2 emitted by them can be approximately offset by the absorption of them, which can be thought of namely carbon neutrality (Chang 2013). This paper adopts the average value of coefficient of methane emissions from paddy field of Yunnan Province as 0.194 tCH_4/ha year (Gao 2007). Taking into account that the Global warming potential (GWP) value of CH_4 is 21 times as CO_2, there is equivalently 4.074 tons CO_2 emissions from a hectare of paddy fields per year

As to the carbon emissions from animal husbandry, according to the instruction of the Intergovernmental Panel on Climate Change (IPCC) Guidelines for National

Table 1 Coefficients of greenhouse gas emissions of livestock (unit: kg/each year)

Livestock	Coefficient of CH_4		Coefficient of N_2O	Conversion coefficient: tCO_2/(each year)
	Enteric fermentation	Feces		
Pig	1	0.940	0.53	0.20504
Cattle	55	1.414	1.34	1.60009
Sheep	5	0.150	0.33	0.21045
Poultry	0	0.012	0.02	0.00645

Greenhouse Gas Inventories (Eggelston et al. 2006), Kaiyuan should calculate the carbon emissions from pigs, cattle, sheep, and poultry farming. This part of the carbon emissions mainly involves the following aspects: methane from the enteric fermentation of livestock and nitrous oxide and methane from fecal treatment for livestock. As the GWP values of methane and nitrous oxide are, respectively, 21 and 310, the animal husbandry generates vast greenhouse gas emissions contributing to global warming. According to “The People's Republic of China National Greenhouse Gas Inventory” (Gao 2007) and the report of Food and Agriculture Organization (FAO) in 2004, namely “Livestock's Long Shadow” (Steinfield et al. 2006) that carries out a study of nitrous oxide emissions from livestock in China, the default values of greenhouse gas emissions from livestock in Kaiyuan have been taken as Table 1.

2.1.3 Carbon Accounting for Energy Consumption

It is critical for this paper to establish the carbon emission model for the energy consumption field. It comprehensively covers the carbon emissions from consumption of primary energy used in urban transportation, construction, production, and daily life in Kaiyuan. To demonstrate the relationship between the corresponding carbon emissions and consumption of different energies, the formula has been expressed in Eq. (1).

$$CO_2 = K * E \tag{1}$$

where CO_2 is the carbon dioxide emissions; E stands for different types of energy use, which can be uniformly converted by coefficients into standard coal consumption; K refers to carbon intensity or coefficient of carbon emissions that differs among different countries or regions with discriminating technical conditions and energy structure. In this study, we assume a constant coefficient K which namely indicates that the adjustment of energy structure has not been taken into account. As K value is constant, the problem of low-carbon research has turn into an issue of energy utilization.

In particular, coal takes high proportion in energy structure with low energy efficiency in Kaiyuan. It is reported by “Summary of first industrial pollution source

Table 2 Indicators for fossil energy conversion

Fuel (kg)	Q_{dw} (kJ/kg)	Emission factor ($kgCO_2/TJ$)	Oxygenation efficiency	Carbon emission coefficient (tCO_2e/t)	Standard coal coefficient
Raw coal	20,908	87,300	1	1.83	0.7143
Coke	28,435	95,700	1	2.72	0.9714
Diesel	42,652	72,600	1	3.10	1.4571

Note Q_{dw} refers to lower heating value of the fuels

census of Kaiyuan" in 2007 that raw coal accounts for 90 % energy use in Kaiyuan. Combining with the energy conversion coefficients of some fossil energy (Table 2) provided by "China energy statistical yearbook 2010" (2011), we calculate the coefficient of CO_2 emission per standard coal using the weighted average method and find that it should be 2.38 tons (CO_2/E).

2.2 Scenario Analysis

Scenario analysis refers to the whole process of predicting how the situations take place and comparing what influence they will, respectively, bring. Its basic view is that future is full of uncertainty, but some situations in the future can be predicted (Zhu et al. 2011). Based on various key assumptions in the major evolutions of economy, industry, or technology, possible solutions in the future can be formulated by deliberate reasoning and description. Combined with related mathematical models, some quantitative analysis and evaluation can be carried out for these possible trends and, on this basis, control measures of adverse trends or promotion actions of favorable trend can be put forward.

In this paper, scenario analysis is the basis for the prediction and evaluation of the development of low-carbon city for Kaiyuan, with uncertain and periodic characteristics. It includes six steps: (1) Determination of focal point. On the basis of deep understanding of low-carbon city development, the current core problem for Kaiyuan is the balance of CO_2 emissions along with economic development. (2) System evaluation. Based on the importance degree, make out the selected factors for further assumptions. In our paper, the economic development, policy of local government, carbon intensity, and technology would be included in different solutions. (3) Determine the alternatives. Setup three alternatives and scene descriptions for each alternative in this paper. (4) Modeling. Based on the alternative solutions, carry out the quantitative estimations with the aid of quantitative models. (5) Scenario Testing. Compare the implications from different scenarios and discuss the feasible solutions. (6) Solution selection. According to the result of prediction and combined with the assumptions to achieve these goals, analyze the prediction results.

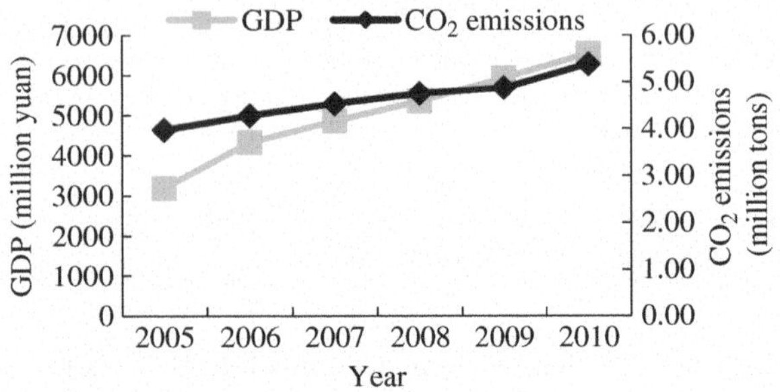

Fig. 1 GDP and CO_2 emissions of Kaiyuan (all time-series data associated with economic variables in this paper are in 2005 constant price)

3 Current Situation of Kaiyuan in the Development of Low-Carbon City

Based on the carbon accounting model demonstrated in Sect. 2, the overall CO_2 emissions and intensity can be estimated with the help of data associated with economic scale, energy use and electricity consumption, farming, and animal husbandry development of Kaiyuan.

The trend of economy scale and CO_2 emissions of Kaiyuan from 2005 to 2010 is shown in Fig. 1. By authors' calculation, there are 3.96 million tons CO_2 emissions in 2005 and 5.38 million tons in 2010 for Kaiyuan, with an average growth rate at 6.31 %. As can be seen that with the economy development, CO_2 emissions consistently and proportionally expanded with GDP. Despite the fact that the growth rate in GDP is slightly higher than that of CO_2 emissions for past few years, the latter has been observed an obvious rise since 2009, with a 10.47 % increase in carbon emissions in 2010. It is a significant dilemma facing Kaiyuan at current development stage how to balance the goals of low-carbon city with sustainable economic growth.

From the perspective of carbon emissions per capita, the abatement situation is far more than unoptimistic. In recent years, Kaiyuan has made significant improvement in its economic development and residents' living standard (Fig. 2).

Fig. 2 GDP per capita and CO_2 emissions per capita

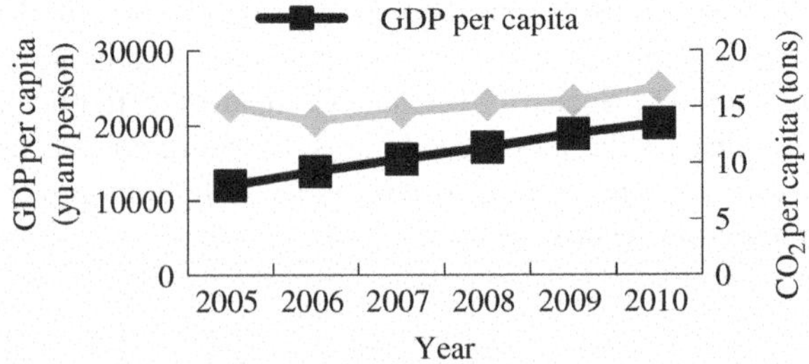

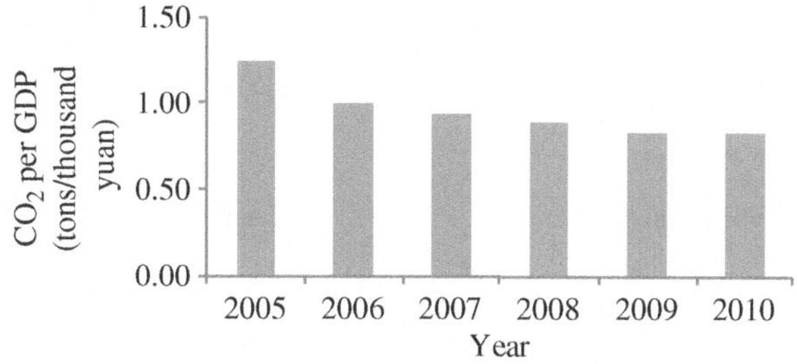

Fig. 3 CO_2 emissions per GDP of Kaiyuan

GDP per capita has increased from 12,016 in 2005 up to 20,242 in 2010; however, per capita emissions have not been fundamental improved. Since 2006, carbon emissions per capita of Kaiyuan have kept stable growth with an average annual growth rate of 4.97 %. In 2010, CO_2 emissions per capita of Kaiyuan have reached up to 16.66 tons, far more than that of the cities with the same city size. Judged from the current situation, the carbon emissions per capita may continue to rise in the forthcoming years, which is in urgent need to focus on transition in the high-carbon lifestyle of residents and the coal-based energy structure of the industry.

To the credit of the past few years, a profound conservation and emissions reduction work has made definite achievements, mainly reflecting in the carbon intensity index. The CO_2 emissions per GDP have decreased from 1.245 tons per thousand yuan in 2005 to 0.823 tons in 2010, with a drop of 34.10 % (see Fig. 3). However, as the energy and carbon intensive industries take high proportions in the industrial system of Kaiyuan, namely with a relative high-carbon emissions at the base period, great efforts should be made to guarantee the coordinated development of the environment and the economy.

4 The Scenario Prediction of Carbon Emissions in Kaiyuan from 2010 to 2020

In order to determine the low-carbon development path of Kaiyuan, this paper lays out three scenarios of carbon dioxide emissions in Kaiyuan from 2010 to 2020: one in which development continues as usual without any policy interventions (BAU), one in which development continues according to current government policies (BP), and the other one which assumes a low-carbon growth path (LC). As the above analysis, the carbon emissions in Kaiyuan are mainly from fossil energy use, agriculture, and livestock husbandry. But the carbon emissions from agriculture and livestock husbandry are not taken into account in the three scenarios due to its low proportion and soggy growth. Specifically, BAU scenario is set on the basis of the pattern of current economic development and carried out in accordance with the traditional way, in which the future carbon emissions are calculated by the constant carbon intensity as the base year without any low-carbon policy requirement; BP

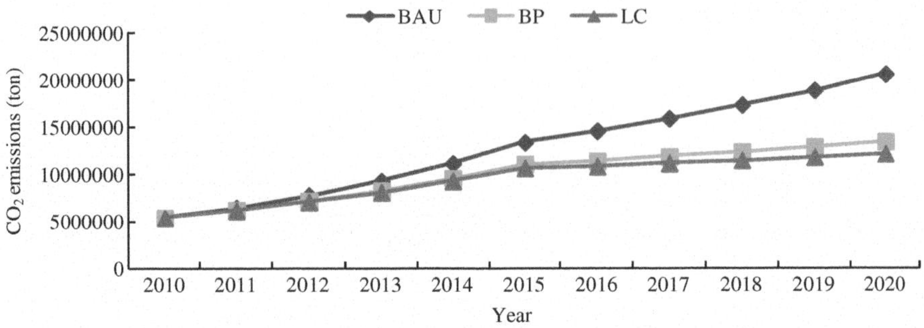

Fig. 4 Carbon emissions of Kaiyuan over 2010–2020

scenario is set with the constraint indices of reducing resource consumption consulting national, provincial, and municipal targets in the context of the policy implementation during the "Twelfth Five-year Plan" period (2011–2015) by such measures as the adjustment of economic structure, the application of energy-conserving technology, the popularization of cleaner production, the use of renewable energy, and the establishment of urban ecological transportation and building systems; LC scenario not only considers all issues in BP scenario but also assumes that the low-carbon construction is strengthened to the greatest extent, which involves specific carbon reduction measures and low-carbon technology projects in critical areas (see Table 4). The scenario analysis with the three scenarios attempts to verify the low-carbon economic feasibility and affordability in Kaiyuan.

Figure 4 shows the predictions of carbon emissions in the three scenarios of Kaiyuan in the next decade, with 2010 as the base year. It is obvious that the difference among the three scenarios will increase as time progresses.

In BP scenario, since all emission-reduction measures may not be implemented in a short term, two-stage targets are established: the first stage is the design and layout one for low-carbon development from 2011 to 2015, the target of which is to reduce carbon emissions 5.7153 million tons fewer than BAU scenario; the second stage is the expansion one for low-carbon development from 2016 to 2020, the target of which is to reduce carbon emissions 25.3142 million tons fewer than BAU scenario.

In LC scenario, as the related departments do their utmost to reduce carbon emissions from all sides, it is estimated that the carbon emissions will be reduced 6.694 million tons fewer than BAU scenario in the first stage (2011–2015) and 29.8813 million tons fewer than BAU scenario in the second stage (2016–2020).

An appreciable distinction of carbon intensity of Kaiyuan among the various scenarios appears in Table 3. The prediction values of carbon intensity in BP scenario will be cut to 0.681 tons per thousand RMB of GDP in 2015 and 0.536 in 2020 in constant prices, decreasing from 2010 level by 17.26 and 34.82 %, respectively, which in LC scenario will drop to 0.657 in 2015 and 0.485 in 2020, decreasing from 2010 level by 20.17 and 41.06 %, respectively.

Table 3 Prediction of carbon intensity of Kaiyuan (unit: tons/thousand RMB)

Year	GDP (thousand RMB)	BAU	BP	LC
2010	6,532,090	0.823	0.823	0.823
2015	16,253,930	0.823	0.681	0.657
2020	25,008,690	0.823	0.536	0.485

In reality, because the national, provincial, and municipal Twelfth Five-year Plans have explicitly set goals to reduce carbon emissions and China's five-year plans can generally be fulfilled, the carbon emissions in Kaiyuan may probably change in terms of BP scenario at least during the "Twelfth Five-year Plan" period (2011–2015). Meanwhile, some negative factors could cause a diversion of carbon emissions from BP scenario to BAU scenario, such as introduction of high-energy-consuming and high-emission industries, path dependence of simple reproduction or expansion of reproduction, and unbridled energy consumption. Correspondingly, some positive factors could lead the society to LC scenario, such as introduction of advanced low-carbon technologies, concern about energy conservation of the whole society, and development of renewable energy resources.

Kaiyuan possesses certain advantages in the aspects of economic power and industrial technologies in Yunnan province, which provide conditions to achieve the balance between economic growth and carbon reduction by 2020. In the next decade, Kaiyuan needs to focus on and encourage low-carbon development, not only seizing the opportunity of relatively low international energy prices to promote economic growth but also tackling key technical problems on energy conservation and emission reduction, adjusting and upgrading the industrial structure, and struggling for the initiative and the power of discourse in the future low-carbon competition.

5 Policy Recommendations

It needs every department including those in social, economic, and environmental domains to innovate their perspectives and means for the comprehensive management in the planning and construction of low-carbon city. Through effective arrangement in the rights, obligations, responsibilities, and procedures, "market failure" could be solved by macroeconomic management and indirect regulation of government meanwhile "government failure" could be overcome through the establishment of comprehensive decision-making and management mechanism and promoting public participation, which creates favorable environment for implementation of low-carbon planning. Specifically in Kaiyuan, support conditions and countermeasure, involving planning organization, responsibility system, supporting policies, financial support, technical support, human resources, statistics and monitoring system, and market mechanism as well as propaganda and communication, should be fit out for the following measures in corresponding departments (see in Table 4).

Table 4 Carbon reduction measures for Kaiyuan

Department	Measure
Energy	The development and utilization of hydropower, wind power, biomass energy, methane, and other clean energy; process agricultural and forestry wastes into biomass-coal, power generation by biomass gasification, and gasification of biomass for centralized gas supply
Industry	Adopt corresponding energy-saving technologies and products, develop industrial power saving technology, support the construction of key energy-saving projects, introduce advanced energy-saving equipment, improve the policies of the cogeneration projects, and orderly develop of cogeneration projects
Construction	In the progress of building construction, reconstruction and residence, implement the standard of 65 % energy saving by adopting energy-saving technology, equipment and materials, improving the performance of heat preservation and heat insulation and the efficiency of heating, air conditioning and refrigeration heating system, strengthening the operation management of the energy system, and using renewable energy (wind and solar) to ensure the quality of indoor thermal environment and lighting and supply of hot water
Traffic	Optimize traffic space layout; advocate the public to choose public transport, cycling, and online shopping; the improvement of tax measures to guide the transition in vehicle structure; strengthen the management of traffic demand, developing intelligent transportation information system and promoting the construction of ecotype highway
Residents living	Promote the use of clean energy and energy-saving appliances, encourage residents autonomous energy-saving behavior, and advocate the residents to use biomass fuel
Farmland	Improve the way of paddy planting by adopting the measured soil fertilizer, and reasonable irrigation technology to reduce emissions of methane and enhance carbon sink in soil
Livestock breeding	Concentrated treatment in livestock feces treatment, utilization technology of dry wet depart, rain sludge separation and biogas fermentation; using planting and raising mode that combines agriculture with animal husbandry
Forestry	Artificial afforestation, restoration of damaged ecosystems, building an agroforestry system, strengthening sustainable management of the forest, improvement in carbon uptake, and carbon sequestration of the forest
Garbage disposal	Reduction and classification in living garbage, decrease in the proportion of waste incineration, gas recycling, and disposal of the methane from landfill

Acknowledgments Financial support: China National Natural Sciences Foundation (Grant reference number: 71073007, 71273022); Special Research of Generalized Virtual Economy [GX2011-1014 (M)]; Beijing planning project of philosophy and social sciences (12JGB061); Beijing planning project of educational Science (priority funding, AAA12007).

References

Akhmat G, Zaman K, Shukui T, Sajjad F (2014) Does energy consumption contribute to climate change? Evidence from major regions of the world. Renew Sustain Energy Rev 36:123–134

Chang SJ (2013) Solving the problem of carbon dioxide emissions. Forest Policy Econ 35:92–97

Eggelston S, Buendia L, Miwa K, Ngara T, Tanabe K (2006) IPCC guidelines for national greenhouse gas inventories. The Institute for Global Environmental Strategies (IGES) for the IPCC

Energy statistics division of the national bureau of statistics of the People's Republic of China (2011) China energy statistical yearbook 2010. China Statistics Press, Beijing (in Chinese)

Gao G (2007) The People's Republic of China national greenhouse gas inventory. China Environmental Science Press, Beijing (in Chinese)

Hu H, Zhang X, Lin L (2014) The interactions between China's economic growth, energy production and consumption and the related air emissions during 2000–2011. Ecol Ind 46:38–51

Khanna N, Fridley D, Hong L (2014) China's pilot low-carbon city initiative: a comparative assessment of national goals and local plans. Sustain Cities Soc 12:110–121

Steinfield H, Gerber P, Wassenaar T, Castel V, Rosales M (2006) Livestock's long shadow: environmental issues and options. Food and Agriculture Organization of the UN (FAO)

Streets DG (1984) Acid rain control and United States energy policy. Energy 9(11–12):1049–1063

Xia Y, Xu M (2012) A 3E model on energy consumption, environment pollution and economic growth-an empirical research based on panel data. Energy Procedia 16(C):2011–2018

Zhang N, Lior N, Jin H (2011) The energy situation and its sustainable development strategy in China. Energy 36(6):3639–3649

Zhu Z, Bai H, Xu H, Zhu T (2011) An inquiry into the potential of scenario analysis for dealing with uncertainty in strategic environmental assessment in China. Environ Impact Assess Rev 31(6):538–548

Erratum to: Analysis on Urban Crisis Conduction Under the Background of Big Data

Yan Li and Yun Lu

Erratum to: Chapter "Analysis on Urban Crisis Conduction Under the Background of Big Data" in: S. Feng et al. (eds.), *Low-carbon City and New-type Urbanization*, Environmental Science and Engineering, DOI 10.1007/978-3-662-45969-0_28

The original acknowledgement of this chapter was unfortunately wrong. The correct acknowledgement should be as below:

This study is supported by the National Natural Science Foundation of China (Nos.71402097); Youth Foundation Program in Shanghai International Studies University, "Financial Crisis Control System in the information age" (Nos. Kx181102); and the Major Research Program of Shanghai International Studies University (Nos. 2013114ZD001).

The online version of the original chapter can be found under
DOI 10.1007/978-3-662-45969-0_28

Y. Li (✉) · Y. Lu
College of International Business, Shanghai International Studies University,
550 Daling Road, Shanghai 200083, China
e-mail: liyan0417@shisu.edu.cn

S. Feng et al. (eds.), *Low-carbon City and New-type Urbanization*,
Environmental Science and Engineering, DOI 10.1007/978-3-662-45969-0_36

Retraction Note to: A Fast Time Series Shapelets Data Mining Algorithm

Zheng Zhang

Retraction Note to: A Fast Time Series Shapelets Data Mining Algorithm in: S. Feng et al. (eds.), *Low-carbon City and New-type Urbanization*, Environmental Science and Engineering, DOI 10.1007/978-3-662-45969-0_32

The article titled "A Fast Time Series Shapelets Data Mining Algorithm" published in Low-carbon City and New-type Urbanization, pp. 357–366, DOI: 10.1007/978-3-662-45969-0_32, needs to be retracted at the request of editorial team of Chinese Low-carbon City Development International Conference, since 69 % of its content overlaps with a previously published paper in Proceedings of the 2013 SIAM International Conference on Data Mining: "Fast Shapelets: A Scalable Algorithm for Discovering Time Series Shapelets" (2013).

The online version of the original chapter can be found at DOI 10.1007/978-3-662-45969-0_32

Z. Zhang (✉)
Information Center, GanSu Party School, Jianning Road 199#, Anning District, Lanzhou 730070, People's Republic of China
e-mail: 630533797@qq.com

Z. Zhang
Complex Data Research Center, BeiHang University, Beijing 100083, People's Republic of China

S. Feng et al. (eds.), *Low-carbon City and New-type Urbanization*, Environmental Science and Engineering, DOI 10.1007/978-3-662-45969-0_37

Author Index

S. Feng et al. (eds.), *Low-carbon City and New-type Urbanization*,
Environmental Science and Engineering, DOI 10.1007/978-3-662-45969-0

The manufacturer's authorised representative in the EU is Springer Nature Customer Service Centre GmbH, Europaplatz 3, 69115 Heidelberg, Germany. If you have any concerns regarding our products, please contact ProductSafety@springernature.com

Printed and bound by CPI Group (UK) Ltd, Croydon, CR0 4YY
23/09/2026
02221535-0007